Encyclopaedia of Mathematical Sciences
Volume 140

Low-Dimensional Topology

Subseries Editors:
R.V. Gamkrelidze V.A. Vassiliev

Springer
Berlin
Heidelberg
New York
Barcelona
Hong Kong
London
Milan
Paris
Tokyo

To my mom, with love and gratitude

Preface

Homology 3–spheres are closed oriented 3–manifolds having the homology of the 3–sphere. These objects may look rather special but they have played a prominent role in manifold topology for a long time. This book is a survey of various ideas and constructions used in their study, from the classical Rokhlin invariant through Casson's theory and its numerous generalizations up to the most recent gauge theoretical invariants. It is hardly possible to account for all the ramifications of this theory in one book. While trying to give as complete an account as possible, I was largely guided by my personal research interests in deciding which topics to cover in more detail.

Some results in the book have never appeared in monograph form before while others are rather well documented in the literature. The state of affairs in low dimensional topology by the early 1980s is described in Mandelbaum's survey [203]. The book [275] can be recommended as a gentle introduction to the Casson invariant aimed at graduate students. Akbulut and McCarthy's book [5] is a more advanced text on the topic. Walker's and Lescop's extensions of the Casson invariant are described in their respective books [302] and [182]. I highly recommend the surveys [139] and [164] for the gauge theoretical approach to the Casson invariant. The book [70] is a standard reference for Floer homology.

While working on this book, I greatly benefitted from conversations with Jim Bryan, Christopher Herald and Sławomir Kwasik. During the course of my work I held visiting positions in the Mathematics Departments at Duke, Princeton and Tulane Universities, and at Max Planck Institut für Mathematik in Bonn. I am thankful to these institutions for their hospitality and support. This work was also partially supported by NSF Grant DMS-0196523.

Coral Gables, Florida *Nikolai Saveliev*

May 2002

Table of Contents

1 Homology 3–Spheres

Let R be a commutative ring with an identity element. By an *R–homology 3–sphere* we will mean a closed oriented 3–manifold M such that $H_*(M, R) = H_*(S^3, R)$. In dimension three, the concepts of topological, smooth, and PL manifolds coincide, as do the concepts of diffeomorphism, homeomorphism and PL–homeomorphism, see [26] and [222]. Therefore, we can work in all three categories interchangeably. Usually, $\mathbb{Z}$–homology spheres are referred to as *integral homology spheres*, and $\mathbb{Q}$–homology spheres as *rational homology spheres*. Homology 3–spheres exist in abundance. We begin with some examples and constructions.

1.1 Integral Homology 3–Spheres

Among closed 3–manifolds, integral homology spheres M are characterized by the property that $H_1(M, \mathbb{Z}) = 0$. This property can be easily checked if one knows the fundamental group of M: the abelianization of $\pi_1(M)$ is isomorphic to $H_1(M, \mathbb{Z})$.

1.1.1 Homotopy 3–Spheres

An obvious example of a homology 3–sphere is S^3. A homology 3–sphere which is simply–connected is called a *homotopy 3–sphere*; it is always homotopy equivalent to S^3. It is not known if there exist homotopy spheres not homeomorphic to S^3. The famous and still unsolved *Poincaré conjecture* asserts that every homotopy 3–sphere is homeomorphic to S^3.

1.1.2 Poincaré Homology Sphere

The icosahedral group I is the group of isometries of a regular icosahedron (or dodecahedron) in $\mathbb{R}^3$ centered at the origin. It is naturally a subgroup of $SO(3)$. The universal cover of $SO(3)$ is the group $SU(2)$. Algebraically, the projection $SU(2) \to SO(3)$ is the quotient by the center $\{\pm 1\}$ of $SU(2)$, and topologically, it is the map $S^3 \to \mathbb{R}P^3$. The lift of I to $SU(2)$ is denoted I^* and is called the *binary icosahedral group*. It has presentation

$$I^* = \langle\, a, b \mid (ab)^2 = a^3 = b^5 \,\rangle.$$

The group I^* has order 120. Its center has order 2 and is generated by a^3. The abelianization of I^* is trivial, hence, the quotient $P = S^3/I^*$ is an integral homology sphere. It is called the *Poincaré homology sphere*. It is oriented as the quotient of S^3 with the standard orientation. Several equivalent descriptions of P will be given later in this chapter, see also [161].

In fact, it is known that the only finite non-trivial group that can occur as the fundamental group of an integral homology 3–sphere is I^*, see Section 1.2.1. However, it is still not known if P is the only homology 3–sphere with fundamental group I^* (of course, it depends in part on the Poincaré conjecture).

1.1.3 Brieskorn Homology Spheres

Let p, q and r be pairwise relatively prime positive integers. The *Brieskorn homology sphere* $\Sigma(p,q,r)$ is defined as the link of Brieskorn singularity,

$$\Sigma(p,q,r) = \{\, x^p + y^q + z^r = 0 \,\} \cap S^5 \ \subset \mathbb{C}^3. \tag{1.1}$$

The (singular) complex surface given by the equation $x^p + y^q + z^r = 0$ has a canonical orientation. It induces a canonical orientation of the link $\Sigma(p,q,r)$ by the rule "outside normal first". The link $\Sigma(2,3,5)$ is orientation preserving homeomorphic to the Poincaré homology sphere. If any of p, q or r is equal to 1, the manifold $\Sigma(p,q,r)$ is homeomorphic to S^3. If $1/p + 1/q + 1/r < 1$ then $\pi_1\Sigma(p,q,r)$ is infinite. Brieskorn homology spheres are a special case of Seifert fibered homology spheres.

1.1.4 Seifert Fibered Homology Spheres

Let $a_1, \dots, a_n$ be positive integers, $n \geq 3$. Let $B = (b_{ij})$ be an $(n-2) \times n$–matrix of complex numbers such that each of the maximal minors of B is non-zero. Then the variety

$$V_B(a_1, \dots, a_n) = \{\, b_{i1} z_1^{a_1} + \ldots + b_{in} z_n^{a_n} = 0,\ i = 1, \dots, n-2 \,\} \subset \mathbb{C}^n$$

is a complex surface which is non–singular except perhaps at the origin. The link of singularity at the origin,

$$\Sigma(a_1, \dots, a_n) = V_B(a_1, \dots, a_n) \cap S^{2n-1}, \tag{1.2}$$

with the induced orientation, is a smooth 3–manifold which does not depend on B up to a homeomorphism. The manifold $\Sigma(a_1, \dots, a_n)$ is an integral homology sphere if and only if $a_1, \dots, a_n$ are pairwise relatively prime integers, see [242], in which case $\Sigma(a_1, \dots, a_n)$ is called a *Seifert fibered homology sphere*.

Seifert fibered homology sphere $\Sigma(a_1,\ldots,a_n)$ can be described topologically as follows. Let $F = S^2 \setminus \operatorname{int}(D_1^2 \cup \ldots \cup D_n^2)$ be the n–punctured 2-sphere and consider an S^1–bundle $W \to F$ with Euler number b and with fixed trivialization over ∂F. The boundary of W consists of n tori, $(\partial D_k^2) \times S^1$. Given n pairs of relatively prime integers, (a_k, b_k), paste n solid tori $D_k^2 \times S^1$ into W in such a way that the homology class $a_k(S^1 \times \{1\}) + b_k(\{1\} \times S^1)$ in the kth boundary component of W is null-homologous in $D_k^2 \times S^1$ after pasting. In other words, perform rational Dehn surgery on the link shown in Figure 1.1, see Section 1.1.6.

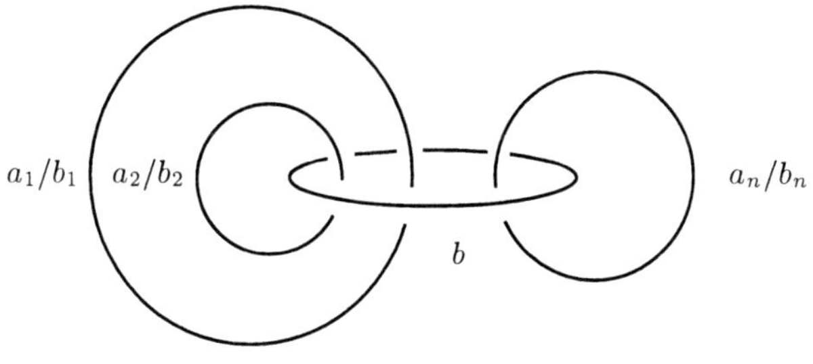

Fig. 1.1.

This construction results in a closed manifold $M(b; (a_1, b_1), \ldots, (a_n, b_n))$. The integers b, $(a_1, b_1), \ldots, (a_n, b_n)$ are called *Seifert invariants* of M. The fundamental group of M has presentation

$$\langle\, x_1, \ldots, x_n, h \mid h \text{ central},\ x_k^{a_k} = h^{-b_k},\ x_1 \cdots x_n = h^{-b}\,\rangle, \qquad (1.3)$$

hence the first integral homology of M is isomorphic to the cokernel of the linear map $A : \mathbb{Z}^{n+1} \to \mathbb{Z}^{n+1}$ with the matrix

$$A = \begin{pmatrix} a_1 & 0 & \ldots & 0 & b_1 \\ 0 & a_2 & & 0 & b_2 \\ \vdots & & \ddots & & \vdots \\ 0 & 0 & & a_n & b_n \\ 1 & 1 & \ldots & 1 & b \end{pmatrix}. \qquad (1.4)$$

The manifold M is an integral homology sphere if and only if $\det A = \pm 1$, or, as a little calculation shows, $a_1 \cdots a_n \cdot (\sum (b_k/a_k) - b) = \pm 1$. Suppose that M is an integral homology sphere, then choosing between $+1$ and -1 amounts to fixing an orientation on M. Fix the orientation by choosing $+1$ so that

$$a_1 \cdots a_n \cdot \sum_{k=1}^{n} b_k/a_k = 1 + b \cdot a_1 \cdots a_n. \qquad (1.5)$$

This equation implies in particular that the a_k are pairwise relatively prime. Moreover, by reducing (1.5) modulo a_k we see that the b_k are uniquely determined modulo a_k for each k. An easy argument using Kirby calculus shows that two integral homology spheres $M(b;(a_1,b_1),\dots,(a_n,b_n))$ and $M(b';(a_1,b'_1),\dots,(a_n,b'_n))$ as above are orientation preserving homeomorphic if and only if $b'_k = b_k$ modulo a_k for all k. Thus the invariants a_k together with equation (1.5) determine a unique integral homology sphere. It is orientation preserving homeomorphic to $\Sigma(a_1,\dots,a_n)$, see [242]. Also note that one can always choose invariants b_k of $M(b;(a_1,b_1),\dots,(a_n,b_n))$ so that $b = 0$; in this case the Seifert invariants are called *unnormalized*.

Another description of the homology spheres $\Sigma(a_1,\dots,a_n)$ is as follows. A closed oriented 3–manifold M admitting a fixed point free action of S^1 is called a *Seifert fibered manifold*, see [242]. For any Seifert fibered manifold M, its S^1–quotient is a closed Riemann surface F, not necessarily oriented. Note that the projection map $M \to F$ induces an epimorphism $H_1(M,\mathbb{Z}) \to H_1(F,\mathbb{Z})$, therefore, a necessary condition for M to be an integral homology sphere is that $F = S^2$.

All Seifert fibered homology spheres $\Sigma(a_1,\dots,a_n)$ are Seifert fibered manifolds. The S^1–action on $\Sigma(a_1,\dots,a_n)$ thought of as the link (1.2) is given by the formula

$$t(z_1,\dots,z_n) = (t^{q_1} z_1,\dots,t^{q_n} z_n),$$

where $t \in S^1$ and $q_k = a_1 \cdots a_n / a_k$. This action is fixed point free. The only points of $\Sigma(a_1,\dots,a_n)$ which may have non-trivial isotropy are those with one zero coordinate, say, $\{ z_k = 0 \}$. At such a point the isotropy group has order a_k. The fiber through such a point corresponds to the core circle $\{0\} \times S^1 \subset D^2_k \times S^1$ in the topological description of $\Sigma(a_1,\dots,a_n)$. If $a_k \geq 2$ this fiber is called a *singular fiber* of degree a_k. All other fibers are called *non–singular.*

Conversely, if M is a Seifert fibered manifold and an integral homology sphere at the same time then $M = \pm\,\Sigma(a_1,\dots,a_n)$ for some pairwise relatively prime $a_1,\dots,a_n$, see [242].

To finish with various descriptions of $\Sigma(a_1,\dots,a_n)$, we mention that they can also be viewed as homogeneous spaces, see Milnor [218] for $n = 3$ and Neumann–Raymond [242] for general case.

Let m be the number of parameters a_k in an integral homology sphere $\Sigma(a_1,\dots,a_n)$ such that $a_k \geq 2$. If $m \leq 2$ the origin is not a singular point of V_B and we immediately conclude that $\Sigma(a_1,\dots,a_n) = S^3$. If $m \geq 3$ the manifold $\Sigma(a_1,\dots,a_n)$ is never simply-connected. In the special case of $n = 3$ we get back the Brieskorn homology spheres $\Sigma(p,q,r)$, see Section 1.1.3.

1.1.5 Dehn Surgery on Knots

By a *knot* k in a closed oriented 3–manifold M we mean a smoothly embedded circle in M. Given a knot $k \subset M$, let $N(k)$ be its (closed) tubular neighborhood. By cutting M open along the 2–torus $\partial N(k)$ we obtain two manifolds – one is the *knot exterior* K which is the closure of $M \setminus N(k)$, and the other is the solid torus $N(k)$ which we identify with $D^2 \times S^1$. The group $\pi_1(K)$ is called the *knot group.*

One can use an arbitrary homeomorphism $h : \partial D^2 \times S^1 \to \partial K$ to glue $D^2 \times S^1$ back in K. The space resulting from this construction is a closed oriented manifold $Q = K \cup_h (D^2 \times S^1)$, in general different from M. We say that Q is obtained from M by *Dehn surgery.* The homeomorphism h maps the circle $\partial D^2 \times \{1\}$ into a simple closed (unoriented) curve $c \subset \partial K$. The curve c uniquely determines manifold Q once the knot k is fixed.

Suppose that k is a knot in S^3, then the knot exterior K has integral homology of the circle. Any simple closed curve m on the torus ∂K whose image in K generates $H_1(K, \mathbb{Z}) = \mathbb{Z}$ is called a *meridian* of k. Up to isotopy, there is a unique non–separating simple closed curve ℓ on ∂K which is homologically trivial in K. It can be represented as $\ell = F \cap \partial K$ where F is a Seifert surface of k in S^3. The curve ℓ is called a *canonical longitude* . The pair (m, ℓ) is quasi-oriented, that is, oriented up to a simultaneous change of orientations of both m and ℓ, by the requirement that $\mathrm{lk}(m, \ell) = 1$. Here lk is the linking number. If the knot k is oriented then m and ℓ are oriented by the convention that $\mathrm{lk}(m, k) = 1$. The curves m and ℓ form a basis in $H_1(\partial K, \mathbb{Z})$ so that every element in $H_1(\partial K, \mathbb{Z})$ is of the form $c = p \cdot m + q \cdot \ell$ with $p, q \in \mathbb{Z}$.

Let $h : \partial D^2 \times S^1 \to \partial K$ be a homeomorphism which sends $\partial D^2 \times \{1\}$ to a curve $c = p \cdot m + q \cdot \ell$. The manifold $Q = K \cup_h (D^2 \times S^1)$ is said to be obtained by *p/q–surgery* of S^3 along k and is denoted by $S^3 + (p/q) \cdot k$. A p/q–surgery is called *integral* if $q = \pm 1$ or $q = 0$; otherwise, it is called *rational.* Note that the pairs (p, q) and $(-p, -q)$ define the same manifold, as suggested by the notation, since orientation of c is of no importance to us. We also allow $q = 0$ with the understanding that in this case $p = 1$ and the result of $1/0$–surgery along any knot $k \subset S^3$ is S^3. Such surgery is called *trivial.* One can easily check that $|H_1(S^3 + (p/q) \cdot k, \mathbb{Z})| = \mathbb{Z}/|p|$. In particular, $S^3 + (1/q) \cdot k$ is always an integral homology sphere.

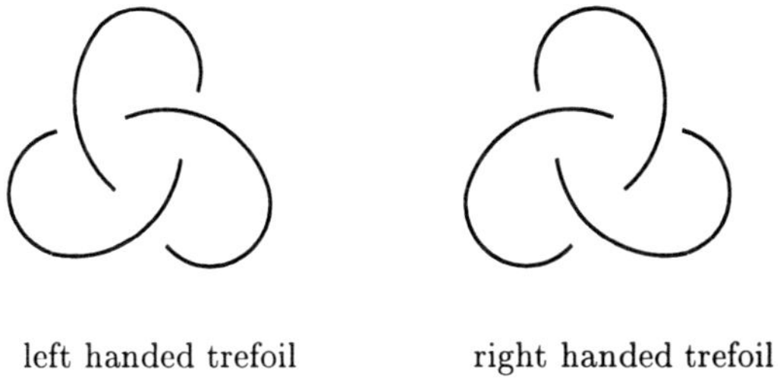

Fig. 1.2.

Example 1.1. Poincaré homology sphere $\Sigma(2, 3, 5)$ is obtained by (-1)–surgery on the left handed trefoil. The same manifold with opposite orientation is obtained by $(+1)$–surgery on the right handed trefoil, while (-1)–surgery on the right handed trefoil yields $\Sigma(2,3,7)$, see Figure 1.2.

Example 1.2. Brieskorn homology spheres $\Sigma(p,q,pqn-1)$ can be obtained by $(-1/n)$–surgery on the left–handed (p,q)–torus knot, and $\Sigma(p,q,pqn+1)$ by $(-1/n)$ surgery on the right–handed (p,q)–torus knot, where $n \geq 1$, see [231]. The integral homology 3-spheres $\Sigma(p,q,r,s)$ with $qr - ps = \pm 1$ are obtained by ± 1 surgery on the connected sum of the right handed (q,r)– and left-handed (p,s)–torus knots.

Example 1.3. The previous example describes all integral homology spheres that can be obtained by surgery on a torus knot – they all are Brieskorn homology spheres $\Sigma(p,q,pqm \pm 1)$. If a knot in S^3 is not torus then it is either hyperbolic or satellite, most knots being hyperbolic, see for instance [7] or [295]. According to Thurston [295] all but finitely many surgeries on a hyperbolic knot result in hyperbolic manifolds. In particular, all but finitely many $(1/q)$–surgeries, $q \in \mathbb{Z}$, result in hyperbolic homology spheres.

Example 1.4. A knot surgery presentation need not be unique. For example, any Brieskorn homology sphere $\Sigma(2,3,6m+1)$ can also be obtained by $+1$ surgery on a *twist knot* of type $(2n+2)_1$ shown in Figure 1.3 (when $n = 1$ this is the *figure eight knot*). Note that the knots $(2n+2)_1$, $n \geq 1$, are hyperbolic. Several infinite families of hyperbolic knots certain surgeries on which are Seifert fibered manifolds can be found in [65].

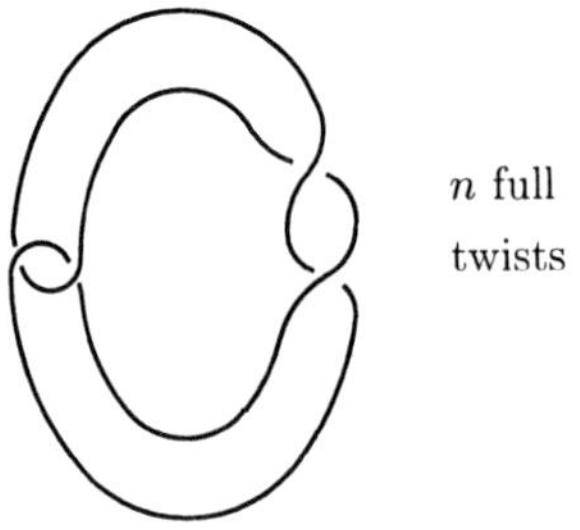

Fig. 1.3.

Not every integral homology sphere can be obtained by surgery on a knot in S^3, either integral or rational: according to Gordon and Luecke [131], any integral homology sphere obtained by Dehn surgery on a knot in S^3 is irreducible. First irreducible example was constructed by David Auckly in [16] by using Taubes' periodic end theorem [293]. He exhibited an integral homology sphere which is hyperbolic and hence irreducible, and which cannot be obtained by either integral or rational surgery on a knot in S^3. It is not known if every Brieskorn homology sphere can be obtained by surgery on a knot in S^3.

Gordon and Luecke [131] have shown that non-trivial Dehn surgery on a non-trivial knot in S^3 never yields S^3. Culler, Gordon, Luecke and Shalen [60] have shown that at most one, either $+1$ or -1, nontrivial surgery on a knot in S^3 can produce a homotopy sphere.

A knot k in S^3 is said to have *Property P* if non-trivial Dehn surgery on k does not yield a homotopy sphere. Many knots have been shown to have Property P, see Section 3.6.3. However, it is still not known if all non-trivial knots have this property. Proving this would be equivalent to showing that a counter-example to the Poincaré conjecture can not be obtained by surgery on a knot in S^3.

The surgery theory extends from knots in S^3 to knots in integral homology 3–spheres with only minor adjustments. The concepts of canonical meridians and longitudes etc. are valid in any integral homology 3–sphere.

Example 1.5. Let k_n be the singular fiber of degree a_n in a Seifert fibered homology sphere $\Sigma(a_1, \dots, a_n)$ then $(-1/m)$–surgery on k_n yields Seifert fibered homology sphere $\Sigma(a_1, \dots, a_n + m \cdot a)$ where $a = a_1 \cdots a_{n-1}$ is the product of the first $(n-1)$ Seifert invariants.

Example 1.6. A knot k in an integral homology sphere Σ is called *trivial* if it can be represented as a trivial knot in S^3 in a connected sum decomposition $\Sigma = \Sigma \# S^3$. Surgery along k yields the manifold $\Sigma \# \Sigma'$ where Σ' is obtained by surgery on k in S^3.

Example 1.7. Let k be a knot in an integral homology sphere Σ and let Σ' be the integral homology sphere obtained by (± 1)–surgery of Σ along k. Denote by k^* the image in Σ' of the canonical longitude of k. Then (∓ 1)–surgery of Σ' along k^* gives back Σ.

1.1.6 Surgery on Links

A *link* $\mathcal{L}$ in a closed oriented 3–manifold M is a collection of pairwise disjoint knots, $\mathcal{L} = k_1 \cup \ldots \cup k_m$. Performing surgery on a link $\mathcal{L}$ means performing Dehn surgery on each of the knots $k_1, \dots, k_m$. Suppose that $\mathcal{L}$ is a link in an integral homology sphere Σ then each of the knots $k_1, \dots, k_m$ has canonical meridian and longitude, and surgery on $\mathcal{L}$ can be described by an m–tuple $(p_1/q_1, \dots, p_m/q_m)$ where each of the p_i/q_i is a rational number or $1/0$. Such an m–tuple is referred to as a *framing* of $\mathcal{L}$. If all the p_i/q_i are integers or $1/0$, the surgery is called *integral*, otherwise it is called *rational*. The manifold resulting from Σ by doing surgery along $\mathcal{L}$ is denoted $\Sigma + (p_1/q_1) \cdot k_1 + \ldots + (p_m/q_m) \cdot k_m$ or $\chi_\Sigma(\mathcal{L})$. The following result is due to Lickorish [197] and Wallace [303].

Theorem 1.8. *Every closed oriented 3-manifold can be obtained by integral surgery on a link $\mathcal{L}$ in S^3. Moreover, the link $\mathcal{L}$ can be chosen so that all its individual components are unknotted and framed by ± 1.*

Proof. Every closed oriented 3–manifold M admits a Heegaard splitting $M = M_1 \cup_f M_2$ where M_1 and M_2 are handlebodies of genus g and $f : F \to F$ is an automorphism of $F = \partial M_1 = \partial M_2$. Every automorphism of the Riemann surface F is known to be a product of finitely many Dehn twists along simple closed curves, see Section 1.1.13. Multiplying the gluing automorphism f by a Dehn twist τ_c corresponds to performing an integral Dehn surgery on the knot $c \subset M$. Starting with a Heegaard splitting of S^3, the manifold M can be obtained by performing finitely many Dehn twists, thus performing integral surgery on a link in S^3. The second statement of the theorem is an exercise in the Kirby calculus.

Any presentation of a 3–manifold M by surgery on a framed link is called a *surgery presentation* of M. There are many surgery presentations of the same manifold. Any two integral surgery presentations of the same manifold can be related by a sequence of *Kirby moves* [159], consisting of blow ups, blow downs and handle slides, although finding such a sequence may be a complicated task. Any two rational surgery presentations are related by a sequence of generalized Kirby moves, see [77] and [263].

Let $\mathcal{L} = k_1 \cup \ldots \cup k_m$ be an *oriented* framed link in an integral homology sphere Σ whose ith component is framed by p_i/q_i, and let lk denote the linking number. The symmetric matrix $A(\mathcal{L}) = (\ell_{ij})$, $i, j = 1, \ldots, m$, with the entries

$$\ell_{ij} = \begin{cases} p_i/q_i, & \text{if } i = j, \\ \operatorname{lk}(k_i, k_j), & \text{if } i \neq j, \end{cases}$$

is called the *linking matrix* of $\mathcal{L}$. The matrix $F(\mathcal{L}) = (q_i \cdot \ell_{ij})$ is called a *framing matrix* of $\mathcal{L}$. The latter is a presentation matrix for $H_1(\chi_\Sigma(\mathcal{L}), \mathbb{Z})$, therefore, $\chi_\Sigma(\mathcal{L})$ is an integral homology sphere if and only if $\det F(\mathcal{L}) = \pm 1$ or, equivalently, $\det A(\mathcal{L}) = \pm 1/q_1 \cdots q_m$.

Theorem 1.8 can be ramified in several directions. Every closed oriented 3–manifold can be obtained by integral surgery on a link in S^3 all of whose framings are even, see Milnor [214] and Kaplan [156]. A link $\mathcal{L} = k_1 \cup \ldots \cup k_m$ in an integral homology sphere Σ is called *algebraically split* if $\operatorname{lk}(k_i, k_j) = 0$ for all $i \neq j$. Every integral homology sphere can be obtained by surgery on an algebraically split link in S^3 all components of which are framed by ± 1. This result is based on the observation that all elementary transformations needed to diagonalize the linking matrix can be realized by the Kirby moves.

A stronger result in this direction was proved by Hilden [143]. He showed that every integral homology 3-sphere can be obtained by surgery on a framed link $\mathcal{L} \subset S^3$ of the following special type: $\mathcal{L}$ has $2n$ components $k_1, \ldots, k_{2n}$, where for each $i = 1, \ldots, n$, the components $k_{2i-1} \cup k_{2i}$ are the boundary curves of an orientable surface F_i of genus 1, and also the F_i are pairwise disjoint. One of the boundary components of each of the F_i is framed by 1, the other by -1. Matveev [210] improved this result by showing that a 3-dimensional manifold is an integral homology sphere if and only if it can be obtained from S^3 by surgery on a link whose components are framed by ± 1 and bound pairwise disjoint Seifert surfaces of genus 1.

1.1.7 Connected Sums and Splicing

Given two oriented integral homology spheres, Σ_0 and Σ_1, their *connected sum* $\Sigma_0 \# \Sigma_1$ is also an integral homology sphere which inherits the same orientation from each of Σ_0 and Σ_1. A surgery presentation of $\Sigma_0 \# \Sigma_1$ is a disjoint union of those of Σ_0 and Σ_1. If an integral homology sphere Σ is homeomorphic to a connected sum $\Sigma_0 \# \Sigma_1$ such that neither summand is S^3, we say that Σ is *reducible*. Otherwise, we say that Σ is *irreducible*. According to [220], every integral homology sphere is a connected sum of irreducible homology spheres other that S^3 in an essentially unique way.

Let $k_0 \subset \Sigma_0$ and $k_1 \subset \Sigma_1$ be oriented knots in oriented integral homology spheres, and let $N(k_0)$ and $N(k_1)$ be their tubular neighborhoods. Denote by $m_0, \ell_0 \subset \partial N(k_0)$ and $m_1, \ell_1 \subset \partial N(k_1)$ canonical meridians and longitudes. Form

$$\Sigma = (\Sigma_0 \setminus \operatorname{int} N(k_0)) \cup (\Sigma_1 \setminus \operatorname{int} N(k_1)) \tag{1.6}$$

pasting along boundaries by matching m_0 to ℓ_1 and ℓ_0 to m_1. An easy Mayer–Vietoris sequence argument shows that Σ is an integral homology sphere. Since the pasting map reverses orientation, Σ inherits the same orientation from each of Σ_0 and Σ_1. The manifold Σ is called the *splice* of Σ_0 and Σ_1 along knots k_0 and k_1.

Splicing can be described in terms of surgery presentations as follows. Suppose that Σ_0 and Σ_1 are given by surgery on oriented framed links $\mathcal{L}_0$ and $\mathcal{L}_1$ in S^3 so that their linking matrices have determinant 1 (if the determinant is -1 or if either of the links is empty, blow it up once). Assign zero framing to $k_0 \subset \Sigma_0$ (if the knot k_0 is drawn inside the surgery presentation $\mathcal{L}_0$, it will be framed by the integer a found from the condition that the linking matrix of $\mathcal{L}_0 \cup k_0$ has zero determinant). Frame k_1 similarly as a knot inside the surgery presentation $\mathcal{L}_1$. The surgery presentation of the splice Σ is then obtained from a disjoint union of framed links $\mathcal{L}_0 \cup k_0$ and $\mathcal{L}_1 \cup k_1$ by linking the oriented knots k_0 and k_1 geometrically once so that their linking number is 1, see Figure 1.4.

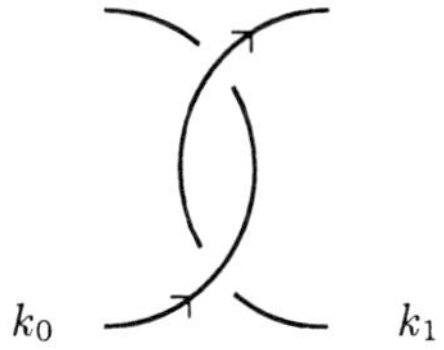

Fig. 1.4.

Connected sum is a special case of splicing. Namely, given two integral homology spheres Σ_0 and Σ_1, their splice along trivial knots is $\Sigma_0 \# \Sigma_1$. Another curious relation between connected sum, splicing, and Dehn surgery is as follows. Let k_0 and k_1 be oriented knots in integral homology spheres Σ_0 and Σ_1, and k_0^* and k_1^* the images of their canonical longitudes in the surgered manifolds $\Sigma_0 - k_0$ and $\Sigma_1 - k_1$, respectively. Then the splice of Σ_0 and Σ_1 along the knots k_0 and k_1 is the result of $(+1)$-surgery of $(\Sigma_0 - k_0)\#(\Sigma_1 - k_1)$ along the knot $k_0^* \# k_1^*$. This is readily seen from the description of splicing given above. Another proof can be found in [109].

The splicing is a special case of a more general operation on integral homology spheres, when the exteriors of knots $k_0 \subset \Sigma_0$ and $k_1 \subset \Sigma_1$ are glued together along an arbitrary orientation reversing homeomorphism of their boundary tori. In the canonical meridian-longitude bases (m_0, ℓ_0) and (m_1, ℓ_1), such a homeomorphism is given by the matrix

$$A = \begin{pmatrix} q & u \\ p & v \end{pmatrix} \tag{1.7}$$

with $\det A = -1$. Our convention is that the first column of A represents the image of ℓ_0 and the second column the image of m_0 in terms of ℓ_1 and m_1. The resulting manifold is an integral homology sphere if and only if $p = \pm 1$; the splicing corresponds to $p = u = 1$ and $q = v = 0$.

1.1.8 Splice Decomposition

A modern point of view on integral homology 3-spheres as presented for example in [76] is that every such manifold can be obtained from certain canonical pieces with the help of just two operations, which are connected sum and splicing. This approach is briefly explained in this section.

For the purposes of this section, a *link* $\mathcal{L} = (\Sigma, K) = (\Sigma, k_1 \cup \ldots \cup k_m)$ will be a pair consisting of an integral homology sphere Σ and a collection K of disjoint oriented knots $k_1, \ldots, k_m$ in Σ. Empty links are also allowed; they are just homology spheres. If $\mathcal{L} = (\Sigma, K)$ and $\mathcal{L}' = (\Sigma', K')$, we can form the disjoint sum $\mathcal{L} + \mathcal{L}' = (\Sigma \# \Sigma', K \cup K')$ by taking connected sum of Σ and Σ' along discs which do not intersect K and K'. A link is called reducible if it can be expressed as a disjoint sum such that neither summand is an empty link in S^3; otherwise, it is called *irreducible*. Because of Milnor's unique factorization theorem [220], for most purposes no generality is lost by considering only irreducible links.

Given links $\mathcal{L} = (\Sigma, K)$ and $\mathcal{L}' = (\Sigma', K')$, choose components $k \subset K$ and $k' \subset K'$. Splicing Σ and Σ' along k and k' as described in (1.6) results in an integral homology sphere Σ''. The link $(\Sigma'', (K \setminus k) \cup (K' \setminus k'))$ is called the splice of $\mathcal{L}$ and $\mathcal{L}'$ along k and k'. A link $(\Sigma(a_1, \ldots, a_n), k_1 \cup \ldots \cup k_m)$ where the link components $k_1, \ldots, k_m$ are fibers in $\Sigma(a_1, \ldots, a_n)$ is said to be a *Seifert link*. Note that a general Seifert link may include singular fibers as well as some non-singular fibers. A *simple* link is defined to be an irreducible link $\mathcal{L} = (\Sigma, K)$ with the property that any incompressible torus in $\Sigma \setminus \operatorname{int} N(K)$ is boundary parallel. By Thurston's hyperbolization theorem [295] this is equivalent to saying that the link complement $\Sigma \setminus K$ admits a complete hyperbolic structure of finite volume, except possibly if K is empty and Σ is not sufficiently large.

The splitting theorem of Jaco and Shalen [152] and Johannson [153] implies that any irreducible link $\mathcal{L} = (\Sigma, K)$ can be expressed as a result of splicing together a collection of Seifert links and simple links, and moreover the minimal way of doing this is essentially unique.

More precisely, Eisenbud and Neumann show in [76] the following. Given a link $\mathcal{L}'' = (\Sigma'', K'')$ and an embedded torus $T^2 \subset \Sigma'' \setminus K''$, one can "desplice" $\mathcal{L}''$ along T^2, that is, represent $\mathcal{L}''$ as the result of splicing links (Σ, K) and (Σ', K') along knots k and k'. This operation is defined uniquely up to reversing the orientations of both k and k'. Moreover, if $\Sigma'' = S^3$ then $\Sigma = \Sigma' = S^3$ as well. The splitting theorem of Jaco-Shalen and Johannson can now be interpreted as follows. Given an irreducible link $\mathcal{L} = (\Sigma, K)$, there exists a collection of disjoint incompressible non-boundary-parallel embedded tori in the link exterior $\Sigma \setminus \operatorname{int} N(K)$ such that if one "desplices" as above along all these tori, each resulting link is either a Seifert link or a simple link. Furthermore, a minimal such collection of tori is included up to isotopy in any such collection, and is therefore unique.

Any link which can be obtained from a finite number of Seifert links by splicing is called a *graph link*. In other words, graph links admit splicing decomposition without simple components. Empty graph links are called *graph homology spheres*. Graph links with one component are called *graph knots*. All irreducible graph links are classified as in [76] by their splice diagrams.

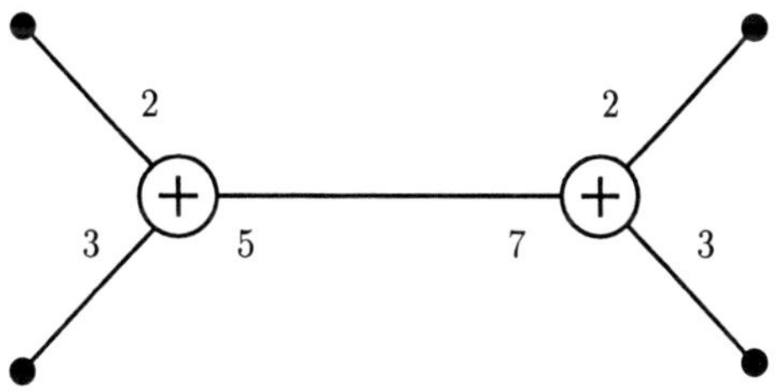

Fig. 1.5.

A *splice diagram* is a finite tree with vertices of three types. Vertices with at least three adjacent edges are called *nodes*. Each node with n adjacent edges corresponds to a Seifert fibered homology sphere $\Sigma(a_1, \ldots, a_n)$. The edges adjacent to the node correspond to the singular fibers and are weighted respectively by $a_1, \ldots, a_n$. The node itself carries a plus sign if $\Sigma(a_1, \ldots, a_n)$ is oriented as the link of singularity (1.2) and minus sign otherwise. The other two types of vertices have only one adjacent edge, and they are either boundary vertices or arrowheads. The former represent a singular fiber in a Seifert manifold, and the latter – components of the graph link.

Example 1.9. The manifold obtained by splicing together $\Sigma(2,3,5)$ and $\Sigma(2,3,7)$ along singular fibers of degrees 5 and 7 is described by the splice diagram shown in Figure 1.5.

Example 1.10. The splice of Seifert links may be a Seifert link again. For any j such that $2 \le j \le n-2$, the homology sphere $\Sigma(a_1,\dots,a_n)$ can be obtained by splicing $\Sigma(a_1,\dots,a_j,p)$ and $\Sigma(q,a_{j+1},\dots,a_n)$ along the singular fibers of degrees $p = a_{j+1}\cdots a_n$ and $q = a_1\cdots a_j$, see [76].

1.1.9 Plumbing

Graph homology spheres defined in Section 1.1.8 can be conveniently described by plumbing. Let Γ be a *plumbing graph*, that is a graph with no cycles, not necessarily connected, each of whose vertices x_i carries an integer weight e_i, $i = 1,\dots,s$. Associate to each vertex x_i the D^2–bundle $Y(e_i)$ over S^2 with the Euler number e_i. If the vertex x_i has d_i edges connected to it in the graph Γ, choose d_i disjoint discs in the base of $Y(e_i)$ and call the disc bundle over the jth disc B_{ij}, so that $B_{ij} = D^2 \times D^2$. When two vertices, x_i and x_k, are connected by an edge, identify B_{ij} with $B_{k\ell}$ by exchanging the base and fiber coordinates. This pasting operation is called *plumbing*, and the resulting smooth 4–manifold $P(\Gamma)$ is said to be obtained by *plumbing according to* Γ.

The 4–manifold $P(\Gamma)$ is simply connected. The group $H_2(P(\Gamma),\mathbb{Z})$ has a natural basis represented by the zero-sections of the plumbed bundles. We note that all these sections are embedded 2-spheres, and that they can be oriented in such a way that the intersection form on the second homology of $P(\Gamma)$, see Section 2.1, is given by the matrix $A(\Gamma) = (a_{ij})_{i,j=1,\dots,s}$ with the entries

$$a_{ij} = \begin{cases} e_i, & \text{if } i = j; \\ 1, & \text{if } i \text{ is connected to } j \text{ by an edge;} \\ 0, & \text{otherwise.} \end{cases} \tag{1.8}$$

The 3–manifold $M = \partial P(\Gamma)$ is an integral homology sphere if and only if the matrix $A(\Gamma)$ is unimodular, that is, $\det A(\Gamma) = \pm 1$. In this situation M is called a *plumbed* (integral) homology sphere.

The manifold $P(\Gamma)$ has a very simple surgery description. Every vertex x_i of the plumbing graph Γ should be replaced by an unknot framed by e_i, and the unknots corresponding to every two vertices connected by an edge should be linked geometrically once, see Figure 1.6. Since Γ has no cycles, it does not matter which of the two possible ways to link the circles we may choose. We end up with a link in S^3 which, if properly oriented, has linking

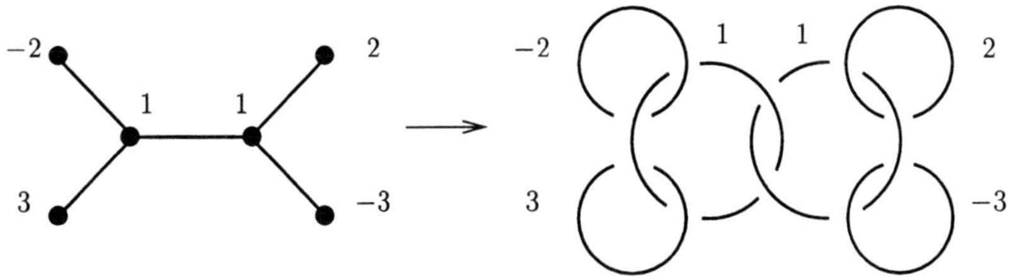

Fig. 1.6.

matrix $A(\Gamma)$. The manifold $P(\Gamma)$ is obtained by attaching 2–handles to the 4–ball according to this link, and its boundary $\partial P(\Gamma)$ is obtained by Dehn surgery on the same link.

Example 1.11. Every Seifert fibered homology sphere $\Sigma(a_1, \dots, a_n)$ can be obtained by plumbing on a star–shaped graph shown in Figure 1.7. The integer weights in the graph are found as follows. We first find a collection of Seifert invariants b, (a_1, b_1),..., (a_n, b_n) from the equation

$$a_1 \cdots a_n \cdot \left(-b + \sum_{i=1}^{n} b_i/a_i \right) = 1.$$

The weights t_{ij} are then found from the continued fractions $[t_{i1}, \dots, t_{im_i}] = a_i/b_i$ where $i = 1, \dots, n$ and we use the continued fraction notation

$$[t_1, \dots, t_k] = t_1 - \cfrac{1}{t_2 - \cfrac{1}{\cdots - \cfrac{1}{t_k}}} \tag{1.9}$$

The equivalence of this description of $\Sigma(a_1, \dots, a_n)$ to the one in Figure 1.1 should be clear from the surgery description of the plumbing given above.

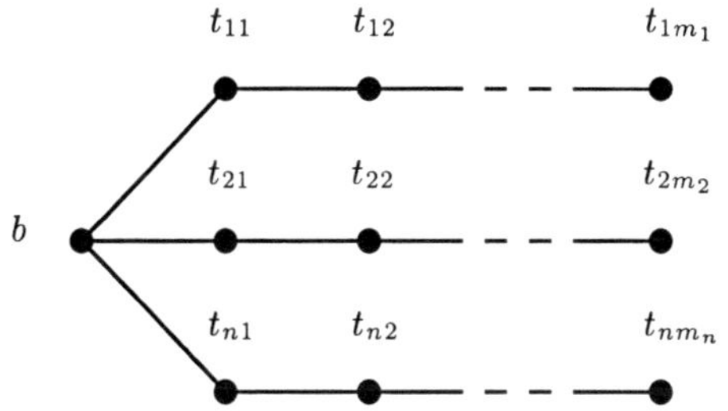

Fig. 1.7.

The same manifold M can be described by various plumbing graphs Γ. One can use the Kirby moves to move between any two plumbing graphs describing M. In fact, the full strength of the Kirby calculus is not needed in this situation: any two plumbing graphs Γ_1 and Γ_2 such that $\partial P(\Gamma_1) = \partial P(\Gamma_2)$ can be connected by a sequence of the simpler moves shown in Figure 1.8, see for example [241].

1a. Delete a component of Γ_1 consisting of an isolated vertex with weight ± 1.

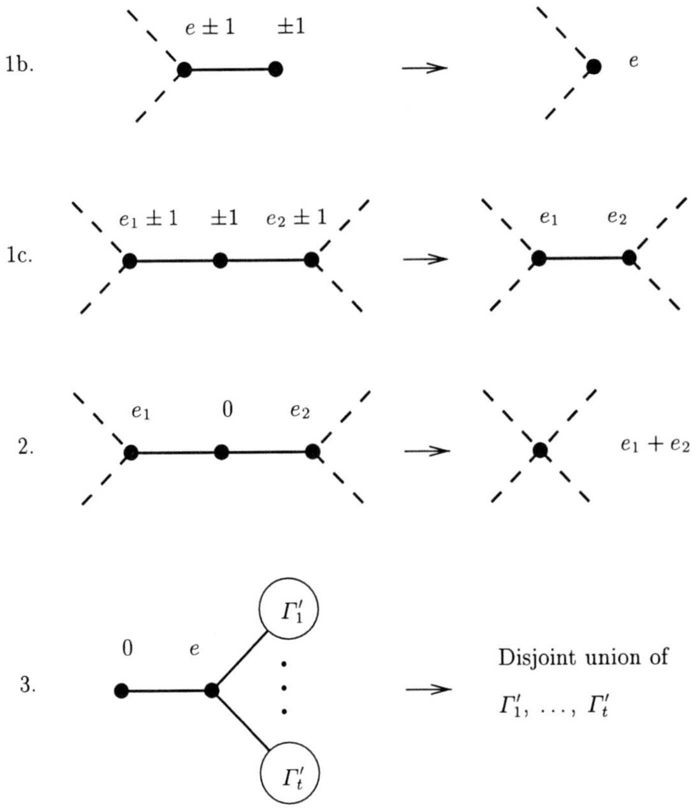

Fig. 1.8.

Example 1.12. Brieskorn homology sphere $\Sigma(3,4,7)$ can be obtained by plumbing along either of the graphs shown in Figure 1.9. For the graph on the left, we used $b = 0$, $b_1 = b_2 = 1$, $b_3 = -4$ and continued fractions $3/1 = [3]$, $4/1 = [4]$, $-7/4 = [-2,-4]$. For the graph on the right, $b = -2$,

$b_1 = -2$, $b_2 = -3$, $b_3 = -4$, and $-3/2 = [-2,-2]$, $-4/3 = [-2,-2,-2]$, $-7/4 = [-2,-4]$.

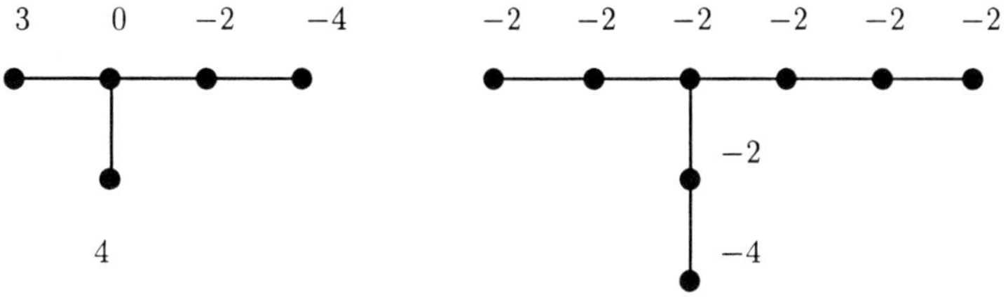

Fig. 1.9.

The splicing operation for plumbed homology spheres can be described as follows. Let $k \subset \Sigma$ be a graph link, see Section 1.1.8. It can be represented by a plumbing diagram with arrowhead vertices attached to the boundary vertices corresponding to the link components.

Example 1.13. Figure 1.10 shows Seifert link (S^3, k) where k is the right handed trefoil, corresponding to a regular fiber in a Seifert fibration of S^3. Figure 1.11 shows Seifert link $(\Sigma(2,3,7), k)$ with k representing the singular fiber of degree 7.

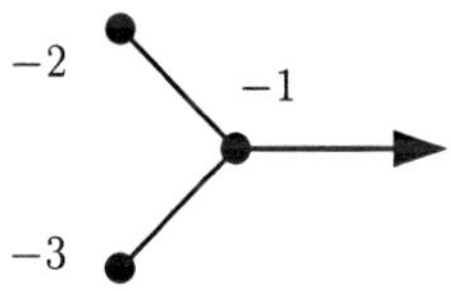

Fig. 1.10.

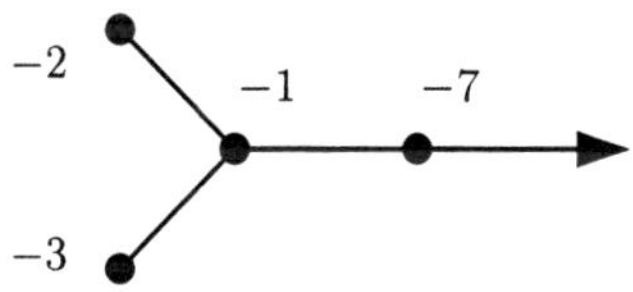

Fig. 1.11.

Suppose we are given two graph links represented by their plumbing diagrams Γ and Γ' as shown in Figure 1.12 with arrows attached to vertices weighted by e_n and e'_m respectively. The plumbing diagram for the spliced link is shown in Figure 1.13.

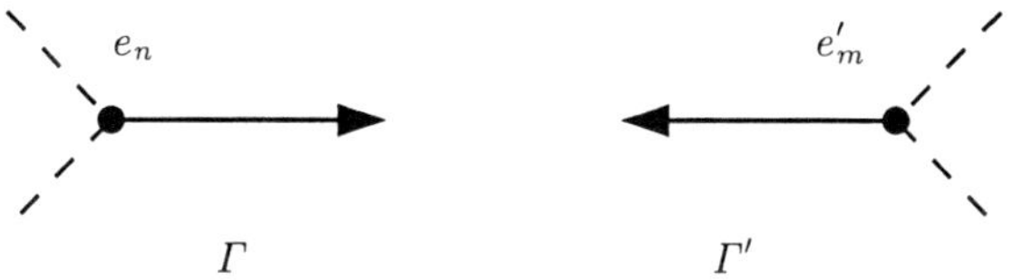

Fig. 1.12.

Here, $a = \det A(\Gamma_0)/\det A(\Gamma)$ where Γ_0 is the portion of the plumbing graph Γ obtained by removing the nth vertex weighted by e_n and all adjacent edges, and similarly for a'.

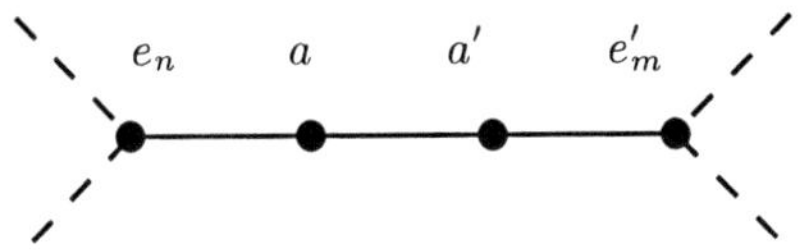

Fig. 1.13.

Example 1.14. The plumbed manifold shown in Figure 1.6 is obtained by splicing together Brieskorn homology spheres $\Sigma(2,3,5)$ and $\Sigma(2,3,7)$ along singular fibers of degrees 5 and 7. This follows after applying the rules of splicing described above and simplifying the resulting manifold with the help of the Kirby calculus. The same manifold is described by the splice diagram shown in Figure 1.5.

Given a plumbing graph Γ, we call a vertex of Γ a *node* if there are three or more edges of the graph meeting at that vertex. Splicing k Seifert fibered homology spheres along Seifert fibers produces a plumbing manifold with at most k nodes, compare with Example 1.10. Conversely, any plumbed homology sphere with k nodes can be obtained by splicing together k Seifert fibered homology spheres along Seifert fibers. Precise description of this "desplicing" operation can be found in [76]. Thus Seifert fibered homology spheres can be thought of as building blocks for plumbed homology spheres. This proves in particular that, for integral homology spheres, the concepts of plumbed and graph homology spheres are equivalent.

1.1.10 Links of Singularities

Given m analytic functions $f_1, \ldots, f_m : (\mathbb{C}^n, 0) \to (\mathbb{C}, 0)$ let $X = \{ f_1 = 0, \ldots, f_m = 0 \}$ be such that the matrix

$$J(p) = \left(\frac{\partial f_i}{\partial z_j} \right) (p)$$

has rank $n-2$ at all points $p \in X$ except perhaps at 0. If rank of $J(0)$ is also equal to $n-2$, we say that $(X,0)$ is smooth, that is, analytically isomorphic to $(\mathbb{C}^2, 0)$. Otherwise, $(X,0)$ is called an *isolated surface singularity.* We will assume that $(X,0)$ is irreducible and normal. For example, if $f : (\mathbb{C}^3, 0) \to (\mathbb{C}, 0)$ satisfies $\nabla f(p) \neq 0$ for any p in a neighborhood of the origin, then $(X,0) = (\{ f = 0 \}, 0)$ is a normal surface singularity.

Fix an arbitrary embedding of $(X, 0)$ into $(\mathbb{C}^n, 0)$. Then there exists ε_0 such that for any $0 < \varepsilon \leq \varepsilon_0$ the $(2n-1)$–dimensional sphere $S_\varepsilon^{2n-1} = \{\, z \in \mathbb{C}^n \mid |z| = \varepsilon \,\}$ intersects $(X, 0)$ transversally. In particular, for such an ε, the intersection

$$\Sigma = X \cap S_\varepsilon^{2n-1}$$

is a closed 3-dimensional real manifold. It does not depend on the embedding $(X, 0) \subset (\mathbb{C}^n, 0)$ and on ε. It is canonically oriented as the boundary of a (singular) complex surface by the rule "outside normal first". The manifold Σ is called the *link of singularity* of $(X, 0)$.

Example 1.15. If $(X, 0)$ is smooth, its link is S^3. All Seifert fibered homology spheres are links of normal surface singularities, see (1.1) and (1.2). The link of singularity at the origin of the polynomial $f(x, y, z) = x^2 + y^2 + z^n$ is the lens space $L(n, n-1)$ with $H_1(L(n, n-1), \mathbb{Z}) = \mathbb{Z}/n$ and in particular is not an integral homology sphere.

Given a normal isolated surface singularity $(X, 0)$, let Σ be its link. Choose a "good resolution" $(\tilde{X}, E) \to (X, 0)$, that is one for which the exceptional curve $E = \cup E_i$ is a union of smooth curves intersecting transversally. Then Σ is obtained by plumbing according to the *resolution dual graph* Γ of E. The link Σ is an integral homology sphere if and only if all E_i are rational, the graph Γ has no loops, and the intersection matrix $A(\Gamma)$ has determinant ± 1.

Thus all links of singularities which are integral homology spheres are plumbed, or graph, homology spheres. Conversely, an integral plumbed homology 3–sphere is a link of singularity if and only if it can be obtained by plumbing on a plumbing graph Γ whose intersection matrix $A(\Gamma)$ is negative definite and unimodular.

Example 1.16. Poincaré homology sphere is the link of Brieskorn singularity $x^2 + y^3 + z^5 = 0$. It is the boundary of the 4–manifold $P(\Gamma)$ obtained by plumbing on the graph Γ shown in Figure 1.14. The matrix $A(\Gamma)$ is commonly denoted by E_8. It is negative definite and unimodular and it is non–diagonalizable over the integers.

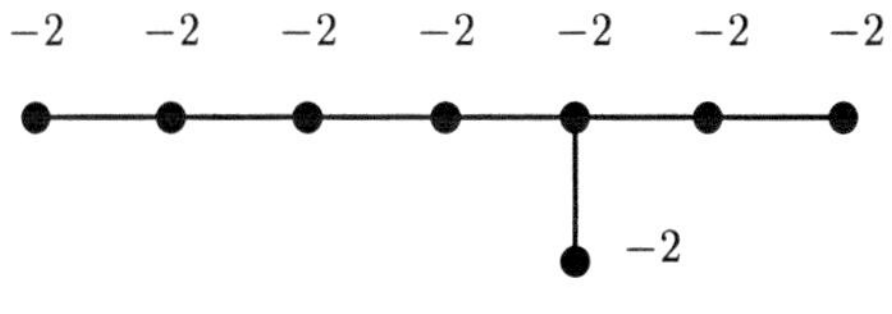

Fig. 1.14.

Example 1.17. In general, the resolution dual graph of the singularity (1.2), whose link is Seifert fibered homology sphere $\Sigma(a_1, \dots, a_n)$, is of the form shown in Figure 1.7. The weights t_{ij} are uniquely determined by the conditions $t_{ij} \le -2$ and $[t_{i1}, \dots, t_{im_i}] = a_i/b_i$ where

$$-a_i < b_i < 0, \quad \text{and} \quad a_1 \cdots a_n \cdot b_i/a_i = 1 \bmod a_i, \tag{1.10}$$

so that

$$b = -1/a_1 \dots a_n + \sum_{i=1}^{n} b_i/a_i \le -1. \tag{1.11}$$

Plumbing on any graph as in Figure 1.7 gives a 4–manifold with boundary $\Sigma(a_1, \dots, a_n)$; the plumbing graph with weights defined by (1.10) and (1.11) is special in that its intersection matrix $A(\Gamma)$ is negative definite and it is minimal (i.e. admits no (-1)–blow downs).

Example 1.18. The splice of Brieskorn homology spheres $\Sigma(p, q, r)$ and $\Sigma(p', q', r')$ along the singular fibers of degrees r and r' is the link of a singularity if and only if $rr' > pp'qq'$.

Example 1.19. Let $g(x, y, z) = f(x, y) + z^n$ define an analytic map $g : (\mathbb{C}^3, 0) \to (\mathbb{C}, 0)$ with an isolated singularity at 0 and let Σ be its link. Then Σ is an integral homology sphere if and only if the plane curve $f(x, y) = 0$ has only one branch at 0 and n is prime to each entry of each essential Puiseaux pair for this plane curve, see [76], page 49. In this case Σ can be explicitly described by its splicing diagram, see [243], Proposition 2.1.

1.1.11 Mutations

Let F be a closed oriented surface of genus two, and $\tau : F \to F$ an involution which makes F into a two-fold branched cover of S^2 with branch set consisting of six points, see Figure 1.15. Given an embedding of F into an oriented 3–manifold M, one can construct another 3-manifold M^τ by cutting M open along F and regluing two boundary components using τ. We say that M^τ is obtained from M by *mutation* along F, or that M^τ is a *mutant* of M. If M is an integral homology sphere then so is its mutant M^τ since $\tau_* : H_1(F, \mathbb{Z}) \to H_1(F, \mathbb{Z})$ is minus the identity.

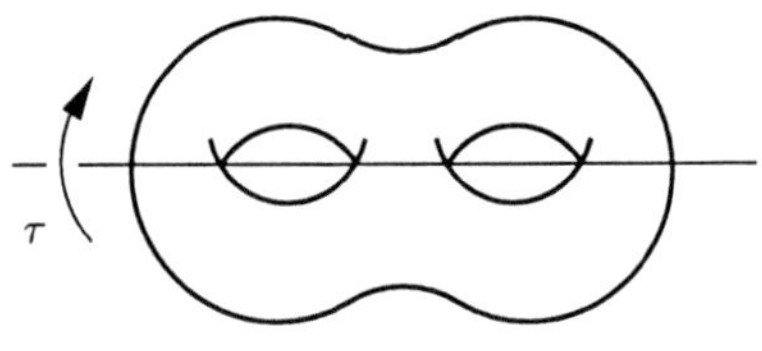

Fig. 1.15.

Mutation was introduced in [265] as an analogue for closed 3-manifolds of the mutation of knots described by Conway. If $k \in S^3$ is a knot and $S^2 \subset S^3$ is a sphere intersecting k in four points, then a *mutant* of k is a knot k^τ obtained by cutting S^3 open along S^2 and regluing using one of the three orientation preserving involutions shown in Figure 1.16 (the sphere in Figure 1.16 is punctured at the intersection points with k). The sphere S^2 is sometimes called the *Conway sphere.*

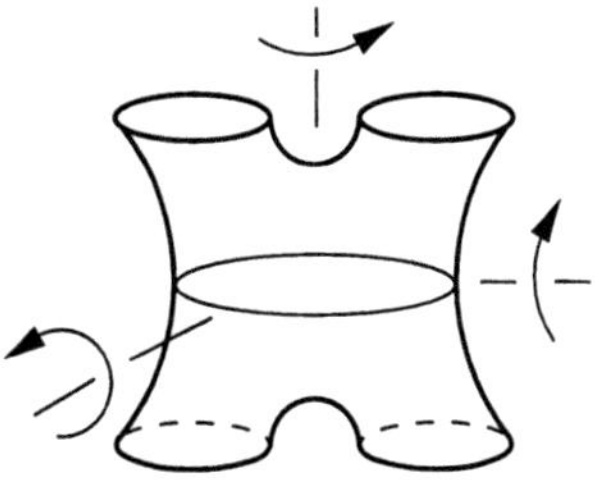

Fig. 1.16.

Observe that a specific choice of τ gives a pair of S^0 s on the knot; each S^0 is preserved by τ. A mutation τ is called *unlinked* if these S^0 s are unlinked on the knot k. In the special case of unlinked mutation, the manifold obtained by p/q–surgery on k^τ is the mutant of the manifold obtained by p/q–surgery on k. In this case the genus two surface is the union of the Conway sphere with four punctures and two tubes which follow the two arcs of k that lie outside of the Conway sphere.

There are no general methods for determining whether a 3–manifold differs from its mutant as most known invariants of M are preserved under mutation, see for instance [264], [265], and also Proposition 2.16 and Theorems 3.20, 4.11 and 6.68. It is known that if a homology sphere M is homeomorphic to S^3 then also $M^\tau = S^3$. However, in general M^τ need not be homeomorphic to M. Examples of integral homology spheres M such that $M \neq M^\tau$ can be found in [163].

1.1.12 Branched Covers

Let Σ be an integral homology sphere, k a knot in Σ, and $K = \Sigma \setminus \operatorname{int} N(k)$ its exterior. An n–fold *cyclic branched cover* of Σ with branch set k is the 3–manifold

$$\Sigma_n = \tilde{K}_n \cup_h N(k),$$

where $\tilde{K}_n$ is the n–fold regular (unbranched) cover of K with $\pi_1(\tilde{K}_n)$ the kernel of the abelianization homomorphism $\pi_1(K) \to H_1(K, \mathbb{Z}) = \mathbb{Z}$ followed by the mod n reduction homomorphism $\mathbb{Z} \to \mathbb{Z}/n$. The boundary of $\tilde{K}_n$ is a torus, and the gluing map $h : \partial\tilde{K}_n \to \partial N(k) = \partial K$ is a homeomorphism such that $h(\tilde{m}) = m$ where $m \subset \partial K$ is a meridian of k and $\tilde{m}$ is its lift to $\partial \tilde{K}_n$.

The manifold Σ_n is closed and oriented. The group $H_1(\Sigma_n, \mathbb{Z})$ is finite if and only if no root of the Alexander polynomial $\Delta_k(t)$ of the knot k is an n th root of unity ζ_i, $i = 1, \dots, n$, and in this case

$$|H_1(\Sigma_n, \mathbb{Z})| = \left| \prod_{i=0}^{n-1} \Delta_k(\zeta_i) \right|, \tag{1.12}$$

see Theorem 8.21 of [44]. Here, the Alexander polynomial $\Delta_k(t)$ should be normalized so that there are no negative powers of t and the constant term is positive. A result usually attributed to Fox, see Lemma 2.10 of [126], says that for any knot k and for any positive integers n and r, there exists an epimorphism $\varphi : H_1(\Sigma_{nr}, \mathbb{Z}) \to H_1(\Sigma_r, \mathbb{Z})$. In particular, if Σ_{nr} is an integral homology sphere then so is Σ_r.

Note that the manifold Σ_n is naturally equipped with an orientation preserving PL homeomorphism $\tau : \Sigma_n \to \Sigma_n$ of order n which acts as a covering transformation on $\tilde{K}_n$ and as a meridional rotation through $(360/n)^\circ$ on $N(k)$ thus fixing the knot k. The quotient map $\pi : \Sigma_n \to \Sigma$ is a one–to–one map on the knot k and a regular n-fold branched cover away from it.

Conversely, given a closed oriented 3–manifold M and an orientation preserving PL homeomorphism $\tau : M \to M$ of order n, the quotient map $\pi : M \to M/\tau$ makes M into an n–fold cyclic cover of M/τ, branched or not. If M is an integral homology sphere then M/τ is always a rational homology sphere. If the fixed point set of τ is non-empty then M/τ is in fact an integral homology sphere. Note that the requirement that τ be PL is essential. Even though any homeomorphism of a closed 3–manifold is isotopic to a PL homeomorphism, it is no longer true equivariantly, see for instance [179]. If we admit group actions which are not PL we may end up with wildly knotted fixed point sets.

Example 1.20. For any integer $n \geq 1$, let $\tau : S^3 \to S^3$ be a $(360/n)^\circ$ rotation about an axis in S^3, which is viewed as a one-point compactification of $\mathbb{R}^3$. It defines a $\mathbb{Z}/n$–action on S^3 whose quotient space is S^3 thus making S^3 into an n–fold cyclic branched cover of itself with branch set a trivial knot. The corresponding projection map $\pi : S^3 \to S^3$ can be thought of as a double suspension of the map $S^1 \to S^1$ of the unit circle in the complex plane given by $z \mapsto z^n$.

Example 1.21. Suppose that k is a knot in S^3 with trivial Alexander polynomial, $\Delta_k(t) = 1$, for example, k is an untwisted Whitehead double, see Section 3.4.3. Then any branched cover of S^3 with branch set k is an integral homology sphere.

Example 1.22. Any Brieskorn homology sphere $\Sigma(p, q, r)$ is an r-fold cyclic branched cover of S^3 with branch set a (p, q)–torus knot. If $\Sigma(p, q, r)$ is viewed as the link of singularity (1.1) then the $\mathbb{Z}/r$–action on $\Sigma(p,q,r)$ can be given by the formula $(x,y,z) \mapsto (x,y,e^{2\pi i/r}z)$. A similar action on $\Sigma(a_1,\dots,a_n)$ viewed as the link (1.2) makes it into an a_n–fold cyclic branched cover of $\Sigma(a_1,\dots,a_{n-1})$ with branch set a regular fiber in $\Sigma(a_1,\dots,a_{n-1})$.

Example 1.23. Generalizing the previous example, consider the link Σ of an isolated singularity at the origin of $f(x,y)+z^n$, compare with Example 1.19. Suppose that Σ is an integral homology sphere, then the action $(x,y,z) \mapsto (x,y,e^{2\pi i/n}z)$ on Σ turns it into an n–fold branched cover of S^3 with branch set the link of singularity at 0 of $f(x,y)$. This branch set is an iterated torus knot and therefore a graph knot.

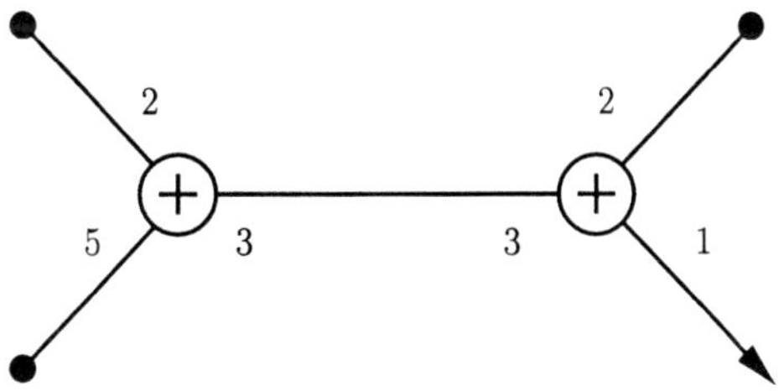

Fig. 1.17.

Example 1.24. In general, branched covers of graph knots are described as follows. Suppose that a graph knot $k \subset \Sigma$ is given by its splice diagram Γ, see Section 1.1.8 and [76]. Every internal edge of Γ has two weights, and every external edge has one. Call a weight on an edge of Γ "near" (resp. "far") if it is on the end of the edge nearest to (resp. furthest from) the arrowhead of Γ representing the knot k. According to [243], Theorem 2.4, the n–fold cyclic branched cover Σ_n of Σ is an integral homology sphere if and only if n is prime to all near weights in Γ, and Σ_n is then given by the splice diagram obtained from Γ by multiplying each far weight by n.

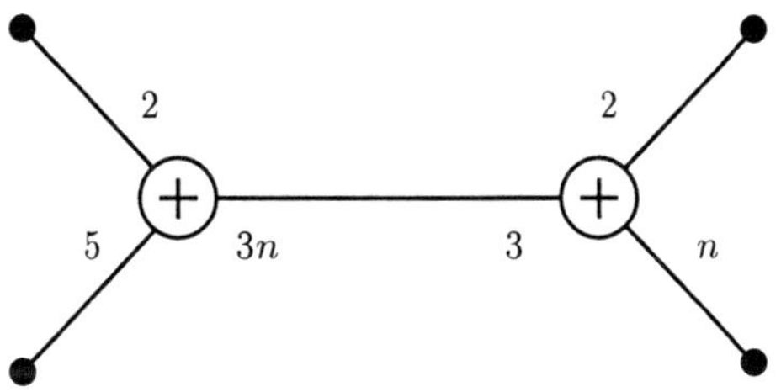

Fig. 1.18.

For example, let n be relatively prime to 2, 3, and 5, then the n–fold branched cover of the graph knot shown in Figure 1.17 is an integral homology sphere given by the splice diagram in Figure 1.18. This is a splice of $\Sigma(2,5,3n)$ and $\Sigma(2,3,n)$ along the singular fibers of degrees $3n$ and 3, respectively.

An ample source of examples of (double) branched covers is Dehn surgery on strongly invertible links, see [225]. A link $\mathcal{L}$ in S^3 is called *strongly invertible* if there is an orientation preserving involution of S^3 which preserves $\mathcal{L}$ and induces on its every component an involution with exactly two fixed points.

Fig. 1.19.

Let M be a 3–manifold obtained by surgery on a strongly invertible link $\mathcal{L}$ in S^3. Represent S^3 as Euclidean space $\mathbb{R}^3$ with an ideal point at infinity. One may assume without loss of generality that there is an axis E in $\mathbb{R}^3$ such that the 180° rotation about E induces in each component of $\mathcal{L}$ an involution with two fixed points. This rotation defines an involution on the exterior of $\mathcal{L}$ which extends to an orientation preserving involution $\tau : M \to M$. One verifies that $M/\tau = S^3$ and that the quotient map $M \to S^3$ is a double branched cover whose branch set k_τ is a link in S^3. If M is an integral homology sphere (or more generally, a $\mathbb{Z}/2$–homology sphere), the branch set is necessarily a knot, as implied by the Smith theory. In either case, the branch set k_τ can be described as follows.

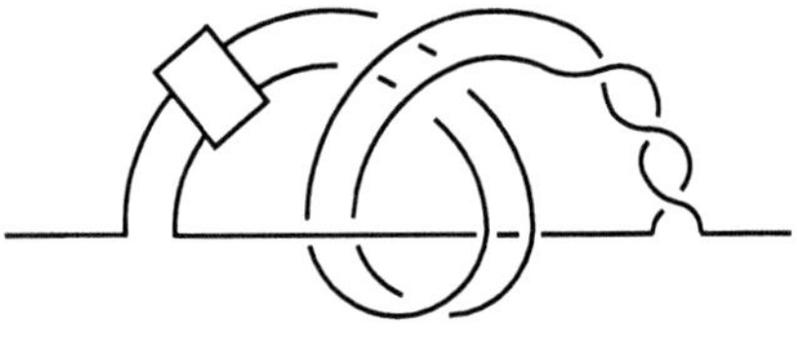

Fig. 1.20.

Assume for the sake of simplicity that $\mathcal{L}$ consists of one component. Consider a tubular neighborhood $N(\mathcal{L})$ and a canonical longitude ℓ on the boundary of $N(\mathcal{L})$. One can always arrange that ℓ is invariant with respect to τ.

Shown in Figure 1.19 is a tubular neighborhood of a trefoil and its canonical longitude ℓ.

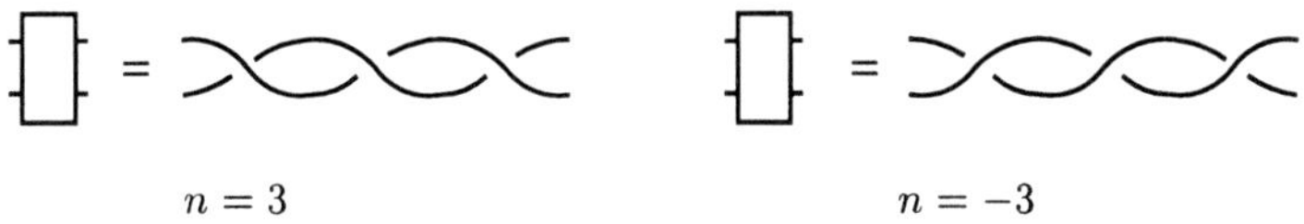

Fig. 1.21.

If M is obtained by n–surgery on $\mathcal{L}$ where n is an integer then the branch set k_τ is obtained as shown in Figure 1.20 (which corresponds to the knot in Figure 1.19; it should be clear how to modify this in general situation). The box in Figure 1.21 contains n half-twists. The twists are right-handed if n is positive, and left-handed if n is negative, see Figure 1.21. The free ends in Figure 1.20 are connected at infinity.

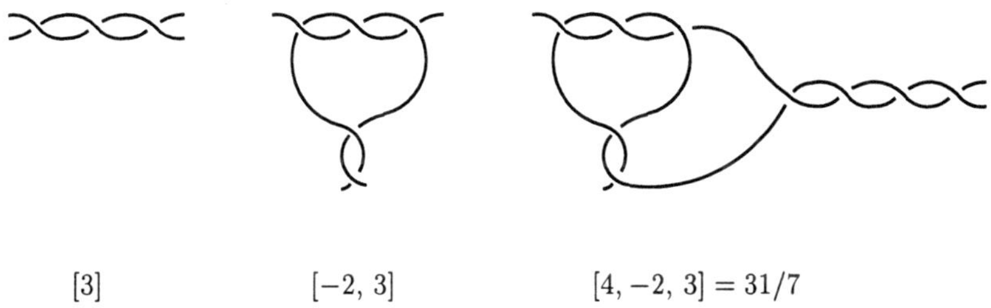

Fig. 1.22.

If M is obtained by a rational surgery on $\mathcal{L}$ with coefficient p/q, the box to insert in Figure 1.20 will contain a *rational tangle* of type p/q. The rational tangle is determined by a continued fraction decomposition $p/q = [t_1, \dots, t_n]$, see (1.9). For example, the rational tangle of type $31/7$ corresponding to the continued fraction $31/7 = [4, -2, 3]$ is shown in Figure 1.22.

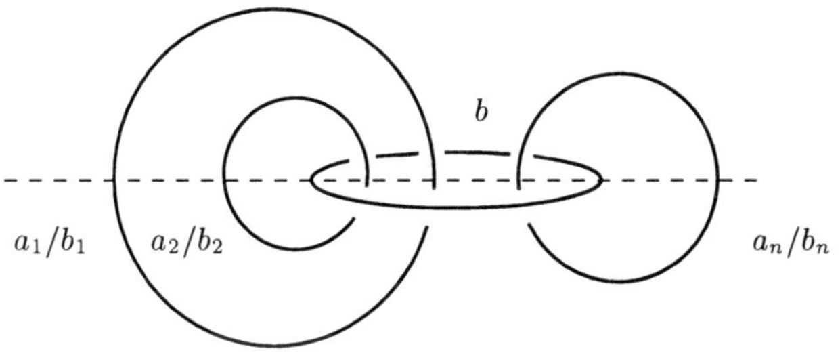

Fig. 1.23.

Example 1.25. Every Seifert fibered homology sphere $\Sigma(a_1, \dots, a_n)$ is a double branched cover of S^3 with respect to the 180° rotation about the axis shown in Figure 1.23. If $\Sigma(a_1, \dots, a_n)$ is viewed as the link of singularity (1.2), this rotation corresponds to the involution $(z_1, \dots, z_n) \mapsto (\bar{z}_1, \dots, \bar{z}_n)$ induced by complex conjugation. The branch set of this covering is shown in Figure 1.24. It is called a *Montesinos knot* of type $(a_1, \dots, a_n)$, see [44] and [226].

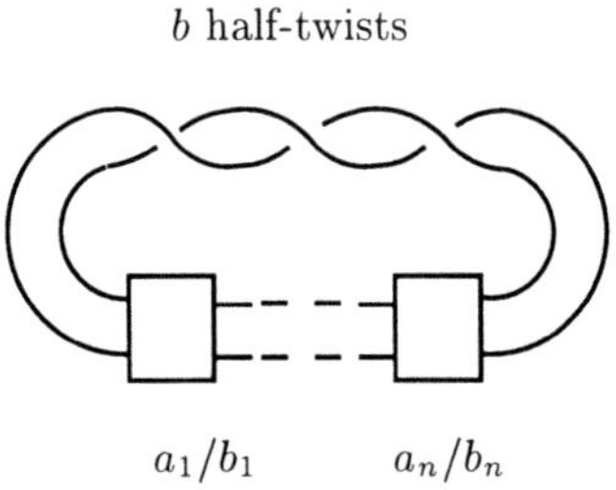

Fig. 1.24.

Example 1.26. Every graph homology sphere can be obtained by splicing Seifert fibered homology spheres $\Sigma(a_1, \dots, a_n)$, see Section 1.1.8. The involution described in Example 1.25 commutes with the splicing operation and extends to the entire class of graph homology spheres. Thus any graph homology sphere Σ can be represented as a double branched cover over S^3 with branch set a knot k. Each splicing component of Σ is a Seifert fibered homology sphere with branch set a Montesinos knot, see Figure 1.24. The knot k is obtained from these knots by band splicing, see Figure 1.25. The knot k is called a *knot of plumbing type* or a (generalized) *Montesinos knot.*

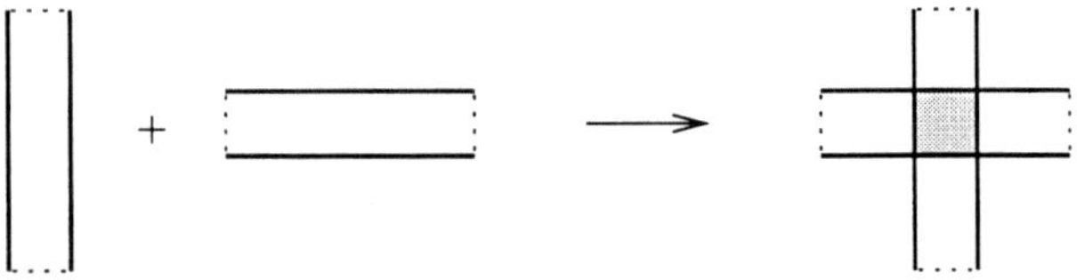

Fig. 1.25.

There may exist many different ways to represent a given manifold Σ as an n-fold cyclic branched cover of S^3. At the same time, not every integral homology sphere is a cyclic branched cover of S^3. In fact, Myers [237] produced an infinite family of pairwise non–homeomorphic irreducible homology 3–spheres each of which admits no PL involution. This implies in particular that these manifolds cannot be double branched covers of S^3. Along the same lines, Flapan [92] produced integral homology spheres without PL homeomorphisms of any finite order. All examples of this nature have been obtained by splicing two copies of S^3 along the knots one of which is non-amphicheiral, the other is non-invertible, and at least one of them has no free cyclic symmetries of finite order.

It has been known since Alexander that every closed orientable 3-manifold can be represented as an *(irregular) branched cover* of S^3, see for example [263]. This result has been improved in work of Hilden [142] and Montesinos [224] who showed that every closed orientable 3-manifold is an irregular 3-fold branch cover of S^3 with branch set a knot. Moreover, if one does not require that the covering be 3-fold, the same knot can be used universally for all 3-manifolds. One of the *universal knots* is the figure-eight knot, see [144].

1.1.13 Heegaard Splittings of Homology Spheres

Every closed oriented 3–manifold M can be represented as a union of two handlebodies, M_1 and M_2, glued along their common boundary, a Riemann surface F of genus g. Such a representation, $M = M_1 \cup_F M_2$, is called a *Heegaard splitting*; any two Heegaard splittings of the same manifold are stably equivalent, see [259] and [287].

For a fixed genus g, the boundaries of handlebodies M_1 and M_2 in a Heegaard splitting are identified via an orientation preserving homeomorphism $f : F \to F$. Isotopic homeomorphisms result in homeomorphic manifolds. The group $\mathcal{M}_g$ of all orientation preserving homeomorphisms of F modulo isotopies is called the *mapping class group*. It is generated by Dehn twists along simple closed curves in F, see Lickorish [197]. *Dehn twist* τ_c along a simple closed curve $c \subset F$ is a self–homeomorphism of F which cuts F along c, twists one end through 360° in one of the possible two directions, and glues the ends back together.

The group $\mathcal{M}_g$ acts naturally on the first homology $H_1(F, \mathbb{Z})$. The subgroup $\mathcal{T}_g \subset \mathcal{M}_g$ which acts trivially on $H_1(F, \mathbb{Z})$ is called the *Torelli group*. If $S^3 = M_1 \cup_F M_2$ is a Heegaard splitting of S^3 such that F has genus g then cutting S^3 along F and regluing M_1 and M_2 by a homeomorphism $f \in \mathcal{T}_g$ always results in an integral homology sphere. Moreover, every integral homology sphere can be obtained by this construction, see [96] and [154]. In fact, to obtain all integral homology spheres, a smaller group $\mathcal{K}_g \subset \mathcal{T}_g$ is sufficient, namely, the group generated by Dehn twists on separating simple closed curves in F, see [230].

1.2 Rational Homology Spheres

A *rational homology 3–sphere* M is a closed oriented 3–manifold such that $H_*(M,\mathbb{Q}) = H_*(S^3,\mathbb{Q})$. Among closed oriented 3–manifolds, rational homology 3–spheres are characterized by the property that $H_1(M,\mathbb{Z})$ is finite.

1.2.1 Spherical Space Forms

A well–known problem in the 3–dimensional topology is to list all finite groups which occur as the fundamental groups of some closed 3–manifold. So far, all known examples come from the finite subgroups $\Gamma \subset SO(4)$ which operate freely and linearly on $S^3 \subset \mathbb{R}^4$. The associated 3–manifolds S^3/Γ are known as the (linear) *spherical space forms*; they all are naturally rational homology 3–spheres. It turns out that, in all known examples, there exists an S^1–action on S^3 which commutes with Γ so that the orbit space S^3/Γ is always a Seifert fibered manifold, see [251]. It should be also noted that the class of spherical space forms coincides with the class of the links of quotient singularities, see for example [218]. The following is a complete list of spherical space forms S^3/Γ.

- $\Gamma = \mathbb{Z}/p = \langle\, x \mid x^p = 1 \,\rangle$. The manifold S^3/Γ is a *lens space* $L(p,q)$, $p,q > 0$, given by the $(-p/q)$-surgery on the unknot in S^3. It is obtained by plumbing on the graph shown in Figure 1.26 with $-p/q = [x_1,\dots,x_n]$ where $[\ ,\ ]$ stands for continued fraction (1.9). Two lens spaces, $L(p,q)$ and $L(p,q')$ are homeomorphic if and only if $qq' = \pm 1 \bmod p$, which was first proved by Reidemeister [258]. It should be noted that $L(p,q)$ and $L(p,q')$ are homotopy equivalent if and only if $qq' = \pm\, m^2 \bmod p$ for some integer m, see Whitehead [304].

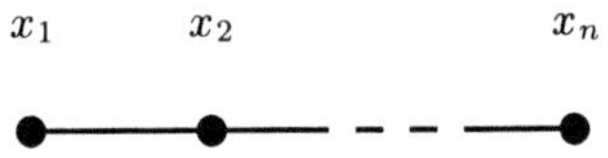

Fig. 1.26.

- $\Gamma = \mathbb{Z}/m \times D^*_{4p}$ where $m,p > 0$, $(m,2p) = 1$ and $D^*_{4p} = \langle\, x, y \mid x^2 = (xy)^2 = y^p \,\rangle$ is the *binary dihedral group* of order $4p$. The manifold S^3/Γ is called a *prism manifold*; it is a Seifert fibered manifold described by the diagram in Figure 1.27, see also Example 1.30. It should be mentioned that the prism manifolds can also be realized as Seifert fibered manifolds over the real projective space $\mathbb{R}P^2$, see Section 1.2.3.

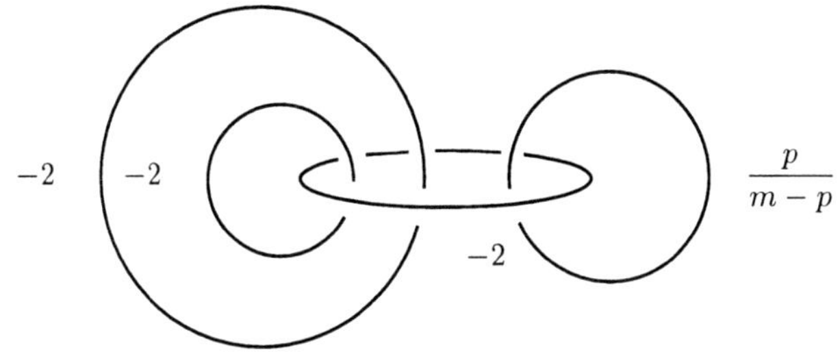

Fig. 1.27.

• $\Gamma = \mathbb{Z}/m \times D'_{2^{k+2}p}$, $m > 0$, $k \geq 0$, $p \geq 3$ odd, and $(m, 2p) = 1$, where $D'_{2^{k+2}p} = \langle\, x, y \mid x^{2^{k+2}} = 1,\ y^p = 1,\ xy^{-1} = yx \,\rangle$. The manifold S^3/Γ is given by the diagram in Figure 1.28. Note that $D'_{4p} = D^*_{4p}$ for odd p.

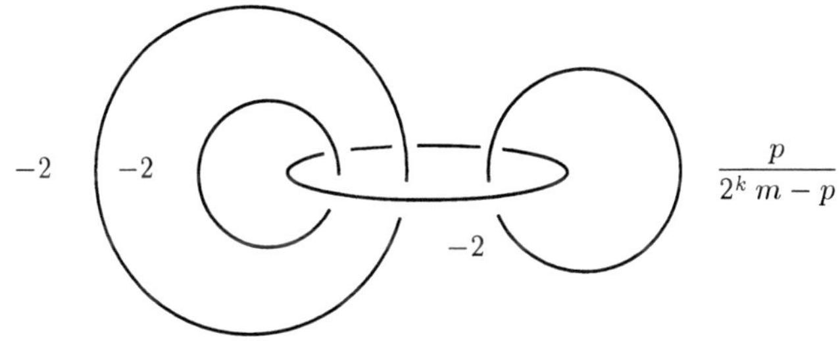

Fig. 1.28.

• $\Gamma = \mathbb{Z}/m \times T^*$ where $m > 0$, $(m, 6) = 1$ and $T^* = \langle\, a, b \mid (ab)^2 = a^3 = b^3 \,\rangle$ is the *binary tetrahedral group* of order 24. The manifold S^3/Γ is realized by surgery on the link shown in Figure 1.29 where p and q are integers such that $(p, 3) = 1$, $(q, 3) = 1$, and $m = -3 + 2p + 2q$.

• $\Gamma = \mathbb{Z}/m \times T'_{8\cdot 3^k}$ where $m > 0$, $(m, 6) = 1$, $k \geq 1$, and $T'_{8\cdot 3^k} = \langle\, x, y, z \mid x^2 = (xy)^2 = y^2,\ zxz^{-1} = y,\ zyz^{-1} = xy,\ z^{3^k} = 1 \rangle$. The manifold S^3/Γ is obtained by surgery on the link shown in Figure 1.29 where p and q are integers such that $(p, 3) = 1$, $(q, 3) = 1$, and $3^{k-1}m = -3 + 2p + 2q$. Note that $T'_{24} = T^*$.

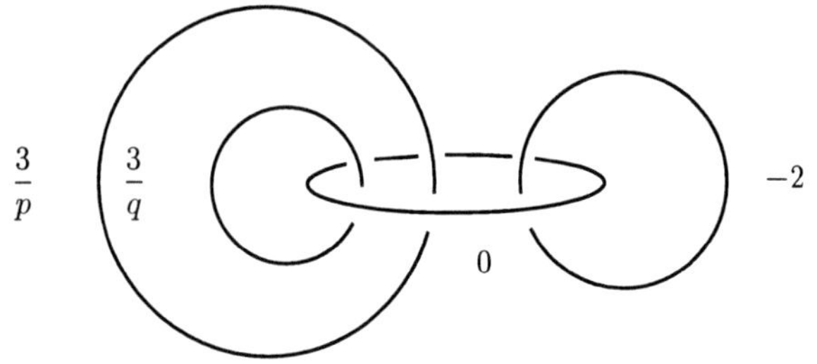

Fig. 1.29.

• $\Gamma = \mathbb{Z}/m \times O^*$ where $m > 0$, $(m, 6) = 1$ and $O^* = \langle\, a, b \mid (ab)^2 = a^3 = b^4 \,\rangle$ is the *binary octahedral group* of order 48. The manifold S^3/Γ is realized by surgery on the link shown in Figure 1.30 where p and q are integers such that $(p, 3) = 1$, q is odd, and $m = -6 + 4p + 3q$.

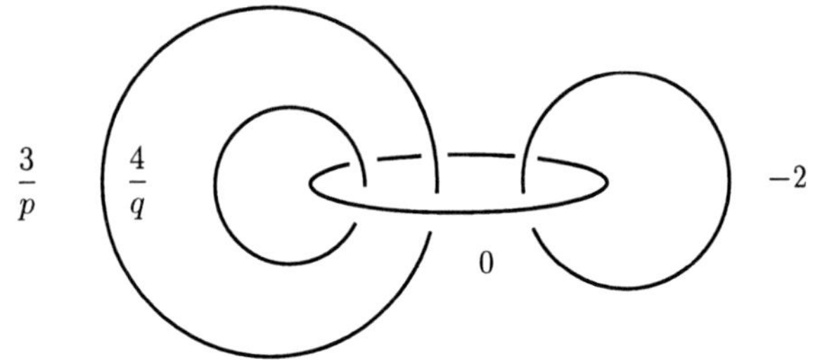

Fig. 1.30.

• $\Gamma = \mathbb{Z}/m \times I^*$ where $m > 0$, $(m, 30) = 1$ and $I^* = \pi_1(\Sigma(2, 3, 5)) = \langle\, a, b \mid (ab)^2 = a^3 = b^5 \,\rangle$ is the *binary icosahedral group* of order 120. The manifold S^3/Γ is realized by surgery on the link shown in Figure 1.31 where p and q are integers such that $(p, 3) = 1$, $(q, 5) = 1$, and $m = -15 + 10p + 6q$. If $m = 1$, this is the Poincaré homology sphere.

It should be noted that apart from the lens spaces whose homeomorphism classification was already mentioned, any two spherical space forms S^3/Γ from the list above are homeomorphic if and only if their fundamental groups are isomorphic.

It is not known whether there exists a smooth free action of any group Γ on S^3, or on a homotopy 3–sphere, not conjugate to one of the linear actions listed above. In 1957, Milnor [215] established that any group Γ acting freely on S^3 or a homotopy 3–sphere must satisfy certain cohomological condition. He compiled a list of all finite groups satisfying all the restrictions known at

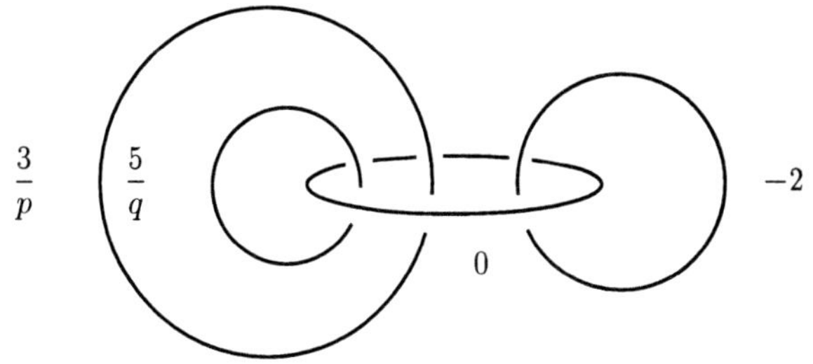

Fig. 1.31.

the time. In addition to the groups listed above, Milnor's list included the following groups.

- The semidirect product $Q(8n,k,\ell)$ of the odd cyclic group $\mathbb{Z}/k\ell$ with the generalized quaternionic group $Q(8n)$. More precisely, $Q(8n,k,\ell) = \langle\, x,y,z \mid x^2 = y^{2n} = (xy)^2,\ z^{k\ell} = 1,\ xzx^{-1} = z^r,\ yzy = z^{-1}\,\rangle$ where n, k and ℓ are odd and pairwise relatively prime integers, $n > k > \ell \geq 1$, and r satisfies $r = -1 \bmod k$, $r = 1 \bmod \ell$.

- Groups $Q(8n,k,\ell)$ with the same presentation as above but with n even.

- An extension $O(48,3^{k-1},\ell)$ of the odd cyclic group $\mathbb{Z}/3^{k-1}\ell$, with ℓ relatively prime to 3, by the binary octahedral group O^*. More precisely, $O(48,3^{k-1},\ell) = \langle\, x,p,q,r,a \mid x^{3^k} = p^4 = a^\ell = 1,\ p^2 = q^2 = r^2,\ pqp^{-1} = q^{-1},\ xpx^{-1} = q,\ xqx^{-1} = pq,\ rxr^{-1} = x^{-1},\ rpr^{-1} = qp,\ rqr^{-1} = q^{-1},\ ap = pa,\ aq = qa,\ rar^{-1} = a^{-1}\,\rangle$.

- The product of any of the groups from the three classes above with a cyclic group of relatively prime order.

In years to follow, the family $Q(8n,k,\ell)$ with n even was ruled out by Ronnie Lee [186] and the family $O(48,3^{k-1},\ell)$ was eliminated by C. B. Thomas [294]. Therefore, we are left with just the groups $Q(8n,k,\ell)$, n odd, and their products with cyclic groups of relatively prime order.

1.2.2 Dehn Surgery

Let k be a knot in an integral homology sphere Σ and let $M = \Sigma + (p/q)\cdot k$ be the result of (p/q)–surgery of Σ along k, see Section 1.1.5. The manifold M is a rational homology sphere if and only if $p \neq 0$, in which case $H_1(M,\mathbb{Z}) = \mathbb{Z}/|p|$.

Example 1.27. For any relatively prime integers p and q, the lens space $L(p,q)$ is obtained by $(-p/q)$–surgery on a trivial knot in S^3. Lens spaces can also be obtained by surgery on other knots. There are essentially two cases: if $k \subset S^3$ is a torus knot then $S^3 + (p/q) \cdot k$ is a lens space for infinitely many integers p, see [231], though when k is non–trivial this occurs at most once for each p, see [35]. If $k \subset S^3$ is not a torus knot then, according to cyclic surgery theorem [60], at most two non–trivial surgeries on k yield lens spaces, and if there are two, they correspond to consecutive integers. Several families of non–torus knots with lens space Dehn surgeries can be found in [25] and [80]. If $L(p,q)$ arises from *integral* surgery on a knot in S^3 then $\pm q$ is a quadratic residue modulo p, see [80]. Note that the latter condition is equivalent to the condition that $L(p,q)$ be homotopy equivalent to $L(p,1)$, see Section 1.2.1.

Example 1.28. A rational homology sphere M is called a *homology lens space* if $H_1(M,\mathbb{Z}) = \mathbb{Z}/p$, $p \geq 1$. A closed oriented 3–manifold M is a homology lens space if and only if there exists a knot k in an integral homology 3–sphere Σ such that $M = \Sigma + (p/q) \cdot k$, see [35]. In the special case when $M = S^3 + (p/q) \cdot k$, the groups $\pi_1(S^3 \setminus k)$ and hence $\pi_1(M)$ have weight one, that is, they are normally generated by one element. Boyer and Lines [35] show that there are infinitely many prime homology lens spaces M such that $\pi_1(M)$ has weight one but M cannot be obtained by surgery on a knot in S^3. The proof relies on their extension of the Casson invariant, which is a special case of the Walker invariant discussed in Chapter 4.

Example 1.29. Every integral homology sphere obtained by Dehn surgery on a knot in S^3 is irreducible, see Section 1.1.5. This is not always the case for rational homology spheres. For example, the result of 6–surgery on the right handed trefoil is a connected sum of two lens spaces, $L(2,1)\#L(3,1)$, see for example [263].

Not all rational homology spheres can be obtained by surgery on a knot, however, every one of them can be obtained by surgery on a link, see Theorem 1.8. Given a framed link $\mathcal{L}$ in an integral homology sphere Σ with linking matrix $A(\mathcal{L})$ and framing matrix $F(\mathcal{L})$, see Section 1.1.6, denote by $\chi_\Sigma(\mathcal{L})$ the result of surgery of Σ along the link $\mathcal{L}$. The manifold $\chi_\Sigma(\mathcal{L})$ is a rational homology sphere if and only if $\det A(\mathcal{L}) \neq 0$, in which case $|H_1(\chi_\Sigma(\mathcal{L}),\mathbb{Z})| = |\det F(\mathcal{L})|$.

The surgery theory can be extended beyond the realm of links in integral homology spheres. Let k be a knot in a closed oriented 3–manifold M and $K = M \setminus \operatorname{int} N(k)$ its exterior. A homology class $c \in H_1(\partial K,\mathbb{Z})$ is called *primitive* if it is not a multiple of any other class in $H_1(\partial K,\mathbb{Z})$; in particular, it is not trivial. A primitive class c can be represented by a non-separating simple closed curve on ∂K, which we again call c. Given a primitive homology class c, consider the manifold

$$K_c = K \cup_h (D^2 \times S^1) \tag{1.13}$$

where $h : \partial D^2 \times S^1 \to \partial K$ is a homeomorphism sending $\partial D^2 \times \{1\}$ to c. We say that K_c is obtained from M by *Dehn surgery* along k, compare with Section 1.1.5.

For a knot in an integral homology sphere M, any primitive class $c \in H_1(\partial K, \mathbb{Z})$ is uniquely described by a rational number p/q, see Section 1.1.5. The situation is a little more complicated for knots in rational homology spheres. Namely, if k is an oriented knot in a rational homology sphere M then every primitive class c can be uniquely described by a pair (p, q), where $p = \mathrm{lk}_M(k, c) \in \mathbb{Q}$ and q is the integer such that c is homologous to $q \cdot k$ in $N(k)$. Since orientation of c is of no importance to us, we may assume that q is non-negative. If k happens to be homologous to zero in M then p is an integer.

The *linking number* $\mathrm{lk}_M(k, c)$ is defined as follows. If k represents zero in $H_1(M, \mathbb{Z})$, choose an oriented Seifert surface F with $\partial F = k$ and define $\mathrm{lk}_M(k, c)$ as the algebraic intersection number of F and c in M. In general, there exists an integer N such that $N \cdot k$ represents zero in $H_1(M, \mathbb{Z})$, and we define $\mathrm{lk}_M(k, c) = \mathrm{lk}_M(N \cdot k, c)/N \in \mathbb{Q}$. Note that the linking number lk_M is symmetric, $\mathrm{lk}_M(k, c) = \mathrm{lk}_M(c, k)$.

A *framed link* in a rational homology sphere M is a link $\mathcal{L} = k_1 \cup \ldots \cup k_m$ each component k_i of which is oriented and equipped with a primitive class c_i specified by a pair (p_i, q_i) with a positive q_i. The manifold obtained by surgery on a framed link $\mathcal{L}$ is denoted by $\chi_M(\mathcal{L})$. The *linking matrix* of a framed link $\mathcal{L}$ is the matrix $A(\mathcal{L}) = (\ell_{ij})$ with

$$\ell_{ij} = \frac{1}{q_j}\,\mathrm{lk}_M(k_i, c_j) = \begin{cases} \mathrm{lk}_M(k_i, k_j), & \text{if } i \neq j, \\ p_i/q_i, & \text{if } i = j. \end{cases}$$

The matrix $F(\mathcal{L}) = (q_j \cdot \ell_{ij})$ is called a *framing matrix* of $\mathcal{L}$. It is a presentation matrix for $H_1(\chi_M(\mathcal{L}), \mathbb{Z})$, and

$$|H_1(M, \mathbb{Z})| \cdot |\det(F(\mathcal{L}))| = |H_1(\chi_M(\mathcal{L}), \mathbb{Z})|.$$

1.2.3 Seifert Fibered Manifolds

By *Seifert fibered manifold* we mean a closed oriented 3–manifold admitting a fixed point free action of S^1, see Section 1.1.4. Let M be a Seifert fibered manifold then $F = M/S^1$ is a closed Riemann surface, not necessarily orientable. The quotient map $M \to F$ induces an epimorphism $H_1(M, \mathbb{Q}) \to H_1(F, \mathbb{Q})$. Therefore, M can be a rational homology sphere only if F is either the 2–sphere or the real projective plane $\mathbb{R}P^2$.

If $F = S^2$ then M is homeomorphic to a Seifert fibered manifold $M(b; (a_1, b_1), \ldots, (a_n, b_n))$ whose surgery presentation is shown in Figure 1.1. The first integral homology of M equals the cokernel of the map (1.4) hence M is a rational homology sphere if and only if $\det A = a_1 \ldots a_n \cdot (\sum (b_k/a_k) - b) \neq 0$, in which case $|H_1(M, \mathbb{Z})| = |\det A|$. The manifolds M with $n \leq 2$ are lens spaces.

Any Seifert fibered manifold M with the quotient $M/S^1 = \mathbb{R}P^2$ has surgery presentation shown in Figure 1.32, where $a_k \geq 1$ and $b_k \neq 0$, see for example [223]. It is always a rational homology sphere with $|H_1(M,\mathbb{Z})| = 4 \cdot a_1 \cdots a_n$.

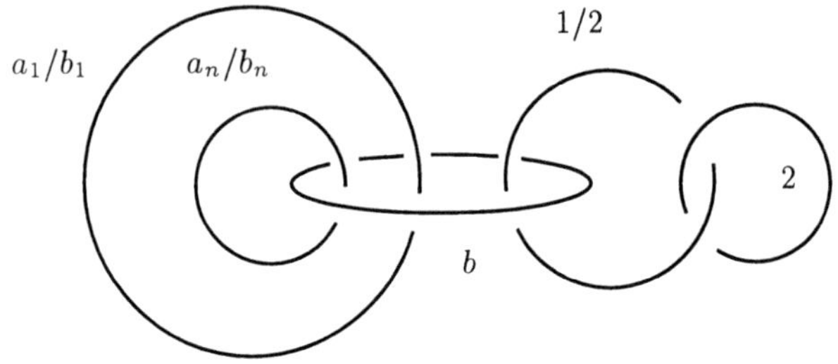

Fig. 1.32.

1.2.4 Links of Singularities

The link of an isolated surface singularity $(X,0) \subset (\mathbb{C}^n,0)$ was defined in Section 1.1.10 as the canonically oriented closed 3–manifold $\Sigma = X \cap S^{2n-1}_\varepsilon$ where ε is a sufficiently small positive number. Choose a "good resolution" $(\tilde{X},E) \to (X,0)$, that is one for which the exceptional curve $E = \cup E_i$ is a union of smooth curves intersecting transversally. Then Σ is obtained by plumbing according to the resolution dual graph Γ of E. The link Σ is a rational homology sphere if and only if all E_i are rational and the graph Γ has no loops. Then $|H_1(\Sigma,\mathbb{Z})| = |\det A(\Gamma)|$, where $A(\Gamma)$ is the matrix (1.8).

Thus all links of singularities which are rational homology spheres are plumbed manifolds. Conversely, a rational plumbed homology 3–sphere is a link of singularity if and only if it can be obtained by plumbing on a plumbing graph Γ whose intersection matrix $A(\Gamma)$ is negative definite.

Example 1.30. For any integer $n \geq 3$ the singularity at zero of the complex polynomial $x^2\, y + y^n + z^2$ has resolution dual graph shown in Figure 1.33. Its link is a prism manifold, compare with Figure 1.27.

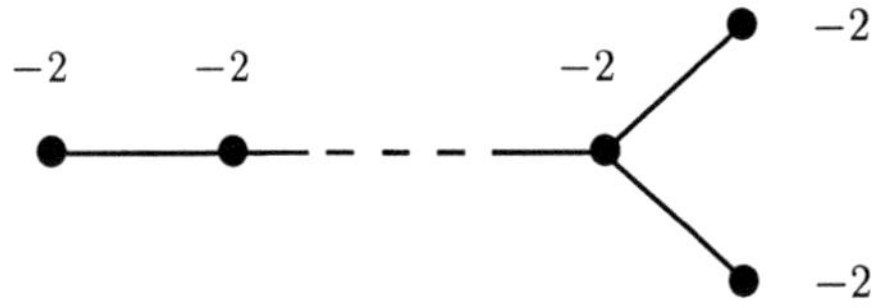

Fig. 1.33.

An important class of examples are the links $\Sigma(a_1,\dots,a_n)$ given by equation (1.2). Each of the $\Sigma(a_1,\dots,a_n)$ admits a fixed point free action of the circle,

$$t \cdot (z_1, \dots, z_n) = (t^{q_1} z_1, \dots, t^{q_n} z_n) \quad \text{where} \quad q_j = (\operatorname{lcm}_k a_k) / a_j$$

so that $\Sigma(a_1, \dots, a_n)$ is a Seifert fibered manifold. Its S^1–quotient is a Riemann surface of genus g, and its unnormalized Seifert invariants are $(0; s_1(t_1, b_1), \dots, s_n(t_n, b_n))$, where $s_j(t_j, b_j)$ means (t_j, b_j) repeated s_j times and

$$t_j = (\operatorname{lcm}_k a_k) / (\operatorname{lcm}_{k \neq j} a_k), \qquad s_j = (\prod_{k \neq j} a_k) / (\operatorname{lcm}_{k \neq j} a_k),$$

$$g = \frac{1}{2} \left(2 + (n-2) (\prod_k a_k) / (\operatorname{lcm}_k a_k) - \sum_j s_j \right),$$

and b_j are determined (up to equivalence of Seifert invariants) by the equation

$$\sum_j s_j b_j / t_j = (\prod_k a_k) / (\operatorname{lcm}_k a_k)^2,$$

see [242]. All links $\Sigma(a_1, \dots, a_n)$ such that $g = 0$ are rational homology spheres. Links $\Sigma(a_1, \dots, a_n)$ with pairwise relatively prime $a_1, \dots, a_n$ are integral homology spheres.

Example 1.31. For any integer $n \geq 2$, the link $\Sigma(2, 2, n)$ is orientation preserving homeomorphic to the lens space $L(n, n-1)$. Its plumbing graph is the graph with $(n-1)$ vertices shown in Figure 1.34.

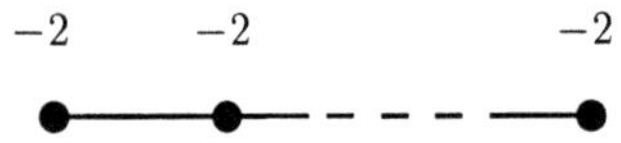

Fig. 1.34.

1.2.5 Branched Covers

An n–fold cyclic branched cover M of an integral homology 3–sphere Σ with branch set a knot k is a rational homology sphere if and only if no root of the Alexander polynomial $\Delta_k(t)$ is an n th root of unity, and in this case the order of $H_1(M, \mathbb{Z})$ is given by the formula (1.12). Here, the Alexander polynomial $\Delta_k(t)$ should be normalized so that there are no negative powers of t and the constant term is positive. Any regular (unbranched) finite sheeted cover of a rational homology sphere is a rational homology sphere.

Conversely, if a rational homology sphere M admits an orientation preserving PL homeomorphism $\tau : M \to M$ of order n then M is an n–fold cyclic cover of M/τ, branched or not. A transfer argument shows that M/τ is always a rational homology sphere. Most results of Section 1.1.12 hold for rational homology spheres (and for more general 3–manifolds). The most essential difference is however that the branch set of $M \to M/\tau$ need not be connected. We illustrate this point for double branched covers of S^3.

Suppose that M is a closed oriented 3–manifold and $\tau : M \to M$ is an involution which makes it into a double branched cover of S^3. The branch set of this cover is in general a link $\mathcal{L}$ of several components. Turaev [297] calls every unordered pair

$$\{\text{ an orientation of } \mathcal{L}, \text{ the opposite orientation of } \mathcal{L}\,\}$$

a *quasiorientation* of $\mathcal{L}$, and establishes a one-to-one correspondence between quasiorientations of $\mathcal{L}$ and spin structures on M. Spin structures on M are classified by the group $H^1(M, \mathbb{Z}/2)$ hence $\mathcal{L}$ is a knot if and only if M is a $\mathbb{Z}/2$–homology sphere.

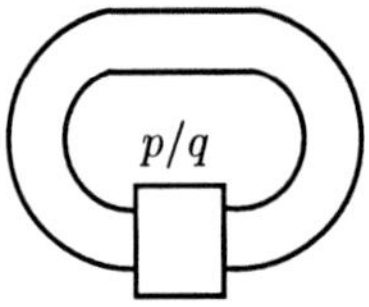

Fig. 1.35.

Example 1.32. For any relatively prime integers $p, q \neq 0$ the lens space $L(p,q)$ is a rational homology sphere obtained by $(-p/q)$–surgery on the unknot in S^3. The natural involution on the unknot with two fixed points makes $L(p,q)$ into a double branched covering of S^3. The branch set of this covering is the link $\mathcal{L}$ shown in Figure 1.35, where the box represents a rational tangle of type p/q, see Section 1.1.12. The link $\mathcal{L}$ is called a *two bridge link*, see [44]. It has one component if p is odd and two components if p is even.

2 Rokhlin Invariant

In the early 1950's, Rokhlin [260] announced a theorem asserting that the signature of a smooth closed spin 4–manifold is divisible by 16. This result, known as the Rokhlin Theorem, has played a distinguished role in the 4-dimensional topology, see [132] for a survey. Among other things, it gave rise to the Rokhlin invariant, see [75] or [146], whose properties are related to the most fundamental questions of the manifold theory, such as triangulability of topological manifolds. Attempts at refining the Rokhlin invariant led to introduction of the Casson invariant and its numerous ramifications which are discussed later in this book.

2.1 The Rokhlin Theorem

Let X be a smooth compact oriented 4–dimensional manifold, with or without boundary. The intersection number of cycles $a \cdot b$ induces a bilinear form

$$Q : H_2(X,\mathbb{Z})/\operatorname{tor} \otimes H_2(X,\mathbb{Z})/\operatorname{tor} \to \mathbb{Z} \tag{2.1}$$

called the *intersection form* of X. Its signature, that is the number of positive eigenvalues minus the number of negative eigenvalues, is called the *signature* of X and is denoted $\operatorname{sign}(X)$. The intersection form Q is called *even* if $Q(u,u) = 0 \bmod 2$ for all u. The manifold X is called *spin* if its second Stiefel-Whitney class $w_2(X) \in H^2(X,\mathbb{Z}/2)$ vanishes. The intersection form of a spin manifold is even. If X has no 2–torsion in its homology then X is spin if and only if its intersection form is even. All 4-dimensional manifolds X with $H_1(X,\mathbb{Z}) = 0$, including all simply connected manifolds, have no 2–torsion.

Theorem 2.1. *If X is a smooth closed oriented spin 4-manifold then* $\operatorname{sign}(X) = 0 \bmod 16$. *In particular, if X is a smooth closed oriented 4–manifold without 2-torsion whose intersection form is even, then* $\operatorname{sign}(X) = 0 \bmod 16$.

This result is known as the *Rokhlin Theorem*, see [260] and also [132]. It has many different proofs: a geometric proof can be found in [99], a homotopy-theoretical in [221], see also [203] and [275]. Still, the shortest proof of Theorem 2.1 is arguably the one based on the Atiyah–Singer index theorem [10].

Proof. Since X is a spin manifold, there is a well defined Dirac operator with even index. According to the index theorem, this index equals $-p_1(X)/24$, where $p_1(X)$ is the first Pontryagin number of X. This implies that $p_1(X)$ is divisible by 48. By the index theorem again, $\operatorname{sign}(X) = p_1(X)/3$, which completes the proof.

Several remarks are due. First, the intersection form of a closed 4–manifold is always unimodular, and the signature of an even unimodular form is always divisible by 8 by purely algebraic reasons, see [282]. The Rokhlin theorem improves this result by a factor of two for the intersection forms of *smooth* spin 4-manifolds. Second, Theorem 2.1 does not hold for topological manifolds, see Example 3.35, or manifolds with boundary, see Example 1.16. Finally, there exists a closed oriented smooth 4-manifold X, constructed by Habegger [135], with even intersection form whose signature equals 8 (the manifold X is not spin). Another description of Habegger's manifold can be found in [86].

Theorem 2.1 can be sharpened as follows, see [158]. For a closed smooth 4–manifold X, we say that $w \in H_2(X,\mathbb{Z})$ is *characteristic* if the modulo 2 reduction of its Poincaré dual equals the Stiefel–Whitney class $w_2(X)$.

Example 2.2. Let X be a simply-connected closed smooth 4–manifold and represent it as $X = X_0 \cup D^4$ where $X_0 = X \setminus \operatorname{int} D^4$ is a manifold with boundary S^3. Suppose that X_0 is the result of a 4–dimensional surgery of D^4 on a framed link $\mathcal{L} = k_1 \cup \ldots \cup k_n$ in S^3, that is X_0 is obtained from D^4 by gluing 2-handles along the components of $\mathcal{L}$. A basis in $H_2(X,\mathbb{Z})$ can then be generated by the surfaces $F_i = F_i' \cup D^2$ where F_i' is a Seifert surface of the knot k_i (with its interior pushed radially inside D^4), and D^2 is the axial disc of the corresponding 2–handle. The intersection form Q in this basis is isomorphic to the linking matrix of $\mathcal{L}$. A class $w \in H_2(X,\mathbb{Z})$ is characteristic if and only if

$$w \cdot x = x \cdot x \bmod 2 \quad \text{for all} \quad x \in H_2(X,\mathbb{Z}). \tag{2.2}$$

Equation (2.2) is equivalent to the linear system $Q\,w = \operatorname{diag} Q \bmod 2$ where $\operatorname{diag} Q$ is the column of the diagonal elements of Q. This system has a unique solution over $\mathbb{Z}/2$ since $\det Q = \pm 1$.

Theorem 2.3. *(Kervaire–Milnor [158]) Let X be a smooth closed oriented 4-manifold whose characteristic class can be represented by a smoothly embedded 2-sphere F in X. Then* $\operatorname{sign}(X) - F \cdot F = 0 \bmod 16$.

If a characteristic class $w \in H_2(X,\mathbb{Z})$ is represented by an embedded surface F of arbitrary genus, one can associate with F a quadratic form

$$\tilde{q} : H_1(F,\mathbb{Z}/2) \to \mathbb{Z}/2$$

and its *Arf–invariant* $\operatorname{Arf}(X,F) \in \mathbb{Z}/2$, see [99] and Example 2.5. Then the following generalized *Rokhlin theorem* holds, see [99], [208] and [261].

Theorem 2.4. *Let X be a smooth closed oriented 4–manifold whose characteristic class is represented by a smoothly embedded surface F in X. Then* $(\operatorname{sign}(X) - F \cdot F)/8 = \operatorname{Arf}(X, F) \bmod 2$.

We will identify $\operatorname{Arf}(X, F)$ and illustrate Theorem 2.4 only for a special class of examples which will be important in Section 2.3.1.

Example 2.5. Suppose that an integral homology sphere Σ is embedded in X and separates the characteristic surface F into two pieces, $F = F' \cup D^2$, where $F' \subset \Sigma$ is a Seifert surface of a knot $k \subset \Sigma$. This is the case, for example, when X is obtained by surgery on a 4–manifold W with $\partial W = \Sigma$ by attaching a 2–handle along the knot $k \subset \Sigma$, and capping it off to get a closed manifold. Then the quadratic form $\tilde{q} : H_1(F, \mathbb{Z}/2) \to \mathbb{Z}/2$ is isomorphic to the quadratic form q of the knot k defined by the formula $q(x) = S(x, x) \bmod 2$. Here, S is the Seifert form associated with F'. Therefore, $\operatorname{Arf}(X, F)$ equals $\operatorname{Arf}(q)$, the *Arf–invariant* of k, and the latter equals $(1/2) \cdot \Delta_k''(1)$, see for instance [275]. Here, $\Delta_k''(t)$ is the second derivative with respect to t of the Alexander polynomial of k normalized so that $\Delta_k(1) = 1$ and $\Delta_k(t) = \Delta_k(t^{-1})$.

Example 2.6. Let $X = \mathbb{C}P^2$ and let $\eta \in H_2(\mathbb{C}P^2, \mathbb{Z}) = \mathbb{Z}$ be a generator represented by the embedded sphere $S^2 = \mathbb{C}P^1$. A class $s \cdot \eta \in H_2(\mathbb{C}P^2)$ is characteristic if and only if s is odd. The algebraic curve

$$C = \{\, [x_0 : x_1 : x_2] \mid x_0 x_1^{s-1} + x_2^s \,\} \subset \mathbb{C}P^2$$

is homeomorphic to S^2 and represents the class $s \cdot \eta$. It is smoothly embedded in $\mathbb{C}P^2$ except possibly at the point $[1 : 0 : 0]$. Let B be a 4–ball of radius $\varepsilon > 0$ centered at $[1 : 0 : 0]$. In the affine plane $x_0 = 1$ the intersection $\partial B \cap C$ is given by the equations $x_1^{s-1} + x_2^s = 0$, $|x_1|^2 + |x_2|^2 = \varepsilon^2$. Therefore, $\partial B \cap C \subset \partial B = S^3$ is an $(s, s-1)$–torus knot $k_{s,s-1}$. Let F' be a Seifert surface in ∂B with boundary the curve $\partial B \cap C$, then the surface $F = (C \setminus (C \cap \operatorname{int} B)) \cup F'$ represents the class $s \cdot \eta$. An easy calculation with Alexander polynomials using Examples 2.5 and 2.12 verifies Theorem 2.4 in this particular instance:

$$\begin{aligned} \operatorname{Arf}(\mathbb{C}P^2, s \cdot \eta) = \operatorname{Arf}(k_{s,s-1}) &= (s^2 - 1)((s-1)^2 - 1)/24 \\ &= (1 - s^2)/8 = (\operatorname{sign} \mathbb{C}P^2 - s\eta \cdot s\eta)/8 \bmod 2. \end{aligned}$$

In fact, this example is a cornerstone in the geometric proof of Theorem 2.4, see [99] and [275], after one reduces the general case to that of characteristic surfaces in $\mathbb{C}P^2$.

Example 2.7. Let X be a smooth simply connected closed 4–manifold. Every homology class $u \in H_2(X, \mathbb{Z})$ can be represented by an embedded surface $F \subset X$. The Rokhlin Theorem can be used to provide a lower bound on the genus of such a surface, see [262]. For example, if η is a generator in $H_2(\mathbb{C}P^2, \mathbb{Z}) = \mathbb{Z}$ then the class 3η cannot be represented by a 2–sphere because $\text{Arf}(\mathbb{C}P^2, 3\eta) = 1 \bmod 2$, see Example 2.6, but had 3η been spherical we would have had $\text{Arf}(\mathbb{C}P^2, 3\eta) = 0 \bmod 2$. In fact, the *Thom conjecture* asserts that the minimal genus of an embedded surface representing $s \cdot \eta$ in $\mathbb{C}P^2$ is $(s-1)(s-2)/2$. This conjecture was proved in [177] and generalized in [253].

Further ramifications of the Rokhlin theorem can be found in [134] and [208] and, for manifolds of higher dimensions, in [41] and [221].

2.2 Definition of the Rokhlin Invariant

Every closed oriented 3-manifold M with a fixed spin structure bounds a smooth compact oriented spin 4-manifold X whose spin structure restricts on the boundary to that of M. Moreover, X can be chosen simply connected, see Milnor [214]. Kaplan [156] proved this fact by exhibiting an explicit handle decomposition of X.

Let M be a closed oriented 3-manifold with a fixed spin structure, and choose a spin 4–manifold X with $\partial X = M$ as above. Then $\operatorname{sign} X \bmod 16$ is independent of the choice of X by the Rokhlin theorem. Indeed, if X' is another spin manifold with boundary M, one can form the closed manifold $X \cup (-X')$ which will be spin, and apply to it Theorem 2.1. Thus $\operatorname{sign}(X) \bmod 16$ is an invariant of M with a fixed spin structure. Note that the spin-structures on M are in one-to-one correspondence with the elements of $H^1(M, \mathbb{Z}/2)$.

Let M be an oriented $\mathbb{Z}/2$–homology sphere then it has a unique spin structure. Choose a smooth compact oriented spin 4–manifold X such that $\partial X = M$ and define the *Rokhlin invariant*

$$\mu(M) = \frac{\operatorname{sign}(X)}{8} \bmod 2.$$

In general, $\mu(M)$ is a rational residue modulo 2, however, $8 \cdot \mu(M)$ is always an integer modulo 16. If M is an integral homology sphere then $\operatorname{sign}(X)$ is divisible by 8 so that $\mu(M)$ is 0 or 1 mod 2. Also note that the requirement that X be smooth in the definition of $\mu(M)$ is essential since every integral homology sphere M bounds a contractible topological 4-manifold, see [98].

This is the Rokhlin invariant that we are going to deal with most of the time. However, it should be mentioned that if M is not a $\mathbb{Z}/2$–homology sphere, its *Rokhlin invariant* $\mu(M, \mathfrak{s})$ still can be defined as above as an invariant of a pair $(M, \mathfrak{s})$ where $\mathfrak{s}$ is a spin structure on M.

Example 2.8. The Poincaré homology sphere $\Sigma(2,3,5)$ bounds a smooth simply connected spin manifold X with intersection form E_8 and $\operatorname{sign}(X) = -8$, see Figure 1.14. Therefore, $\mu(\Sigma(2,3,5)) = 1 \bmod 2$.

In general, one does not have to construct a spin 4-manifold bounding M to compute $\mu(M)$. Suppose that X is a smooth oriented 4–manifold with $\partial X = M$. Assume for simplicity that $H_1(X, \mathbb{Z}) = 0$ but do not assume that X is spin. Suppose that X has a spherical characteristic surface, that is, a smoothly embedded characteristic surface $F \subset X$ of genus 0. Then

$$\mu(M) = (\operatorname{sign} X - F \cdot F)/8 \bmod 2. \tag{2.3}$$

This follows from Theorem 2.3. More generally, if $F \subset X$ is a characteristic surface of arbitrary genus then the Rokhlin invariant can be computed as

$$\mu(M) = (\operatorname{sign} X - F \cdot F)/8 + \operatorname{Arf}(X, F) \bmod 2, \tag{2.4}$$

where $\operatorname{Arf}(X, F)$ is the Arf–invariant associated with F. The formula (2.4) can be easily checked with the help of Theorem 2.4.

Example 2.9. Let n be an odd integer then lens space $L(n,1)$ is a $\mathbb{Z}/2$–homology sphere. By gluing a 2–handle to D^4 along an unknot in $S^3 = \partial D^4$ with framing $-n$, we obtain a smooth simply connected 4-manifold X_n with $\partial X_n = L(n,1)$ and $\operatorname{sign}(X_n) = -1$. The sphere F obtained by gluing together the axial disc of the 2–handle and a genus zero Seifert surface of the unknot is characteristic and $F \cdot F = -n$. Therefore, $\mu(L(n,1)) = (n-1)/8 \bmod 2$. Another approach is as follows. Suppose that n is positive (negative n can be taken care of by reverse of orientation). Lens space $L(n,1) = -L(n, n-1)$ bounds a spin 4–manifold Y obtained by plumbing on the chain shown in Figure 1.26 with $n-1$ vertices all weighted by 2. Then $\mu(L(n,1)) = \operatorname{sign}(Y)/8 = (n-1)/8 \bmod 2$. A general formula for the Rokhlin invariant of lens spaces can be found in Corollary 2.24.

2.3 Properties of the Rokhlin Invariant

The Rokhlin invariant of homology $\mathbb{Z}/2$–spheres has the following elementary properties. It is an invariant of orientation preserving homeomorphism. If $-\Sigma$ is Σ with opposite orientation then $\mu(-\Sigma) = -\mu(\Sigma) \bmod 2$. If Σ_0 and Σ_1 are $\mathbb{Z}/2$–homology spheres then their connected sum $\Sigma_0 \# \Sigma_1$ is a $\mathbb{Z}/2$–homology sphere, and $\mu(\Sigma_0 \# \Sigma_1) = \mu(\Sigma_0) + \mu(\Sigma_1)$. Some deeper properties of the Rokhlin invariant are discussed below.

2.3.1 Surgery Formula for the Rokhlin Invariant

Given a knot k in an integral homology 3–sphere Σ and a reduced fraction p/q, we denote the manifold obtained by (p/q)-surgery of Σ along k by $\Sigma + (p/q) \cdot k$. This manifold is an integral homology sphere if and only if $p = \pm 1$, and a $\mathbb{Z}/2$–homology sphere if and only if p is odd. By $\Delta_k(t)$ we denote the Alexander polynomial of k normalized so that $\Delta_k(1) = 1$ and $\Delta_k(t) = \Delta_k(t^{-1})$.

Theorem 2.10. *Let k be a knot in an integral homology sphere Σ, and $\Delta_k(t)$ the normalized Alexander polynomial of k. Then*

$$\mu(\Sigma + (1/q)\cdot k) = \mu(\Sigma) + (q/2)\cdot \Delta_k''(1) \bmod 2.$$

Proof. The theorem only needs to be proved for $q = 1$ as the remaining cases will follow by induction and by reverse of orientation. Let X and X' be oriented simply-connected smooth 4–manifolds with even intersection forms whose boundaries are $\partial X = \Sigma$ and $\partial X' = \Sigma' = \Sigma + k$ so that $\mu(\Sigma) = \operatorname{sign}(X)/8 \bmod 2$ and $\mu(\Sigma') = \operatorname{sign}(X')/8 \bmod 2$. Let V be the union of the manifold $[0,1]\times\Sigma$ and a 2–handle attached to $\{1\}\times\Sigma$ along k with framing 1. The boundary of V is $-\Sigma \cup \Sigma'$. The union of a Seifert surface of the knot k in $\{0\}\times\Sigma$ with the cylinder $[0,1]\times k \subset [0,1]\times\Sigma$ and with the axial disc of the 2–handle is a closed oriented surface F smoothly embedded in V with homological self-intersection 1. The manifold $W = X \cup V \cup (-X')$ is a closed smooth oriented simply-connected 4–manifold with intersection form $Q_W = Q_X \oplus Q_V \oplus Q_{-X'}$ and with characteristic surface $F \subset W$. According to Theorem 2.4 and Example 2.5,

$$\frac{1}{8}(\operatorname{sign} W - F\cdot F) = \operatorname{Arf}(W,F) = \frac{1}{2}\,\Delta_k''(1) \bmod 2.$$

On the other hand, $\operatorname{sign} W = \operatorname{sign} X + 1 - \operatorname{sign} X'$ and $F\cdot F = 1$, therefore,

$$\mu(\Sigma') - \mu(\Sigma) = \frac{1}{8}(\operatorname{sign} X - \operatorname{sign} X') = \frac{1}{2}\,\Delta_k''(1) \bmod 2.$$

Corollary 2.11. *If k is a knot in S^3 then*

$$\mu(S^3 + (1/q)\cdot k) = (q/2)\cdot \Delta_k''(1) \bmod 2.$$

Example 2.12. The normalized Alexander polynomial of (a,b)–torus knot k in S^3 is given by the formula

$$\Delta_k(t) = t^{-(a-1)(b-1)/2}\cdot\frac{(t^{ab}-1)(t-1)}{(t^a-1)(t^b-1)}$$

so that $(1/2)\cdot\Delta_k''(1) = (a^2-1)(b^2-1)/24$. Therefore, $\mu(\Sigma(a,b,1\pm mab)) = m\cdot(a^2-1)(b^2-1)/24 \bmod 2$ for any $m \in \mathbb{Z}$, compare with Example 1.2. A formula for the Rokhlin invariant of a general Seifert fibered homology sphere $\Sigma(a_1,\ldots,a_n)$ can be found in Section 2.4.

The following is an extension of the surgery formula for the Rokhlin invariant to rational surgeries of integral homology spheres.

Theorem 2.13. *Let k be a knot in an integral homology sphere Σ, and $\Delta_k(t)$ the normalized Alexander polynomial of k. Then for every odd p, the manifold $\Sigma + (p/q)\cdot k$ is a $\mathbb{Z}/2$–homology sphere, and*

$$\mu(\Sigma + (p/q)\cdot k) = \mu(\Sigma) - \mu(L(p,q)) + (q/2p)\cdot\Delta_k''(1) \bmod 2.$$

Recall that our orientation convention is that lens space $L(p,q)$ is obtained by $(-p/q)$–surgery on an unknot in S^3. The invariant $\mu(L(p,q))$ is calculated explicitly in Corollary 2.24. The proof of this theorem is omitted.

More general surgery formulas for the Rokhlin invariant involving surgery on knots and links in rational and $\mathbb{Z}/2$–homology spheres can be obtained by reducing modulo 2 the surgery formula (4.1), see also Corollary 4.5 and Theorem 4.9.

2.3.2 Surgery on Algebraically Split Links

The surgery formula of Theorem 2.10 admits the following interpretation. Let k be a knot in an integral homology sphere Σ and denote $Y = \Sigma + 0\cdot k$ the result of 0–surgery of Σ along k. Since $H^1(Y,\mathbb{Z}/2)=\mathbb{Z}/2$, the manifold Y admits two different spin structures, say, $\mathfrak{s}_1$ and $\mathfrak{s}_2$. Define $\mu'(k)=\mu(Y,\mathfrak{s}_1)+\mu(Y,\mathfrak{s}_2)$.

Proposition 2.14. *For any knot k in an integral homology sphere Σ,*

$$\mu'(k)=\mu(\Sigma+k)+\mu(\Sigma).$$

Proof. The manifold Y can be obtained by 0–surgery on both $\Sigma+k$ and Σ. Let W_1 and W_2 be the traces of these surgeries, that is, smooth 4–manifolds obtained from $[0,1]\times(\Sigma+k)$, respectively, $[0,1]\times\Sigma$, by attaching a two–handle along $\{1\}\times k$ with zero framing. Then W_1 is a spin cobordism between $\Sigma+k$ and $(Y,\mathfrak{s}_1)$, and W_2 is a spin cobordism between Σ and $(Y,\mathfrak{s}_2)$. Since the intersection forms of both W_1 and W_2 are zero, this proves the claim.

The statement of Theorem 2.10 is now equivalent to saying that $\mu'(k)$ equals $(1/2)\cdot\Delta_k''(1)$ mod 2, the Arf–invariant of k, compare with (3.1). This result can be generalized to links of two and more components as follows.

Let us consider a link $k\cup\ell$ of two components in an integral homology sphere Σ such that $\operatorname{lk}(k,\ell)=0$. The manifold $Y=\Sigma+0\cdot k+0\cdot\ell$ has $H^1(Y,\mathbb{Z}/2)=(\mathbb{Z}/2)^2$ hence admits four different spin structures $\mathfrak{s}_i$, $1\le i\le 4$. Define

$$\mu''(k,\ell)=\sum_{i=1}^{4}\mu(Y,\mathfrak{s}_i),$$

then an argument similar to that of Proposition 2.14 shows that

$$\mu''(k,\ell)=\mu(\Sigma+k+\ell)+\mu(\Sigma+k)+\mu(\Sigma+\ell)+\mu(\Sigma), \tag{2.5}$$

compare with (3.2). The invariant $\mu''(k,\ell)$ is also known as (the modulo 2 reduction of) the Sato–Levine invariant, see Section 3.3.5. With a bit of work, the formula (2.5) implies that this invariant vanishes if $k\cup\ell$ is a *boundary link*, that is, if k and ℓ bound disjoint Seifert surfaces in Σ.

Let $k\cup\ell\cup m$ be a link of three components in an integral homology sphere Σ such that $\operatorname{lk}(k,\ell)=\operatorname{lk}(k,m)=\operatorname{lk}(\ell,m)=0$. Then the manifold

$Y = \Sigma + 0 \cdot k + 0 \cdot \ell + 0 \cdot m$ has eight different spin–structures $\mathfrak{s}_i$, $1 \le i \le 8$, and we define

$$\mu'''(k, \ell, m) = \sum_{i=1}^{8} \mu(Y, \mathfrak{s}_i).$$

A formula similar to (2.5) expresses $\mu'''(k, \ell, m)$ in terms of the μ–invariants of the eight homology spheres obtained by respective surgeries on $k \cup \ell \cup m$:

$$\begin{aligned}\mu'''(k, \ell, m) &= \mu(\Sigma + k + \ell + m) + \mu(\Sigma + \ell + m) + \mu(\Sigma + k + m) \\ &+ \mu(\Sigma + k + \ell) + \mu(\Sigma + k) + \mu(\Sigma + \ell) + \mu(\Sigma + m) + \mu(\Sigma), \quad (2.6)\end{aligned}$$

compare with (3.12). Kaplan [156] identifies $\mu'''(k, \ell, m)$ with the Massey product of k, ℓ and m; equivalently, if a, b, and c is a basis in $H^1(Y, \mathbb{Z}/2)$ then $\mu'''(k, \ell, m) = (a \cup b \cup c)[Y] \bmod 2$, compare with Section 3.3.5.

A similar construction for algebraically split links of four or more components leads to μ–invariants which vanish identically, compare with Section 3.3.5.

2.3.3 Splicing and Mutation

The following two properties of the Rokhlin invariant are proved with the help of the surgery formula of Theorem 2.10.

Proposition 2.15. *Let Σ be obtained by splicing two integral homology spheres Σ_0 and Σ_1 along knots $k_0 \subset \Sigma_0$ and $k_1 \subset \Sigma_1$. Then $\mu(\Sigma) = \mu(\Sigma_0) + \mu(\Sigma_1)$.*

Proof. We saw in Section 1.1.7 that $\Sigma = \Sigma_0' \# \Sigma_1' + k_0^* \# k_1^*$ where $\Sigma_i' = \Sigma_i - k_i$ and k_i^* is the image in $\Sigma_i - k_i$ of the canonical longitude of k_i, $i = 0, 1$. Since the knot k_i and its canonical longitude form a boundary link in Σ_i, we have that

$$\Delta_{k_i \subset \Sigma_i}(t) = \Delta_{k_i^* \subset \Sigma_i'}(t).$$

The normalized Alexander polynomial of the connected sum $k_0^* \# k_1^*$ is the product of the normalized Alexander polynomials of the components, hence

$$\Delta''_{k_0^* \# k_1^*}(1) = \Delta''_{k_0^*}(1) + \Delta''_{k_1^*}(1),$$

which together with the surgery formula completes the proof.

A surgery formula based proof of the following result can be found in [163], see also Theorem 3.20.

Proposition 2.16. *Let Σ be an integral homology sphere and Σ^τ be obtained from Σ by mutation, see Section 1.1.11. Then $\mu(\Sigma^\tau) = \mu(\Sigma)$.*

2.3.4 Rokhlin Invariant of Branched Coverings

Let k be a knot in an integral homology sphere Σ. For any real number α between 0 and 1, the *equivariant knot signature* $\operatorname{sign}^\alpha(k)$ is defined as the signature of the Hermitian form

$$(1 - e^{2\pi i\alpha})S + (1 - e^{-2\pi i\alpha})S^t, \tag{2.7}$$

where S is a Seifert matrix of the knot k, and S^t is its transpose. The equivariant signature $\operatorname{sign}^{1/2}(k)$ is usually referred to as the *knot signature* and is denoted by $\operatorname{sign}(k)$.

Theorem 2.17. *Let Σ_n be a $\mathbb{Z}/2$–homology sphere represented as an n–fold cyclic branched covering of an integral homology sphere Σ with branch set a knot k. Then*

$$\mu(\Sigma_n) = n \cdot \mu(\Sigma) + \frac{1}{8} \sum_{m=1}^{n-1} \operatorname{sign}^{m/n}(k) \mod 2.$$

In the special case of $n = 2$, we obtain $\mu(\Sigma_2) = (1/8) \cdot \operatorname{sign}(k) \mod 2$.

Proof. Let X be a smooth compact simply-connected manifold with $\partial X = \Sigma$ and with even intersection form. Let $F \subset \Sigma$ be a Seifert surface of k, and push its interior slightly into X so that it intersects Σ only along k. The n–fold branched cover X_n of X whose branch set is the surface F is a smooth compact spin manifold with $\partial X_n = \Sigma_n$. The signatures of X_n and X are related by the formula

$$\operatorname{sign} X_n = n \cdot \operatorname{sign} X + \sum_{m=1}^{n-1} \operatorname{sign}^{m/n}(k),$$

see Viro [301]. Dividing by 8 and reducing modulo 2, we obtain the desired formula for the Rokhlin invariant.

Every closed oriented 3–manifold M can be represented as an (irregular) 3–fold branched cover of S^3, see Section 1.1.12. A formula expressing $\mu(M)$ in terms of such a representation can be found in [50].

2.3.5 Birman–Craggs Homomorphisms

Let $\mathcal{T}_g$ be the Torelli group, see Section 1.1.13. Birman and Craggs [28] used the Rokhlin invariant to produce a finite collection of homomorphisms from $\mathcal{T}_g$ to $\mathbb{Z}/2$ as follows. Choose an embedding $h : F \to S^3$ and identify F with its image. For any $f \in \mathcal{T}_g$, split S^3 along F and reglue the two pieces by the homeomorphism f. Since f is identity on the first integral homology, the resulting manifold Σ_f is an integral homology sphere, and its Rokhlin invariant

$\mu(\Sigma_f) \in \mathbb{Z}/2$ is well defined. Birman and Craggs showed that, for a fixed embedding h, this defines a homomorphism $\mathcal{T}_g \to \mathbb{Z}/2$. They also showed that, although this homomorphism depends on the choice of h, it is not very sensitive to it, and that in fact only a finite number of distinct homomorphisms can arise from different choices of h. A number of results in [28] and [154] relate the value of $\mu(\Sigma_f)$ to the membership of f in various subgroups of $\mathcal{T}_g$. The definition of the Birman–Craggs homomorphisms naturally extends to all $\mathbb{Z}/2$–homology spheres.

2.3.6 Homology Cobordism Invariance

Let R be a commutative ring with an identity element and let Σ_0 and Σ_1 be R–homology 3–spheres. We say that Σ_0 and Σ_1 are *R–homology cobordant* if there exists a smooth compact oriented 4–manifold W with $\partial W = -\Sigma_0 \cup \Sigma_1$ such that the inclusions $\Sigma_i \to W$ induce isomorphisms in R–homology.

The set of equivalence classes of oriented R–homology 3–spheres under this relation is denoted Θ^3_R. With the operation of connected sum, Θ^3_R is an abelian group, called the *homology cobordism group* of oriented R–homology 3–spheres. The additive inverse in this group is obtained by a reverse of orientation.

Best studied is the group $\Theta^3_{\mathbb{Z}}$ of homology cobordisms of oriented integral homology 3–spheres. The Rokhlin invariant μ is obviously an invariant of integral homology cobordism hence it defines a homomorphism

$$\mu : \Theta^3_{\mathbb{Z}} \to \mathbb{Z}/2$$

called the *Rokhlin homomorphism*. The group $\Theta^3_{\mathbb{Z}}$ and the homomorphism μ have played a distinguished role in the topology of manifolds. This role will be discussed in Section 3.6 and in great detail in Chapter 7.

The Rokhlin invariant μ of $\mathbb{Z}/2$–homology spheres is an invariant of $\mathbb{Z}/2$–homology cobordism so it defines a homomorphism on the group $\Theta^3_{\mathbb{Z}/2}$ taking rational values modulo 2. The latter is also called the Rokhlin homomorphism.

Example 2.18. Poincaré homology sphere $\Sigma(2,3,5)$ is not homology cobordant to S^3 because $\mu(\Sigma(2,3,5)) \neq 0$, see Example 2.8.

Example 2.19. Let M be a $\mathbb{Z}/2$–homology 3–sphere. If M is embeddable in $\mathbb{R}^4$ then $\mu(M) = 0 \bmod 2$ because then M is homology cobordant to S^3 (the latter follows from the Alexander duality). For example, Poincaré homology sphere $\Sigma(2,3,5)$ can not be embedded in $\mathbb{R}^4$.

Example 2.20. According to Whitehead [304], two lens spaces $L(p,q)$ and $L(p,q')$ are homotopy equivalent if and only if either qq' or $-qq'$ is a quadratic residue modulo p. For example, $L(7,1)$ and $L(7,2)$ are homotopy equivalent. However, they are neither homeomorphic (orientation preserving or otherwise) nor $\mathbb{Z}/2$–homology cobordant because $\mu(L(7,2)) = -1/4 \bmod 2$ and $\mu(L(7,1)) = 3/4 \bmod 2$.

Example 2.21. Let W be the 4–manifold obtained by attaching a 2–handle to D^4 along the (-1)–framed left-handed trefoil in $S^3 = \partial D^4$. Then ∂W is the Poincaré homology sphere $\Sigma(2,3,5)$ and $H_2(W,\mathbb{Z}) = \mathbb{Z}$. Let $\eta \in H_2(W,\mathbb{Z})$ be a generator then η can be represented by an embedded torus obtained by gluing together a genus one Seifert surface of the trefoil with the axial disc in the 2–handle. However, η cannot be represented by an embedded sphere: a regular neighborhood N of such a sphere would have boundary S^3, and $W \setminus \operatorname{int}(N)$ would then be a homology cobordism between S^3 and $\Sigma(2,3,5)$, a contradiction. Note that the manifold W is homotopy equivalent to the 2–sphere. For more on such manifolds, called *homotopy 2–spheres*, see [268].

2.4 The Rokhlin Invariant of Seifert Fibered and Graph Homology Spheres

In this section we provide both an algorithm and a closed form formula for calculating the Rokhlin invariant of Seifert fibered and plumbed integral and $\mathbb{Z}/2$–homology spheres.

2.4.1 The Algorithm

Let Σ be a plumbed $\mathbb{Z}/2$–homology sphere. It can be thought of as the boundary of a 4–manifold $P(\Gamma)$ obtained by plumbing on a plumbing graph Γ, see Section 1.1.9. Recall that Σ is a $\mathbb{Z}/2$–homology sphere if and only if $\det A(\Gamma) = 1 \bmod 2$ where $A(\Gamma)$ is the matrix (1.8). The manifold Σ is an integral homology sphere if and only if $\det A(\Gamma) = \pm 1$. To calculate the Rokhlin invariant of Σ we will first compute the signature of the matrix $A(\Gamma)$ and then find a characteristic surface $F \subset P(\Gamma)$.

There is a simple algorithm to diagonalize matrices of the form $A(\Gamma)$, see [73] and [76], which is essentially a graph realization of the Gauss diagonalization process. Let us allow trees weighted by arbitrary rational numbers. Given such a tree, pick a vertex and direct all its edges toward this vertex. The tree can be simplified by performing operations of the two types shown in Figures 2.1 and 2.2 (the vertices can be re-indexed if necessary).

We end up with a finite collection of isolated points weighted by rational numbers $d_1, \dots, d_s$. The diagonal matrix D with diagonal entries $d_1, \dots, d_s$ is a diagonalization of $A(\Gamma)$, that is, $A(\Gamma) = U^t D U$ where $\det U = \pm 1$. In particular, $\det A(\Gamma) = \det D$ and $\operatorname{sign} A(\Gamma) = \operatorname{sign} D$.

Example 2.22. The algorithm described above generalizes the concept of continued fraction: the diagonalization of the matrix $A(\Gamma)$ corresponding to the tree in Figure 1.26 directed toward its left vertex has diagonal entries

$$[x_1, \dots, x_{n-1}, x_n], [x_2, \dots, x_{n-1}, x_n], \dots, [x_{n-1}, x_n], [x_n]$$

where [,] stands for continued fraction, see (1.9). The determinant of $A(\Gamma)$ equals (up to a sign) the numerator of the reduced fraction $[x_1, \dots, x_{n-1}, x_n]$.

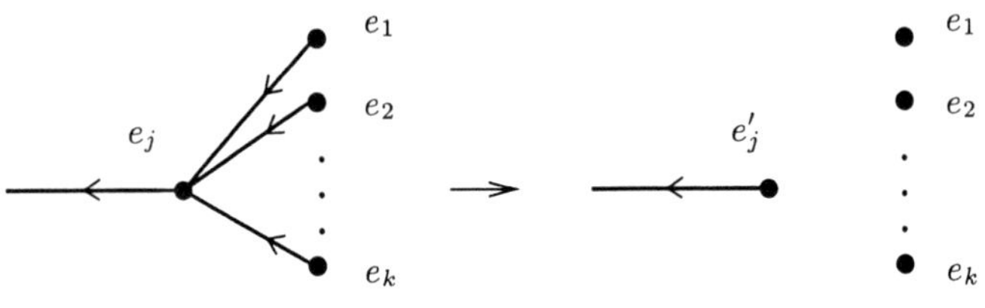

$$e'_j = e_j - \frac{1}{e_1} - \ldots - \frac{1}{e_k}$$

Fig. 2.1.

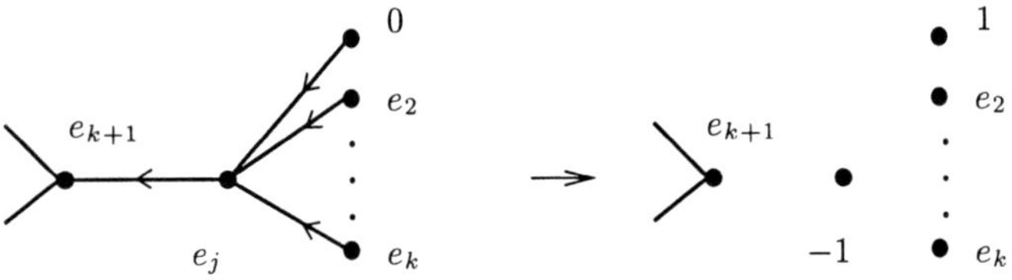

Fig. 2.2.

With this algorithm in place, the signature of $A(\Gamma)$ can be easily computed. Now, to find a characteristic surface F in $P(\Gamma)$, we simply solve the linear system (2.2) over $\mathbb{Z}/2$. In the canonical basis of $H_2(P(\Gamma), \mathbb{Z}/2)$ represented by zero–sections of plumbed bundles, the solution assigns either 0 or 1 to each vertex of the plumbing graph Γ. The vertices with assigned 1 form a subset $\mathcal{S}$ of the vertices of Γ called the *Wu set*. One can show that two vertices connected by an edge in Γ cannot both belong to $\mathcal{S}$. Since the canonical basis is spherical, we connect the 2–spheres corresponding to the vertices in $\mathcal{S}$ by tubes to obtain a spherical characteristic surface F. The Rokhlin invariant is then computed by the formula

$$\mu(\Sigma) = \frac{1}{8}\,(\operatorname{sign} P(\Gamma) - F \cdot F) = \frac{1}{8}\,(\operatorname{sign} A(\Gamma) - \sum_{i \in \mathcal{S}} e_i) \bmod 2.$$

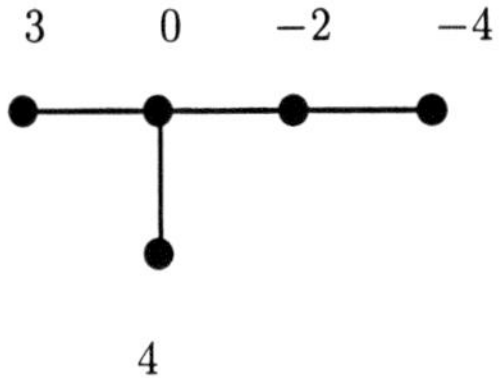

Fig. 2.3.

Example 2.23. Brieskorn homology sphere $\Sigma(3,4,7)$ can be obtained by plumbing on the graph in Figure 2.3. By the algorithm above, $\operatorname{sign} P(\Gamma) = -1$. To find the characteristic surface, consider the matrix

$$A(\Gamma) = \begin{pmatrix} 0 & 1 & 1 & 1 & 0 \\ 1 & 3 & 0 & 0 & 0 \\ 1 & 0 & 4 & 0 & 0 \\ 1 & 0 & 0 & -2 & 1 \\ 0 & 0 & 0 & 1 & -4 \end{pmatrix}$$

and the system $A(\Gamma)\varepsilon = \operatorname{diag} A(\Gamma) \bmod 2$. This system is of the form

$$\begin{aligned} \varepsilon_1 + \varepsilon_2 + \varepsilon_3 \quad &= 0 \\ \varepsilon_0 + \varepsilon_1 \quad &= 1 \\ \varepsilon_0 \quad &= 0 \\ \varepsilon_0 + \quad \varepsilon_4 &= 0 \\ \varepsilon_3 \quad &= 0 \end{aligned}$$

and its only solution is $\varepsilon = (0,1,1,0,0)$. Thus the Wu set $\mathcal{S}$ consists of the vertices weighted by 3 and 4, and $\mu(\Sigma(2,3,7)) = (-1-7)/8 = 1 \bmod 2$.

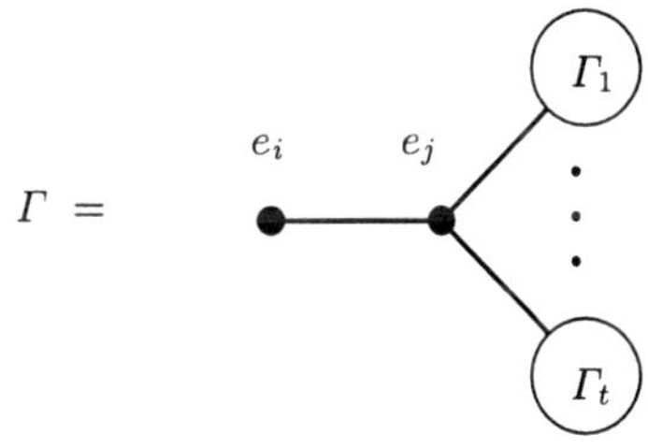

Fig. 2.4.

There is an algorithm which allows to find the Wu set of a plumbing graph without solving big linear systems, see [242]. Given a plumbing graph Γ of the form shown in Figure 2.4, perform one of the following two moves.

- *Move 1.* If e_i is even, replace Γ by the disjoint union Γ' of $\Gamma_1, \dots, \Gamma_t$.
- *Move 2.* If e_i is odd, replace Γ by Γ' shown in Figure 2.5.

Repeating this operation sufficiently many times, we end up with a collection of isolated points. Then $\Sigma = \partial P(\Gamma)$ is a $\mathbb{Z}/2$–homology sphere if and only if all weights on these points are odd. If Σ is a $\mathbb{Z}/2$–homology sphere, define the subset $\mathcal{S}(\Gamma)$ of the vertices of Γ inductively as follows.

- If Γ_0 is a set of isolated points with odd weights, put $\mathcal{S}(\Gamma_0)$ equal to the set of these points.

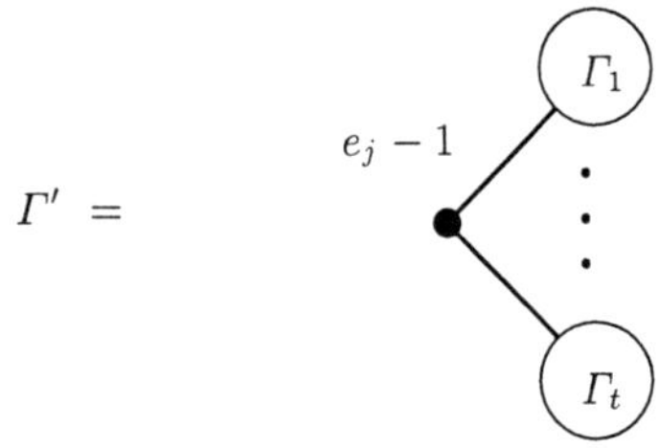

Fig. 2.5.

- If $\mathcal{S}(\Gamma')$ is known and Γ reduces to Γ' by Move 1 above put $\mathcal{S}(\Gamma) = \mathcal{S}(\Gamma') \cup \{ i \}$ or $\mathcal{S}(\Gamma) = \mathcal{S}(\Gamma')$ according as the number of points in $\mathcal{S}(\Gamma')$ adjacent to vertex j is congruent to $e_j - 1$ or e_j modulo 2.
- If Γ reduces to Γ' by Move 2, put $\mathcal{S}(\Gamma) = \mathcal{S}(\Gamma') \cup \{ i \}$ if j is not in $\mathcal{S}(\Gamma')$, and put $\mathcal{S}(\Gamma) = \mathcal{S}(\Gamma')$ if $j \in \mathcal{S}(\Gamma')$.

Then $\mathcal{S}(\Gamma)$ is the Wu set of the graph Γ.

2.4.2 The Formula

The above algorithms can be converted into a closed form formula for the Rokhlin invariant of graph $\mathbb{Z}/2$–homology spheres. We will first state it for Seifert fibered homology spheres.

Let Σ be a Seifert fibered $\mathbb{Z}/2$–homology sphere with unnormalized Seifert invariants $\{ 0;\ (a_1, b_1), \dots, (a_n, b_n) \}$. Then either

(1) all a_i are odd and $\sum b_i$ is odd, or
(2) exactly one of the a_i is even, say a_1.

According to [242], in these two cases the μ-invariant of Σ is given respectively by

$$(1) \quad \mu(\Sigma) = (\sum (c(a_i, b_i) + \operatorname{sign} b_i) + \operatorname{sign} e(\Sigma))/8 \mod 2, \text{ and} \tag{2.8}$$

$$(2) \quad \mu(\Sigma) = (\sum c(a_i - b_i, a_i) + \operatorname{sign} e(\Sigma))/8 \mod 2, \tag{2.9}$$

where $c(a, b)$ is the function described below, $e(\Sigma) = \sum b_i/a_i$, and, in the second case, we are to choose the Seifert invariants b_i so that $(a_i - b_i)$ are odd for all i.

The integers $c(a, b)$ are defined for coprime integer pairs (a, b) with a odd. They are uniquely determined by the recursion:

$$\begin{aligned} &c(a, \pm 1) = 0 \text{ for any odd } a, \\ &c(a \pm 2b, b) = c(a, b), \\ &c(a, b + a) = c(a, b) + \operatorname{sign} b(b + a), \\ &c(a, b) = -c(-a, b) = -c(a, -b). \end{aligned}$$

These recursion formulas, in fact, give the fastest computation of $c(p,q)$ in practice, but various other descriptions of $c(p,q)$ are known, see [242] and also [146] and [147]. For example,

$$\begin{aligned} c(a,b) &= \sum (-1)^i \,\#\, \{\, k \mid 0 < k < b\,,\ i < \frac{ka}{b} < i+1 \,\} \\ &= -\frac{1}{b} \sum_{\eta^b=-1} \frac{(\eta+1)(\eta^a+1)}{(\eta-1)(\eta^a-1)}\,, \\ &= b-1-4\cdot \#\{\, 1 \le i \le \frac{b-1}{2} \mid \frac{b-1}{2} < ai < b \,(\operatorname{mod} b)\,\}, \quad b \text{ odd}. \end{aligned}$$

The following formula is obtained by applying the general formula to a Seifert manifold with one singular fiber.

Corollary 2.24. *Suppose that both p and q are positive and q is odd. If p is odd then $\mu(L(q,p)) = c(p,q)/8$ modulo 2; if p is even then $\mu(L(q,p)) = (c(p-q,p)-1)/8$ modulo 2.*

The above formulas can be used to compute the Rokhlin invariant of all graph homology spheres by using the splicing additivity of μ, see Proposition 2.15.

3 Casson Invariant

Casson defined his λ–invariant for integral homology 3–spheres in a series of lectures at MSRI in 1985. Detailed expositions can be found in [5], [133] and [275]. We first give an axiomatic definition of Casson's invariant and prove its uniqueness and some basic properties. Then we prove existence by providing an explicit construction of λ. Deeper properties of the invariant together with some applications are described in the last three sections of this chapter.

3.1 Definition of the Casson Invariant

Let $\mathcal{S}$ be the class of oriented integral homology 3–spheres modulo orientation preserving homeomorphism. A *Casson invariant* is a map $\lambda : \mathcal{S} \to \mathbb{Z}$ having properties (0), (1) and (2) below.

(0) *$\lambda(S^3) = 0$, and $\lambda(\mathcal{S})$ is not contained in any proper subgroup of $\mathbb{Z}$.*

(1) *For any integral homology sphere Σ, a knot $k \subset \Sigma$, and an integer m, the difference*

$$\lambda\left(\Sigma + \frac{1}{m+1} \cdot k\right) - \lambda\left(\Sigma + \frac{1}{m} \cdot k\right) \tag{3.1}$$

is independent of m.

Therefore, the difference (3.1) is an invariant of the knot $k \subset \Sigma$. It is denoted by $\lambda'(k \subset \Sigma)$ or simply $\lambda'(k)$ and is called a *Casson invariant of a knot.* Let $k \cup \ell$ be a link in a homology sphere Σ with $\operatorname{lk}(k, \ell) = 0$. Then for any integers m and n, the manifold

$$\Sigma + \frac{1}{m} \cdot k + \frac{1}{n} \cdot \ell$$

is an integral homology sphere. The quantity

$$\begin{aligned}&\lambda\left(\Sigma+\frac{1}{m+1}\cdot k+\frac{1}{n+1}\cdot\ell\right)-\lambda\left(\Sigma+\frac{1}{m}\cdot k+\frac{1}{n+1}\cdot\ell\right)\\&-\lambda\left(\Sigma+\frac{1}{m+1}\cdot k+\frac{1}{n}\cdot\ell\right)+\lambda\left(\Sigma+\frac{1}{m}\cdot k+\frac{1}{n}\cdot\ell\right)\\&=\lambda'\left(k\subset\Sigma+\frac{1}{n+1}\cdot\ell\right)-\lambda'\left(k\subset\Sigma+\frac{1}{n}\cdot\ell\right)\\&=\lambda'\left(\ell\subset\Sigma+\frac{1}{m+1}\cdot k\right)-\lambda'\left(\ell\subset\Sigma+\frac{1}{m}\cdot k\right)\end{aligned}\tag{3.2}$$

is independent of both m and n, and is denoted by $\lambda''(k,\ell\subset\Sigma)$ or $\lambda''(k,\ell)$. A two component link $k\cup\ell$ in an integral homology sphere Σ is called a *boundary link* if k and ℓ bound disjoint Seifert surfaces in Σ. If a link $k\cup\ell$ is boundary then $\mathrm{lk}(k,\ell)=0$ but not vice versa.

(2) $\lambda''(k,\ell\subset\Sigma)=0$ *for any boundary link $k\cup\ell$ in a homology sphere Σ.*

The existence and uniqueness of a Casson invariant together with some basic properties are given by the following theorem of Casson.

Theorem 3.1. *There exists a Casson invariant λ which is unique up to sign. Moreover, it has the following properties:*

(3) $\lambda'(\text{trefoil})=\pm1$.

(4) $\lambda'(k\subset\Sigma)=(1/2)\cdot\Delta''_{k\subset\Sigma}(1)\cdot\lambda'(\text{trefoil})$ *for any knot $k\subset\Sigma$.*

(5) $\lambda(\Sigma)=\mu(\Sigma)\bmod 2$ *where μ is the Rokhlin invariant.*

(6) $\lambda(-\Sigma)=-\lambda(\Sigma)$ *where $-\Sigma$ stands for Σ with reversed orientation.*

(7) $\lambda(\Sigma_0\#\Sigma_1)=\lambda(\Sigma_0)+\lambda(\Sigma_1)$.

(8) *If $\lambda(\Sigma)\neq0$ then $\pi_1\Sigma$ has a non-trivial representation in $SU(2)$.*

Here, the Alexander polynomial $\Delta_k(t)$ is normalized so that $\Delta_k(1)=1$ and $\Delta_k(t)=\Delta_k(t^{-1})$. By a trefoil we mean either a left-handed or a right-handed trefoil; by property (4) their λ'–invariants coincide.

Once we prove the above theorem, we will agree to normalize the Casson invariant by setting $\lambda'(\text{trefoil})=+1$, in which case properties (1) and (4) will imply that

$$\lambda\left(\Sigma+\frac{1}{q}\cdot k\right)=\lambda(\Sigma)+\frac{q}{2}\,\Delta''_{k\subset\Sigma}(1).\tag{3.3}$$

This formula is referred to as the *surgery formula* for the Casson invariant.

Remark 3.2. With the above normalization, the Casson invariant of Poincaré homology sphere $\Sigma(2,3,5)$ equals -1 because $\Sigma(2,3,5)$ is obtained by (-1)–surgery on the left-handed trefoil k with $\Delta_k(t)=-1+t+t^{-1}$.

Remark 3.3. Concerning property (5), it should be noted that the Casson invariant lifts the Rokhlin invariant of homology spheres but not the Rokhlin homomorphism $\mu : \Theta^3_{\mathbb{Z}} \to \mathbb{Z}/2$ because λ is not an invariant of homology cobordism. For example, the (-1)–surgery on the square knot k shown in Figure 7.4 results in an integral homology sphere Σ which is homology cobordant to zero, see Example 7.9. On the other hand, an easy calculation shows that $\Delta''_k(1) = 4$, hence $\lambda(\Sigma) = -2$ by the surgery formula.

The proof of Theorem 3.1 is accomplished in several steps. We first prove that (0), (1), (3), and (4) imply uniqueness. This follows from the observation that any integral homology sphere Σ can be obtained by surgery on an algebraically split link framed by ± 1, see Section 1.1.6. More precisely, there exists a link $k_1 \cup \ldots \cup k_n$ in S^3 such that $\Sigma = S^3 + \varepsilon_1 \cdot k_1 + \ldots + \varepsilon_n \cdot k_n$ with all $\varepsilon_i = \pm 1$ and $\operatorname{lk}(k_i, k_j) = 0$. Introduce the notation $\Sigma_i = S^3 + \varepsilon_1 \cdot k_1 + \ldots + \varepsilon_i \cdot k_i$ for $i = 0, 1, \ldots, n$, so that $\Sigma_0 = S^3$ and $\Sigma_n = \Sigma$. Then

$$\begin{aligned}\lambda(\Sigma) &= \lambda(\Sigma_n) \\ &= (\lambda(\Sigma_n) - \lambda(\Sigma_{n-1})) + (\lambda(\Sigma_{n-1}) - \lambda(\Sigma_{n-2})) + \ldots \\ &\quad + (\lambda(\Sigma_1) - \lambda(\Sigma_0)) + \lambda(\Sigma_0),\end{aligned}$$

where $\lambda(\Sigma_0) = \lambda(S^3) = 0$. Properties (1) and (4) imply that $\lambda(\Sigma_i) - \lambda(\Sigma_{i-1}) = \varepsilon_i \cdot \lambda'(k_i \subset \Sigma_{i-1})$. Therefore,

$$\lambda(\Sigma) = \sum_{i=1}^{n} \varepsilon_i \cdot \lambda'(k_i \subset \Sigma_{i-1}) = \left(\sum_{i=1}^{n} \frac{\varepsilon_i}{2} \cdot \Delta''_{k_i \subset \Sigma_{i-1}}(1) \right) \cdot \lambda'(\text{trefoil}), \quad (3.4)$$

so that λ is unique up to the choice of $\lambda'(\text{trefoil}) = \pm 1$.

In our next step, we prove that (0), (1), (3), and (4) imply (5), (6), and (7). Property (5) is immediate from (3.4) and the surgery formula for the Rokhlin invariant, see Theorem 2.10. Properties (6) and (7) follow easily from (3.4) and properties of the Alexander polynomial.

Finally, we show that (0), (1), and (2) imply (3) and (4). This part of the proof is more technically involved. One first uses general position argument to reduce the problem to knots in S^3 by showing that, for any knot k in an integral homology 3–sphere Σ, there exists a knot ℓ in S^3 such that $\lambda'(k \subset \Sigma) = \lambda'(\ell \subset S^3)$ and $\Delta_{k \subset \Sigma}(t) = \Delta_{\ell \subset S^3}(t)$. Next, every knot in S^3 can be made into an unknot by changing crossings in its knot diagram. One uses properties of the Alexander polynomial, property (2), and some Kirby calculus to show that $\lambda'(\ell)$ and $\Delta''_\ell(1)$ change by the same rule as one changes a crossing, see [5], [133] or [275] for details.

This completes the proof of Theorem 3.1 except for the existence part. Existence of a Casson invariant satisfying (0), (1), and (2) will be proved in Section 3.2 by providing an explicit construction. Property (8) will follow immediately from that construction.

3.2 Construction of the Casson Invariant

The Casson invariant is one of the many invariants of a closed 3–manifold M that can be obtained by studying representations of its fundamental group in a certain non-abelian group G. It is one of the characteristic features of $\pi_1(M)$ as a group that its representation variety $\mathcal{R}(M)$ tends to be discrete, in a proper sense. While $\mathcal{R}(M)$ itself only depends on the homotopy type of M, various geometric structures associated with the manifold can be employed to obtain topological invariants via a (creative) count of points in $\mathcal{R}(M)$.

In particular, the Casson invariant of an integral homology sphere M can be obtained by counting representations of $\pi_1(M)$ in $G = SU(2)$. The geometric structure used to obtain a topological invariant is a Heegaard splitting of M and the symplectic geometry associated with it. An alternative gauge-theoretical approach, which will be discussed in Chapter 5, uses flat bundles together with a Riemannian metric on M. The latter construction leads to a refinement of the Casson invariant, the Floer homology, see Chapter 6.

This section is organized as follows. We first review necessary facts about character varieties, and then define the Casson invariant as an intersection number. After that, we check the properties (0), (1) and (2). As we have seen, they imply properties (3) through (7). Property (8) is immediate from the construction.

3.2.1 $SU(2)$–Representation Spaces

Let π be a finitely presented group. The $SU(2)$–representation space of π is the space $R(\pi) = \operatorname{Hom}(\pi, SU(2))$ with the compact open topology, where π is assumed to have the discrete topology. The space $R(\pi)$ can be turned into a real algebraic variety by thinking of $SU(2)$ as a unit sphere in $\mathbb{R}^4$, so that every generator of π is represented by a vector of unit length. The relations in π are then viewed as polynomial equations on these vectors. In particular, $R(\pi)$ is always compact.

The group $SU(2)$ acts on $R(\pi)$ by conjugation. The quotient $\mathcal{R}(\pi) = R(\pi)/SU(2)$ under this action is called the $SU(2)$–*character variety* of π. It is a (compact) real algebraic variety.

A representation $\alpha \in R(\pi)$ is called *irreducible* if and only if its stabilizer coincides with the center of $SU(2)$, that is, with $\{\pm 1\}$. It is called *reducible* otherwise. Reducible representations $\alpha \in R(\pi)$ are precisely representations of π in $U(1)$. In particular, they form a closed subset in $R(\pi)$. The group $SO(3) = SU(2)/\pm 1$ acts freely on the open set of irreducible representations, which we denote by $R^*(\pi)$, with the quotient $\mathcal{R}^*(\pi) = R^*(\pi)/SO(3) \subset \mathcal{R}(\pi)$. Thus $\mathcal{R}^*(\pi)$ may inherit singularities from $R^*(\pi)$, but it does not acquire any quotient singularities.

We will subdivide reducible $SU(2)$ representations into two categories: those with the image in $\{\pm 1\} \subset SU(2)$ will be called *central*, and the rest will be called *abelian*. The central representation θ with image $\{1\}$ is called *trivial*.

The Zariski tangent space $T_\alpha \mathcal{R}(\pi)$ to $\mathcal{R}(\pi)$ at $\alpha \in \mathcal{R}(\pi)$ can be identified as in [123] with $H^1(\pi, \operatorname{ad}\alpha)$, the first group cohomology of π with coefficients in the representation

$$\pi \xrightarrow{\ \alpha\ } SU(2) \xrightarrow{\ \operatorname{ad}\ } \operatorname{Aut}(\mathfrak{su}(2)). \tag{3.5}$$

Correspondingly, the tangent space $T_\alpha R(\pi)$ is identified with the vector space of 1–cocycles of π with coefficients in $\operatorname{ad}\alpha$.

Any compact manifold M of dimension less than or equal to three is homeomorphic to a finite simplicial complex, therefore, its fundamental group is finitely presentable, and the above discussion applies to $\pi = \pi_1 M$. We will use notation $\mathcal{R}(M)$ for $\mathcal{R}(\pi_1 M)$ etc. In particular,

$$T_\alpha \mathcal{R}(M) = H^1(\pi, \operatorname{ad}\alpha) = H^1(M, \operatorname{ad}\alpha).$$

Example 3.4. A handlebody M of genus g has free fundamental group of rank g so that $R(M)$ is a product of g copies of $SU(2)$, and $\mathcal{R}^*(M)$ is a smooth open manifold of dimension $3g - 3$.

Example 3.5. Let F be a closed oriented Riemann surface of genus g and let $F_0 = F \setminus D^2$, where D^2 is an open disc in F. A choice of orientation of F induces an orientation on the curve $\gamma = \partial F_0$ which provides for a well-defined map

$$h : R(F_0) \to SU(2), \quad h(\alpha) = \alpha(\gamma), \tag{3.6}$$

so that $R(F) = h^{-1}(1)$. The group $\pi_1 F_0$ is a free group of rank $2g$, therefore, h can be thought of as a map $h : SU(2)^{2g} \to SU(2)$. The map h is surjective. One can show by an explicit calculation that it is regular at irreducible representations and only at them; in other words, $d_\alpha h : T_\alpha \mathcal{R}(F) \to T_{\alpha(\gamma)} SU(2)$ is onto if and only if α is irreducible. Therefore, by removing reducible representations from $R(F)$, we also get rid of all possible singularities. This shows that $\mathcal{R}^*(F)$ is a smooth open manifold of dimension $6g - 6$.

3.2.2 The Intersection Theory

Let Σ be an integral homology 3–sphere, $M_1 \cup_F M_2$ a Heegaard splitting of Σ, see Section 1.1.13, and $F_0 = F \setminus D^2$. We have the following commutative diagram of fundamental groups:

$$\begin{array}{ccccc}
 & & & \pi_1 M_1 & & \\
 & & \nearrow & & \searrow & \\
\pi_1 F_0 \longrightarrow & \pi_1 F & & & & \pi_1 \Sigma \\
 & & \searrow & & \nearrow & \\
 & & & \pi_1 M_2 & &
\end{array}$$

All homomorphisms in this diagram are surjective, therefore, irreducible representations of $\pi_1\Sigma$ naturally induce irreducible representations of $\pi_1 F_0$, $\pi_1 F$ and $\pi_1 M_k$, $k = 1, 2$. Restricting ourselves to irreducible representations and passing to the free $SO(3)$ quotients, we obtain the following commutative diagram, where $i_k : F \to M_k$ and $j_k : M_k \to \Sigma$ denote inclusions:

$$\begin{array}{ccccc} & & \mathcal{R}^*(M_1) & & \\ & \swarrow^{i_1^*} & & \nwarrow^{j_1^*} & \\ \mathcal{R}^*(F_0) \longleftarrow \mathcal{R}^*(F) & & & & \mathcal{R}^*(\Sigma) \\ & \nwarrow_{i_2^*} & & \swarrow_{j_2^*} & \\ & & \mathcal{R}^*(M_2) & & \end{array}$$

All maps in this diagram are injective, therefore, we can view $\mathcal{R}^*(\Sigma)$ as the intersection of $\mathcal{R}^*(M_1)$ and $\mathcal{R}^*(M_2)$ inside $\mathcal{R}^*(F)$.

Lemma 3.6. *The intersection of open manifolds* $\mathcal{R}^*(M_1) \cap \mathcal{R}^*(M_2) = \mathcal{R}^*(\Sigma)$ *in* $\mathcal{R}^*(F)$ *is compact.*

Proof. First note that any reducible $\alpha : \pi_1\Sigma \to SU(2)$ factors through a copy of $U(1) \subset SU(2)$. The group $U(1)$ is abelian, therefore, we obtain a factorization $\alpha : \pi_1\Sigma \to H_1(\Sigma, \mathbb{Z}) \to U(1) \to SU(2)$. Since $H_1(\Sigma, \mathbb{Z})$ is trivial, we conclude that $\alpha = \theta$, the trivial representation. Thus the only non-compactness in $\mathcal{R}^*(M_1) \cap \mathcal{R}^*(M_2)$ may come from the removal of the trivial representation. This does not really happen because the trivial representation θ is an isolated point in $R(M_1) \cap R(M_2)$.

To prove this, it is sufficient to show that the intersection $R(M_1) \cap R(M_2)$ is transversal at θ. The latter is equivalent to the assertion that $i_1^*\, T_\theta R(M_1) \oplus i_2^*\, T_\theta R(M_2) = T_\theta R(F_0)$. With the identifications $T_\theta R(M_k) = H^1(M_k, \mathfrak{su}(2))$, $k = 1, 2$, and $T_\theta R(F_0) = H^1(F_0, \mathfrak{su}(2)) = H^1(F, \mathfrak{su}(2))$, the space $i_1^*\, T_\theta R(M_1) \oplus i_2^*\, T_\theta R(M_2)$ is the image of the map

$$i_1^* \oplus i_2^* : H^1(M_1, \mathfrak{su}(2)) \oplus H^1(M_2, \mathfrak{su}(2)) \to H^1(F, \mathfrak{su}(2)) \qquad (3.7)$$

which can be included into the Mayer–Vietoris sequence

$$\ldots \to H^1(\Sigma, \mathfrak{su}(2)) \to H^1(M_1, \mathfrak{su}(2)) \oplus H^1(M_2, \mathfrak{su}(2)) \\ \to H^1(F, \mathfrak{su}(2)) \to H^2(\Sigma, \mathfrak{su}(2)) \to \ldots$$

Since Σ is an integral homology sphere the map (3.7) is an isomorphism.

As we know, $\dim \mathcal{R}^*(M_1) = \dim \mathcal{R}^*(M_2) = 3g - 3$ and $\dim \mathcal{R}^*(F) = 6g - 6$. Hence, if the intersection of $\mathcal{R}^*(M_1)$ and $\mathcal{R}^*(M_2)$ in $\mathcal{R}^*(F)$ is transversal, it consists of finitely many points. In practice, both transversal and non-transversal intersections occur. If the intersection is not transversal, a perturbation is needed. We choose an isotopy of $\mathcal{R}^*(F)$ with compact support that carries $\mathcal{R}^*(M_2)$ to $\tilde{\mathcal{R}}^*(M_2)$ where $\tilde{\mathcal{R}}^*(M_2)$ is transversal to $\mathcal{R}^*(M_1)$ so that they intersect in a finite number of points. Given orientations on $\mathcal{R}^*(M_1)$ and $\mathcal{R}^*(M_2)$, see Section 3.2.3, we have an induced orientation on $\tilde{\mathcal{R}}^*(M_2)$. Thus we can define the algebraic intersection of $\mathcal{R}^*(M_1)$ and $\mathcal{R}^*(M_2)$ as the sum

$$\#(\mathcal{R}^*(M_1) \cap \mathcal{R}^*(M_2)) = \sum_{p \in \mathcal{R}^*(M_1) \cap \tilde{\mathcal{R}}^*(M_2)} \varepsilon_p, \tag{3.8}$$

where the number ε_p equals ± 1 depending on whether the orientations of the spaces $T_p\mathcal{R}^*(M_1) \oplus T_p\tilde{\mathcal{R}}^*(M_2)$ and $T_p\mathcal{R}^*(F)$ agree. By a standard homological argument, the number (3.8) is independent of the choice of isotopy, hence the following definition makes sense.

Let $\Sigma = M_1 \cup_F M_2$ be a genus g Heegaard splitting of an integral homology sphere Σ. Its *Casson invariant* is the number

$$\lambda(\Sigma, M_1, M_2) = \frac{(-1)^g}{2}\, \#(\mathcal{R}^*(M_1) \cap \mathcal{R}^*(M_2)). \tag{3.9}$$

It is not immediately clear from this definition that $\lambda(\Sigma)$ is an integer: this will follow from the surgery formula, see Proposition 3.10. The fact that $\lambda(\Sigma, M_1, M_2)$ is actually independent of the choice of Heegaard splitting will be proved in Section 3.2.4.

Remark 3.7. The above definition relies heavily on the fact that Σ is an integral homology sphere: the only reducible representation of its fundamental group is θ. Since θ is isolated, this makes it possible to define an invariant after making a perturbation with compact support. It is technically more difficult but still possible to define an invariant via a similar intersection theory when Σ is a *rational* homology sphere. This generalization, due to Walker, is described in Chapter 4.

3.2.3 Orientations

Let $\Sigma = M_1 \cup_F M_2$ be a Heegaard splitting of genus g and orient the surface F. The intersection form

$$I : H_1(F, \mathbb{R}) \times H_1(F, \mathbb{R}) \to \mathbb{R}, \quad I(a, b) = \langle a \cup b, [F] \rangle,$$

is symplectic and has determinant 1, for any choice of orientation. A choice of symplectic basis for I defines a canonical orientation on $H^1(F, \mathbb{R})$. The Mayer–Vietoris exact sequence implies that both $H^1(M_1, \mathbb{R})$ and $H^1(M_2, \mathbb{R})$ are direct summands of dimension g in the space $H^1(F, \mathbb{R}) = \mathbb{R}^{2g}$. Since the latter is already oriented, an orientation of $H^1(M_1, \mathbb{R})$ defines an orientation of $H^1(M_2, \mathbb{R})$. Fix these orientations.

Orient $SU(2)$ and choose a basis in $H^1(F, \mathbb{R})$ consistent with the orientation to identify $\mathrm{Hom}(\pi_1(F_0), SU(2))$ with $SU(2)^{2g}$. Requiring that the map (3.6) be orientation-preserving orients the manifold $\mathcal{R}(F)$. Choose bases in

$H^1(M_1, \mathbb{R})$ and $H^1(M_2, \mathbb{R})$ consistent with the above orientations and use them to orient $\mathcal{R}^*(M_1)$ and $\mathcal{R}^*(M_2)$, respectively.

These are the orientations which are used to define ε_p in (3.8). One can check that the product $(-1)^g \cdot \varepsilon_p$ is insensitive to various choices made in the construction, including the choice of handlebody to call M_1, and only depends on the orientation of Σ. This makes (3.9) a well defined invariant of a Heegaard splitting.

3.2.4 Independence of Heegaard Splitting

Two Heegaard splittings of Σ are called *equivalent* if there exists an orientation preserving homeomorphism of Σ onto itself taking one splitting into the other. It is obvious that $\lambda(\Sigma, M_1, M_2)$ is the same for equivalent Heegaard splittings. According to a theorem of Reidemeister [259] and Singer [287], any two Heegaard splittings of Σ are *stably equivalent.* The latter means that they become equivalent after adding finitely many unknotted handles to each of them.

In order to show that $\lambda(\Sigma, M_1, M_2)$ only depends on Σ and not on the choice of Heegaard splitting, it suffices to check that it does not change if an unknotted handle is added to $M_1 \cup M_2$. A proof of this fact is a straightforward comparison of the definitions except for the perturbation part, where it gets a little technical. We refer the reader to [5] for a complete proof.

The common value of the $\lambda(\Sigma, M_1, M_2)$ for various Heegaard splittings $\Sigma = M_1 \cup_F M_2$ is denoted by $\lambda(\Sigma)$ and is called the *Casson invariant* of Σ.

3.2.5 Casson Invariant for Knots and Property (1)

To study how the invariant (3.9) changes under surgery of Σ along a knot k, we choose a Heegaard splitting of Σ *compatible* with the knot k.

Lemma 3.8. *There exists a Heegaard splitting $\Sigma = M_1 \cup_F M_2$ such that k is a separating curve on F so that $F = F' \cup F''$. Moreover, one may assume that $M_1 = F' \times I$.*

Proof. Start with a Seifert surface F_0 of $k \subset \Sigma$ and thicken it to get a handlebody $F_0 \times I$. The knot k is a separating curve on the boundary of $F_0 \times I$. The closure K of the complement of $F_0 \times I$ in Σ is not necessarily a handlebody but can be made into one by drilling some tunnels in K. Each of the tunnels adds a handle to $F_0 \times I$. One needs to make sure that the attaching points of each of the new handles are on the same side of the knot k so that it remains separating. To make M_1 equal to $F' \times I$, for each handle that we added, drill a tunnel along its core from $F_0 \times \{1\}$ to $F_0 \times \{0\}$.

Let $\Sigma = M_1 \cup_F M_2$ be a Heegaard splitting of genus g compatible with the knot k and denote by t a Dehn twist of F along k, see Section 1.1.13. For any m, a Heegaard splitting of the manifold $\Sigma + (1/m) \cdot k$ can be obtained from the splitting $\Sigma = M_1 \cup_F M_2$ by composing the gluing map $\partial M_1 \to \partial M_2$ with t^m. Introduce the notations $\mathcal{R}_1 = \mathcal{R}^*(M_1)$ and $\mathcal{R}_2 = \mathcal{R}^*(M_2)$. The map t induces a diffeomorphism $t^* : \mathcal{R}(F) \to \mathcal{R}(F)$ which maps $\mathcal{R}_2$ to $t^*\mathcal{R}_2$, and

$$\lambda\left(\Sigma + \frac{1}{m} \cdot k\right) = \frac{(-1)^g}{2} \#(\mathcal{R}_1 \cap (t^m)^* \mathcal{R}_2) = \frac{(-1)^g}{2} \#(\mathcal{R}_1 \cap (t^*)^m \mathcal{R}_2).$$

Therefore,

$$\begin{aligned} &\lambda\left(\Sigma + \frac{1}{m+1} \cdot k\right) - \lambda\left(\Sigma + \frac{1}{m} \cdot k\right) \\ &= \frac{(-1)^g}{2} \cdot \left[\#(\mathcal{R}_1 \cap (t^*)^{m+1} \mathcal{R}_2) - \#(\mathcal{R}_1 \cap (t^*)^m \mathcal{R}_2)\right] \\ &= -\frac{(-1)^g}{2} \cdot \# \left[(t^*\mathcal{R}_1 - \mathcal{R}_1) \cap (t^*)^{m+1} \mathcal{R}_2\right]. \end{aligned}$$

Thus to prove that (3.9) satisfies property (1) of a Casson invariant we need to show that the number $\#((t^*\mathcal{R}_1 - \mathcal{R}_1) \cap (t^*)^{m+1}\mathcal{R}_2)$ is independent of m. This will be accomplished by showing that the difference $t^*\mathcal{R}_1 - \mathcal{R}_1$ is homologous to a cycle δ such that $t^*\delta = \delta$.

To construct the cycle δ, we first describe in convenient terms the action of t^* on $\mathcal{R}^*(F)$. Since the knot k separates the surface F, the space $R(F) = \operatorname{Hom}(\pi_1 F, SU(2))$ can be identified as

$$R(F) = \{ (\alpha', \alpha'') \mid \alpha'(k) = \alpha''(k) \} \subset R(F') \times R(F'').$$

Suppose that the base point of F belongs to F'. Then the action of t_* on $\pi_1(F)$ is identity on the loops in $\pi_1 F'$, and is given by the formula $t_* x = k^{-1} x k$ on $x \in \pi_1 F''$. Therefore, the induced diffeomorphism $t^* : \mathcal{R}^*(F) \to \mathcal{R}^*(F)$ is given by the formula

$$(\alpha', \alpha'') \mapsto (\alpha', \alpha'(k)^{-1} \cdot \alpha'' \cdot \alpha'(k)).$$

The cycle δ will essentially be a cycle in the space $\mathcal{R}_-(F)$ which we will now define. Let

$$R_-(F) = \{ \alpha : \pi_1 F \to SU(2) \mid \alpha(k) = -1 \} \subset R(F). \qquad (3.10)$$

All representations in $R_-(F)$ are irreducible due to the fact that $0 = [k] \in H_1(F, \mathbb{Z})$ because k is separating. One can think of $R_-(F)$ as the product

$$R_-(F) = \{ (\alpha', \alpha'') \mid \alpha'(k) = -1,\ \alpha''(k) = -1 \} = (h')^{-1}(-1) \times (h'')^{-1}(-1)$$

where $h' : R(F') \to SU(2)$ and $h'' : R(F'') \to SU(2)$ are the maps $h'(\alpha') = \alpha'(k)$ and $h''(\alpha'') = \alpha''(k)$, compare with (3.6). Since $-1 \in SU(2)$ is a regular value for h' and h'', both $\mathcal{R}_-(F') = (h')^{-1}(-1)$ and $\mathcal{R}_-(F'') = (h'')^{-1}(-1)$

are smooth manifolds. Therefore, the space $\mathcal{R}_-(F) = R_-(F)/SO(3)$ gets the structure of a closed smooth manifold of dimension $6g-9$ smoothly embedded in $\mathcal{R}^*(F)$. The map $t^* : \mathcal{R}^*(F) \to \mathcal{R}^*(F)$ restricts to $\mathcal{R}_-(F)$ as the identity map because $t^*(\alpha', \alpha'') = (\alpha', \alpha'(k)^{-1} \cdot \alpha'' \cdot \alpha'(k)) = (\alpha', \alpha'')$.

Lemma 3.9. *There exists a canonical isotopy*

$$H_t : \mathcal{R}^*(F) \setminus \mathcal{R}_-(F) \to \mathcal{R}^*(F) \setminus \mathcal{R}_-(F), \quad t \in [0,1],$$

such that $H_0 = \mathrm{id}$ *and* $H_1 = t^*$. *In other words,* H_t *isotopes* $\mathcal{R}_1$ *to* $t^*\mathcal{R}_1$ *in the complement of* $\mathcal{R}_-(F)$.

Proof. The exponential map $\exp : \mathfrak{su}(2) \to SU(2)$, $A \mapsto e^A$, is a diffeomorphism of the open ball of radius π about 0 in $\mathfrak{su}(2)$ to $SU(2) \setminus \{-1\}$. The natural contraction $X \mapsto t \cdot X$ on the Lie algebra $\mathfrak{su}(2)$ exponentiates to a map $A \mapsto A^t$ of $SU(2) \setminus \{-1\}$ to itself. We define an isotopy $H_t : R(F) \setminus R_-(F) \to R(F) \setminus R_-(F)$ by the formula

$$H_t\,(\alpha', \alpha'') = (\,\alpha',\ (\alpha'(k)^t)^{-1} \cdot \alpha'' \cdot \alpha'(k)^t\,),$$

which makes sense because we removed all representations with $\alpha'(k) = -1$ from $\mathcal{R}^*(F)$. The maps H_t are $SO(3)$–equivariant and thus define the required isotopy.

A transversality argument can now be used to show that the trace of isotopy H_t provides a homology between $t^*\mathcal{R}_1 - \mathcal{R}_1$ and a cycle δ' in a regular neighborhood $\mathcal{N}$ of $\mathcal{R}_-(F)$. The cycle δ' is the trace of $\mathcal{R}_1 \cap \mathcal{N}$ under the isotopy H_t. Therefore,

$$\#((t^*\mathcal{R}_1 - \mathcal{R}_1) \cap (t^*)^{m+1}\mathcal{R}_2) = \#(\delta' \cap (t^*)^{m+1}\mathcal{R}_2) = \#((t^*)^{-(m+1)}\delta' \cap \mathcal{R}_2).$$

In the next section, we will show how to collapse the cycle δ' to the so called difference cycle $\delta \subset \mathcal{R}_-(F)$. Since the action of t^* on $\mathcal{R}_-(F)$ is trivial, $(t^*)^{-(m+1)}\delta = \delta$, and the difference $\#((t^*\mathcal{R}_1 - \mathcal{R}_1) \cap (t^*)^{m+1}\mathcal{R}_2)$ is independent of m. This proves property (1) for the invariant (3.9).

3.2.6 The Difference Cycle

Let $\Sigma = M_1 \cup_F M_2$ be a Heegaard splitting as in Lemma 3.8 with $M_1 = F' \times I$ and $F = \partial(F' \times I) = F' \cup_k F'$. Let $\mathrm{tr} : SU(2) \to [-2, 2]$ be the trace function on the Lie group $SU(2)$, and define the argument function $\arg : SU(2) \to [0, \pi]$ by the formula $\phi(A) = \arccos(\mathrm{tr}(A)/2)$. Then

$$\mathcal{R}_1 = \{\,(\alpha', \alpha') \mid \alpha' \in R^*(F')\,\}/SO(3),$$
$$\mathcal{R}_-(F) = \{\,(\alpha', \alpha'') \mid \alpha', \alpha'' \in R(F'),\ \arg \alpha'(k) = \arg \alpha''(k) = \pi\,\}/SO(3),$$

and a closed regular ε–neighborhood $\mathcal{N}_\varepsilon$ of $\mathcal{R}_-(F)$, for $\varepsilon > 0$ small enough, can be identified as

$$\{ (\alpha',\alpha'') \mid \alpha',\alpha'' \in R(F'),\ \alpha'(k)=\alpha''(k), \pi-\varepsilon \le \arg\alpha'(k) \le \pi \}/SO(3).$$

The cycle δ' is homologous to the cycle

$$\begin{aligned}\delta_\varepsilon &= \bigcup_t \quad H_t(\mathcal{R}_1 \cap \partial\mathcal{N}_\varepsilon),\\ &= \bigcup_t \quad \{ (\alpha', \alpha'(k)^{-t}\cdot\alpha'\cdot\alpha'(k)^t) \mid \arg\alpha'(k) = \pi-\varepsilon \}/SO(3)\\ &= \{ (\alpha', g^{-1}\cdot\alpha'\cdot g \mid g \in SU(2)^\varepsilon \}/SO(3),\end{aligned}$$

where $SU(2)^\varepsilon$ consists of all $g \in SU(2)$ with $0 \le \arg g \le \pi - \varepsilon$. Since the limiting value of $SU(2)^\varepsilon$ as $\varepsilon \to 0$ is the entire group $SU(2)$, we see that the limit as $\varepsilon \to 0$ of the cycles δ_ε is a cycle $\delta \in \mathcal{R}_-(F)$ given by the formula

$$\delta = \{ (\alpha', g^{-1}\cdot\alpha'\cdot g) \mid g \in SU(2) \}/SO(3).$$

The cycle δ has the following interpretation. The product action of $SO(3)\times SO(3)$ on $R_-(F')\times R_-(F')$ defines the projection $\hat{p} : R_-(F')\times R_-(F') \to \mathcal{R}_-(F')\times\mathcal{R}_-(F')$ onto the quotient, with the fiber $SO(3)\times SO(3)$. The map $\hat{p}$ factors through $\mathcal{R}_-(F) = R_-(F')\times R_-(F')/SO(3)$ (where the action of $SO(3)$ is diagonal) to define the projection $p : \mathcal{R}_-(F) \to \mathcal{R}_-(F')\times\mathcal{R}_-(F')$ with the fiber $SO(3)$. Let $\Delta \subset \mathcal{R}_-(F')\times\mathcal{R}_-(F')$ be the diagonal then

$$\begin{aligned}p^{-1}(\Delta) = \{ (\alpha',\alpha'') \mid &\\ &\alpha'' = \mathrm{Ad}_g\,\alpha' \text{ for some } g \in SU(2) \text{ and } \alpha'(k) = \alpha''(k) = -1 \}/SO(3).\end{aligned}$$

Since $SU(2)$ is a regular double cover of $SO(3)$ we see that $\delta = 2\cdot p^{-1}(\Delta)$. This proves in particular the following result.

Proposition 3.10. *Let k be a knot in a homology sphere Σ, then $\lambda'(k) = (-1)^g\,\#(p^{-1}(\Delta)\cap\mathcal{R}_2)$. In particular, $\lambda(\Sigma)$ is an integer for any homology sphere Σ.*

3.2.7 Casson Invariant for Boundary Links and Property (2)

Let $k \cup \ell$ be a boundary link in a homology sphere Σ. Slightly generalizing Lemma 3.8, one can find a Heegaard splitting $\Sigma = M_1 \cup_F M_2$ such that $M_1 = F' \times I$ where F' is a Seifert surface of ℓ, and k is a separating curve in F'. Let τ_k and τ_ℓ be Dehn twists along k and ℓ, see Section 1.1.13, and h_k and h_ℓ their inverses. Then

$$\begin{aligned}\lambda''(k,\ell) &= [\lambda(\Sigma+k+\ell) - \lambda(\Sigma+k) - \lambda(\Sigma+\ell) + \lambda(\Sigma)]\,,\\ &= \frac{(-1)^g}{2}\cdot[\,\#(h_k^*(h_\ell^*\mathcal{R}_1 - \mathcal{R}_1)\cap\mathcal{R}_2) - \#((h_\ell^*\mathcal{R}_1 - \mathcal{R}_1)\cap\mathcal{R}_2)\,].\end{aligned}$$

We have $\delta_\ell = \tau_\ell^* \mathcal{R}_1 - \mathcal{R}_1$ and $\tau_\ell^* \delta_\ell = \delta_\ell$. Therefore, $h_\ell^* \delta_\ell = \delta_\ell$, so $\delta_\ell = h_\ell^* \delta_\ell = \mathcal{R}_1 - h_\ell^* \mathcal{R}_1$. Thus

$$\lambda''(k,\ell) = -\frac{(-1)^g}{2} \cdot \#((h_k^* \delta_\ell - \delta_\ell) \cap \mathcal{R}_2\,).$$

We wish to prove that $h_k^* \delta_\ell - \delta_\ell = 0$ so that the property (2) holds for the invariant defined in (3.9).

We will slightly abuse notations and denote by δ_ℓ both the cycle δ_ℓ and its homology class in $H_*(\mathcal{R}_-(F))$. As we already know, $\delta_\ell = 2 \cdot p^{-1}(\Delta)$. At the level of homology, we have the following commutative diagram (where PD stands for Poincaré duality)

$$\begin{array}{ccc}
H_*(\mathcal{R}_-(F)) & \xrightarrow{\text{PD}} & H^*(\mathcal{R}_-(F)) \\
{\scriptstyle p_*}\big\downarrow & & \big\uparrow{\scriptstyle p^*} \\
H_*(\mathcal{R}_-(F') \times \mathcal{R}_-(F')) & \xrightarrow{\text{PD}} & H^*(\mathcal{R}_-(F') \times \mathcal{R}_-(F'))
\end{array}$$

which implies that $p_*(\delta_\ell) = 2\Delta$ and $\text{PD}(\delta_\ell) = 2 \cdot p^* \, \text{PD}(\Delta)$. The map $h_k^* : \mathcal{R}_-(F') \to \mathcal{R}_-(F')$ induces homomorphisms in both the homology and cohomology of $\mathcal{R}_-(F')$; we call both h_k^*. The following diagram is commutative:

$$\begin{array}{ccc}
H^*(\mathcal{R}_-(F)) & \xrightarrow{h_k^*} & H^*(\mathcal{R}_-(F)) \\
{\scriptstyle p^*}\big\uparrow & & \big\uparrow{\scriptstyle p^*} \\
H^*(\mathcal{R}_-(F') \times \mathcal{R}_-(F')) & \xrightarrow{\tilde{h}_k^*} & H^*(\mathcal{R}_-(F') \times \mathcal{R}_-(F'))
\end{array}$$

The map $\tilde{h}_k^*$ here is induced by $h_k^* : H^*(\mathcal{R}_-(F')) \to H^*(\mathcal{R}_-(F'))$, with the help of the Künneth formula.

Theorem 3.11. *(Newstead [246] and [245]) Let F' be a Riemann surface with connected boundary $\partial F'$, and $h : F' \to F'$ a diffeomorphism such that* $\text{id} = h_* : H_*(F') \to H_*(F')$. *Then h induces the identity map in the cohomology of $\mathcal{R}_-(F')$.*

Since k is a separating curve in F, the homomorphism $(h_k)_* : H_*(F') \to H_*(F')$ is the identity. Theorem 3.11 now implies that $\tilde{h}_k^* = \text{id}$, hence $h_k^* : H^*(\mathcal{R}_-(F)) \to H^*(\mathcal{R}_-(F))$ is the identity map on the image of p^*. In particular, $h_k^*(2 \cdot p^* \, \text{PD}(\Delta)) = 2 \cdot p^* \, \text{PD}(\Delta)$, therefore, $h_k^*(\text{PD}(\delta_\ell)) = \text{PD}(\delta_\ell)$ and $\text{PD}(h_k^*(\delta_\ell)) = \text{PD}(\delta_\ell)$. Since PD is an isomorphism, $h_k^* \delta_\ell = \delta_\ell$.

3.2.8 Casson Invariant of a Trefoil and Property (0)

Let k be a left-handed trefoil in S^3. We will show that $\lambda'(k) = \pm 1$ and thus prove property (0) of the invariant defined in (3.9).

The trefoil k is a fibered knot of genus 1 hence its complement in S^3 is fibered over a circle by Seifert surfaces F' of genus 1, with monodromy f. This fibration provides us with a natural Heegaard splitting $S^3 = M_1 \cup_F M_2$ of genus 2 such that $M_1 = M_2 = F' \times I$ and $F = F' \cup F'$. The corresponding character varieties can be identified as follows:

$$\mathcal{R}_1 = \{(\alpha', \alpha') \mid \alpha' \in R^*(F')\}/SO(3),$$
$$\mathcal{R}_2 = \{(\alpha', f^*\alpha') \mid \alpha' \in R^*(F')\}/SO(3).$$

The homology class $p^{-1}(\Delta) \in H_3(\mathcal{R}_-(F))$ is the inverse image of the diagonal $\Delta \subset \mathcal{R}_-(F') \times \mathcal{R}_-(F')$, where p is the projection $\mathcal{R}_-(F) \to \mathcal{R}_-(F') \times \mathcal{R}_-(F')$. The group $\pi_1 F'$ is a free group on two generators. An easy calculation with matrices shows that $\mathcal{R}_-(F')$ is a point, so that $p^{-1}(\Delta)$ equals (up to a sign) the fundamental class of the closed manifold $\mathcal{R}_-(F)$.

According to Proposition 3.10, we have $\pm\,\lambda'(k) = \#(p^{-1}(\Delta) \cap \mathcal{R}_2)$. Since $p^{-1}(\Delta) = \pm\mathcal{R}_-(F)$, we only need to calculate the intersection $\mathcal{R}_-(F) \cap \mathcal{R}_2$ in $\mathcal{R}^*(F)$. Using the description of $\mathcal{R}_2$ given above, we see that

$$\mathcal{R}_-(F) \cap \mathcal{R}_2 = \{(\alpha', f^*\alpha') \mid \alpha' \in R_-(F')\}/SO(3) \cong \mathcal{R}_-(F') = \{\text{point}\}.$$

To complete the proof, one only needs to check the transversality condition.

3.3 Comments and Ramifications

This section is designed to give a less formal explanation of various points in the construction of the Casson invariant, including the surgery formula and the perturbations.

3.3.1 Pillowcase

Our proof of the surgery formula above can be interpreted nicely in terms of the so called pillowcase picture. The latter will also be useful in many other instances later on. Let $k \subset \Sigma$ be a knot in an oriented integral homology 3–sphere and let K be its exterior. Then K is a compact oriented 3–manifold whose boundary ∂K is a 2–dimensional torus. The inclusion $\partial K \to K$ induces a map of representation varieties $r : \mathcal{R}(K) \to \mathcal{R}(\partial K)$. The variety $\mathcal{R}(\partial K)$ can be described as follows.

We first identify ∂K with the standard 2–torus T^2 by using canonical meridian m and longitude ℓ on ∂K. The corresponding generators of $\pi_1 T^2 = \mathbb{Z} \oplus \mathbb{Z}$ will also be denoted by m and ℓ. Any representation $\alpha : \pi_1 T^2 \to SU(2)$ is conjugate to a representation taking the pair (m, ℓ) to $(e^{i\phi}, e^{i\theta}) \in U(1) \times U(1)$. Conjugating by $j \in SU(2)$ if necessary we may assume that $0 \le \phi \le \pi$. Thus $\mathcal{R}(T^2)$ can be identified with the rectangle $[0, \pi] \times [0, 2\pi]$ whose edges

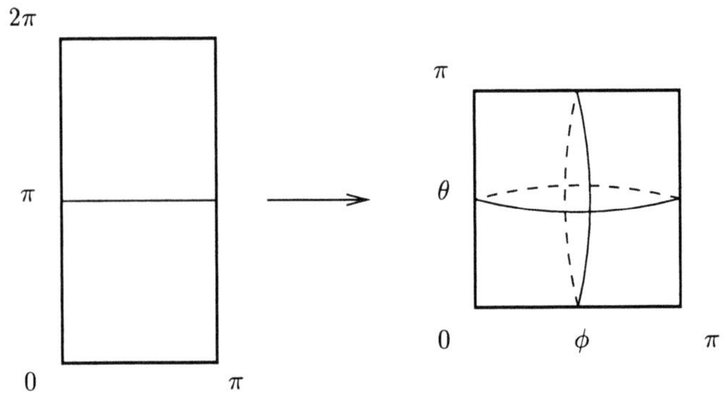

Fig. 3.1.

are glued as shown in Figure 3.1. The resulting variety is topologically a 2–dimensional sphere. It has four singular points, which are the orbits of central representations. This variety is usually referred to as a *pillowcase*.

After a generic perturbation if necessary, the variety $\mathcal{R}(K)$ and its image in $\mathcal{R}(T^2)$ look roughly as follows (for more details, see Theorem 5.14). The variety $\mathcal{R}(K)$ consists of two central orbits, a smooth open arc of abelian orbits with one non-compact end limiting to each of the central orbits, and a smooth 1–dimensional manifold of irreducible orbits with finitely many ends, each limiting to a different point on the abelian arc. The map $r : \mathcal{R}(K) \to \mathcal{R}(T^2)$ is an immersion taking the 1–dimensional strata of $\mathcal{R}(K)$ into the smooth part of $\mathcal{R}(T^2)$.

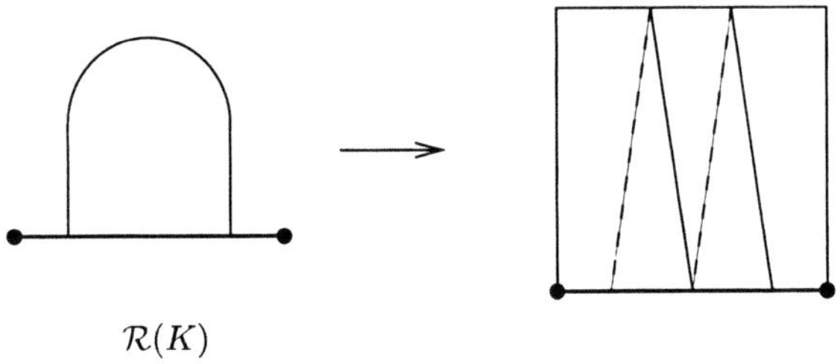

Fig. 3.2.

Example 3.12. Let K be the exterior of a right handed trefoil in S^3. The variety $\mathcal{R}(K)$ is in general position so that no perturbations are needed. It consists of a single arc of irreducible representations limiting to two disjoint points on the abelian arc. The variety $\mathcal{R}(K)$ and its image in the pillowcase are shown in Figure 3.2. The representation variety of the left handed trefoil is the same as that of the right handed trefoil. However, its pillowcase image is different: it is obtained from the one shown in Figure 3.2 by switching solid and dotted lines.

Example 3.13. The representation variety $\mathcal{R}(K)$ of the figure eight knot exterior is in general position, so that no perturbations are needed. It consists of a circle and an abelian arc. The variety $\mathcal{R}(K)$ and its pillowcase image are shown in Figure 3.3.

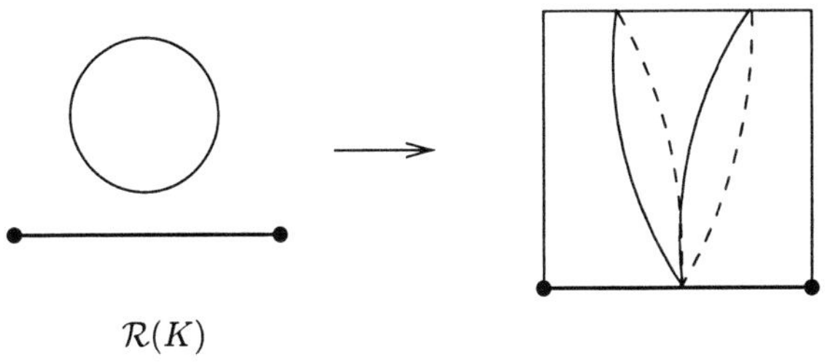

Fig. 3.3.

For explanation to the above examples and for other calculations of character varieties $\mathcal{R}(K)$ and their pillowcase images see [43], [140], [171] and [172].

Let manifold $M = \Sigma + (p/q) \cdot k$ be obtained by Dehn (p/q)–surgery of an integral homology sphere Σ along a knot k. Then $\mathcal{R}(M) \subset \mathcal{R}(K)$ consists of representations $\alpha : \pi_1 K \to SU(2)$ such that $\alpha(m^p \ell^q) = 1$. The image of $\mathcal{R}(M)$ under r in the pillowcase is obtained by intersecting the image of $\mathcal{R}(K)$ with the image of the straight line through the origin with the slope $-p/q$.

Example 3.14. The result of (-1)–surgery along the right-handed trefoil in S^3 is $\Sigma(2,3,7)$. The representation variety $\mathcal{R}(\Sigma(2,3,7))$ has two irreducible representations which can be seen in the pillowcase picture for the right handed trefoil shown on the left in Figure 3.4. The same manifold $\Sigma(2,3,7)$ can be obtained by $(+1)$-surgery along the figure-eight knot, so that $\mathcal{R}(\Sigma(2,3,7))$ can also be seen in the pillowcase picture for the figure-eight knot, on the right in Figure 3.4.

Example 3.15. Let k be a knot in an integral homology sphere Σ and K its exterior, then the image of $\mathcal{R}(\Sigma)$ in the pillowcase consists of the points of $r(\mathcal{R}(K))$ lying on the left edge of the pillowcase (corresponding to $\alpha(m) = 1$). The intersection of $r(\mathcal{R}(K))$ with the upper edge of the pillowcase consists of the images of representations $\alpha : \pi_1 K \to SU(2)$ such that $\alpha(\ell) = -1$. These are representations of $\pi_1 K$ which contribute to the difference cocycle δ, compare with Section 6.5.2.

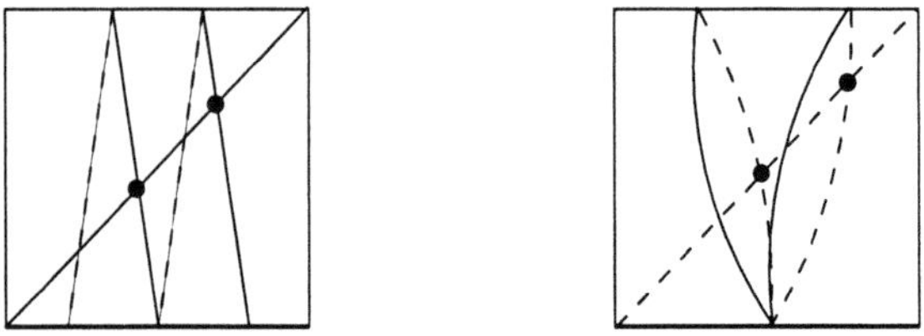

Fig. 3.4.

This last example brings us to a convenient interpretation of the surgery formula (3.3) for the Casson invariant in terms of the pillowcase picture. What follows is just a heuristic explanation of the surgery formula, whose rigorous proof was outlined above. Given a knot k in an integral homology sphere Σ, consider the pillowcase image of $\mathcal{R}(K)$. Its intersection with an appropriate diagonal in the pillowcase coincides with the image of $\mathcal{R}(\Sigma \pm k)$. The isotopy H_t constructed in Lemma 3.9 shows up as a family of curves, see Figure 3.5, interpolating between the diagonal and the corner consisting of the left and upper edges of the pillowcase. Thus the perturbed $\mathcal{R}(\Sigma \pm k)$ splits into a disjoint union of two representation spaces. One is $\mathcal{R}(\Sigma)$, which shows at the left edge of the pillowcase. The other consists of representations of $\pi_1 K$ taking the canonical longitude into -1; a count of these representations results in the term $\lambda'(k \subset \Sigma)$ in the surgery formula. An obvious modification of the above construction explains (3.3) for $q \neq \pm 1$.

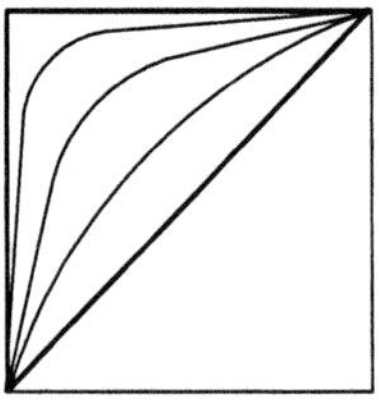

Fig. 3.5.

3.3.2 Perturbations

The Casson invariant was defined in (3.8) via the intersection number of $\mathcal{R}^*(M_1)$ and $\mathcal{R}^*(M_2)$ in $\mathcal{R}^*(F)$ after a perturbation making this intersection transversal. One can avoid perturbations if the intersection happens to be transversal from the beginning.

Lemma 3.16. *Let $\Sigma = M_1 \cup_F M_2$ be a Heegaard splitting of an integral homology sphere Σ. Then the intersection $\mathcal{R}^*(M_1) \cap \mathcal{R}^*(M_2)$ in $\mathcal{R}^*(F)$ is transversal at $\alpha \in \mathcal{R}^*(\Sigma)$ if and only if $H^1(\pi_1 \Sigma, \operatorname{ad} \alpha) = 0$.*

Proof. The group $H^1(\pi_1\Sigma, \operatorname{ad}\alpha)$ can be included in the following Mayer–Vietoris exact sequence in group cohomology, see [42],

$$0 \to H^1(\pi_1\Sigma, \operatorname{ad}\alpha) \to H^1(\pi_1 M_1, \operatorname{ad}\alpha)\oplus$$
$$\oplus\, H^1(\pi_1 M_2, \operatorname{ad}\alpha) \xrightarrow{i_1^*\oplus i_2^*} H^1(\pi_1 F, \operatorname{ad}\alpha) \to \dots \quad (3.11)$$

The intersection in question is transversal if and only if the map $i_1^* \oplus i_2^*$ is surjective. Since α is irreducible, $H^1(\pi_1 M_k, \operatorname{ad}\alpha) = \mathbb{R}^{3g-3}$, $k = 1, 2$, and $H^1(\pi_1 F, \operatorname{ad}\alpha) = \mathbb{R}^{6g-6}$. Therefore, the map $i_1^* \oplus i_2^*$ is surjective if and only if $H^1(\pi_1\Sigma, \operatorname{ad}\alpha) = 0$.

A representation $\alpha : \pi_1(\Sigma) \to SU(2)$ is called *non-degenerate* if $H^1(\pi_1(\Sigma), \operatorname{ad}\alpha) = 0$. A representation variety $\mathcal{R}^*(\Sigma)$ is called *non–degenerate* if all its points are non–degenerate. If $\mathcal{R}^*(\Sigma)$ is non-degenerate then $\lambda(\Sigma)$ can be obtained by an algebraic count of points in $\mathcal{R}^*(\Sigma)$, without having to use perturbations.

Example 3.17. The $SU(2)$ representation variety of a Brieskorn homology sphere $\Sigma(p,q,r)$ is a finite collection of points, see Section 3.5.1. At each $\alpha \in \mathcal{R}^*(\Sigma(p,q,r))$ the cohomology $H^1(\Sigma(p,q,r), \operatorname{ad}\alpha)$ vanishes, see [83]. Therefore, perturbations are not needed, and the Casson invariant of $\Sigma(p,q,r)$ is just an algebraic count of points in $\mathcal{R}^*(\Sigma(p,q,r))$. On the other hand, $\mathcal{R}^*(\Sigma(a_1,\dots,a_n))$ for $n \geq 4$ is not non-degenerate. It contains higher dimensional components, see Section 3.5.2, which are smooth manifolds non–degenerate in the Morse–Bott sense, see Section 5.3.5.

Let $\mathcal{R}_\alpha$ be a connected component of $\mathcal{R}^*(\Sigma)$ which is non-degenerate in the Morse–Bott sense, see Section 5.3.5. Then $\mathcal{R}_\alpha$ is a closed smooth manifold and one can use a Morse function $\phi_\alpha : \mathcal{R}_\alpha \to \mathbb{R}$ to perturb $\mathcal{R}_\alpha$ into a finite collection of non–degenerate points, see Section 5.3.5 for details. In particular, the contribution of $\mathcal{R}_\alpha$ to the Casson invariant will be equal to plus or minus one half of the Euler characteristic of $\mathcal{R}_\alpha$.

An interesting question is whether there exists a representation $\alpha : \pi_1\Sigma \to SU(2)$ which is an isolated point in $\mathcal{R}^*(\Sigma)$ but for which $H^1(\pi_1\Sigma, \operatorname{ad}\alpha) \neq 0$, in other words, an isolated degenerate representation. Nobody has found such a representation so far, and one conjecture states that this is impossible.

3.3.3 The Connected Sum Formula

The property (7) of the Casson invariant asserts that $\lambda(\Sigma_0 \# \Sigma_1) = \lambda(\Sigma_0) + \lambda(\Sigma_1)$. It was proved in Section 3.1 by using the surgery formula. Another proof is as follows.

Let us assume for the sake simplicity that $\mathcal{R}^*(\Sigma_0)$ and $\mathcal{R}^*(\Sigma_1)$ are non–degenerate. Since $\pi_1(\Sigma_0 \# \Sigma_1) = \pi_1(\Sigma_0) * \pi_1(\Sigma_1)$ is a free product, any pair (α_0, α_1) of representations $\alpha_0 : \pi_1\Sigma_0 \to SU(2)$ and $\alpha_1 : \pi_1\Sigma_1 \to SU(2)$

defines a representation $\alpha_0 * \alpha_1$ of $\pi_1(\Sigma_0 \# \Sigma_1)$, and vice versa. Therefore, the space $\mathcal{R}(\Sigma_0 \# \Sigma_1)$ consists of the following three types of (the conjugacy classes of) representations: first, $\alpha_0 * \theta$ where $\alpha_0 \in \mathcal{R}^*(\Sigma_0)$, second, $\theta * \alpha_1$ where $\alpha_1 \in \mathcal{R}^*(\Sigma_1)$, and finally, a copy of $SO(3)$ for each pair $(\alpha_0, \alpha_1) \in \mathcal{R}^*(\Sigma_0) \times \mathcal{R}^*(\Sigma_1)$. Representations of the third type appear in $SO(3)$–families because $\alpha_0 * (g \cdot \alpha_1 \cdot g^{-1})$ is conjugate to $\alpha_0 * \alpha_1$ if and only if $g = \pm 1$.

The representations of the first two types are non–degenerate and contribute to $\lambda(\Sigma_0 \# \Sigma_1)$ with the same signs as they contributed to $\lambda(\Sigma_0)$ and $\lambda(\Sigma_1)$, respectively. The representations of the third type are non–degenerate in the Morse–Bott sense. Since the Euler characteristic of $SO(3)$ is zero, they do not contribute to $\lambda(\Sigma_0 \# \Sigma_1)$ at all. This shows that $\lambda(\Sigma_0 \# \Sigma_1) = \lambda(\Sigma_0) + \lambda(\Sigma_1)$.

3.3.4 The Integrality of $\lambda(\Sigma)$

A somewhat mysterious feature of Casson's construction of $\lambda(\Sigma)$ is that it results in an integer valued invariant, see Proposition 3.10. If $\mathcal{R}(\Sigma)$ is non–degenerate this means that the number of the conjugacy classes of irreducible representations $\pi_1(\Sigma) \to SU(2)$ is even. An explanation of this phenomenon was proposed by Kronheimer, Larsen and Scherk in [174]. They proved the following result: the number of conjugacy classes of irreducible $SU(2)$–representations of *any* finitely presented group π such that $H_1(\pi, \mathbb{Z}) = 0$ is even if it is finite. The proof is purely algebraic and is relying on the theorem of Hasse and Hilbert which states that, for a quaternionic algebra over a global field, the number of places at which ramification occurs is even and finite.

3.3.5 Casson Invariant of Algebraically Split Links

In this section we describe certain properties of the Casson invariant which can be viewed as an integral lift of the properties of the Rokhlin invariant described in Section 2.3.2.

Let Σ be an integral homology 3–sphere. For any knot $k \subset \Sigma$, formula (3.1) defines an integer $\lambda'(k)$ which we showed to be equal to $(1/2) \cdot \Delta''_k(1)$. The modulo 2 reduction of this formula expresses $\mu'(k)$ as the Arf–invariant of k.

For any link $k \cup \ell$ in Σ such that $\operatorname{lk}(k, \ell) = 0$, formula (3.2) defines an integer $\lambda''(k, \ell)$. We proved that $\lambda''(k, \ell)$ vanishes if $k \cup \ell$ is a boundary link. In general, Hoste [148] showed that $\lambda''(k, \ell)$ equals the coefficient a_1 of the Conway polynomial

$$\nabla(z) = z\,(a_0 + a_1 z^2 + \ldots + a_p z^{2p})$$

of link $k \cup \ell$. Another way to identify $\lambda''(k, \ell)$ is as follows. Let F_k and F_ℓ be Seifert surfaces of k and ℓ, respectively. Assume that F_k and F_ℓ are in general position and let c be their (possibly disconnected) intersection. Let c' be a curve parallel to c with respect to the trivialization of a tubular neighborhood of c given by F_k and F_ℓ. Then $\lambda''(k,\ell) = -\operatorname{lk}(c,c')$, see [182]. The integer $\operatorname{lk}(c,c')$ is also known as the *Sato–Levine invariant* of $k \cup \ell$, see for instance [190]. It is an integral lift of the invariant (2.5).

Let $k \cup \ell \cup m$ be a link in Σ such that $\operatorname{lk}(k,\ell) = \operatorname{lk}(k,m) = \operatorname{lk}(\ell,m) = 0$ and define

$$\lambda'''(k,\ell,m) = \lambda''\left(k,\ell \subset \Sigma + \frac{1}{r+1}\cdot m\right) - \lambda''\left(k,\ell \subset \Sigma + \frac{1}{r}\cdot m\right) \quad (3.12)$$

where r is any integer. An argument similar to that of (3.2) can be used to show that $\lambda'''(k,\ell,m)$ is well defined and symmetric with respect to k, ℓ, and m. The invariant $\lambda'''(k,\ell,m)$ equals the coefficient a_1 of the Conway polynomial

$$\nabla(z) = z^2\,(a_0 + a_1 z^2 + \ldots + a_p z^{2p})$$

of the link $k \cup \ell \cup m$, see [148]. It can also be identified with the square of the Massey product of k, ℓ, and m, and with $((a \cup b \cup c)[Y])^2$ where a, b, c is a basis in the first integral cohomology of $Y = \Sigma + 0\cdot k + 0\cdot\ell + 0\cdot m$, see [182]. Reducing the above modulo 2, we obtain a formula for the invariant (2.6).

By iterating the above procedure, one can define λ–invariants for algebraically split links of four and more components. All these invariants vanish identically, see Hoste [148].

3.4 Properties of the Casson Invariant

In addition to the properties (0) through (8) listed in Section 3.1, the Casson invariant has other interesting properties and features which will be discussed in this section.

3.4.1 Splicing Additivity

The splicing of integral homology spheres was defined in Section 1.1.7. The following proposition asserts that the Casson invariant is additive with respect to this operation.

Proposition 3.18. *Let Σ be the splice of integral homology 3–spheres Σ_0 and Σ_1 along knots k_0 and k_1. Then $\lambda(\Sigma) = \lambda(\Sigma_0) + \lambda(\Sigma_1)$; in particular, $\lambda(\Sigma)$ is independent of the choice of k_0 and k_1.*

Proof. The proof of a similar statement for the Rokhlin invariant, see Proposition 2.15, works for the Casson invariant if one does not reduce mod 2.

Suppose that an integral homology sphere Σ is obtained by pasting together the exteriors of knots k_0 and k_1 in integral homology spheres Σ_0 and Σ_1, respectively, according to the matrix

$$A = \begin{pmatrix} q & u \\ p & v \end{pmatrix}$$

with $\det A = -1$ and $p = \pm 1$, see (1.7). The following result is a generalization of the splicing additivity, and it can be proved by similar methods. For a complete proof see [109].

Proposition 3.19. *If Σ is obtained from Σ_0 and Σ_1 by the above operation then $\lambda(\Sigma) = \lambda(\Sigma_0) + \lambda(\Sigma_1) - pv \cdot \lambda'(k_0 \subset \Sigma_0) + pq \cdot \lambda'(k_1 \subset \Sigma_1)$.*

Connected sum and splicing are examples of operations on 3–manifolds which involve gluing along Riemann surfaces of genus 0 and genus 1, respectively. Next section describes how the Casson invariant is affected by mutation, which is a "genus two" operation. Partial results about higher genus operations and their effect on the Casson invariant can be found in [184], see also Section 4.4.

3.4.2 Mutation Invariance

Mutation of 3-manifolds along embedded genus two surfaces was defined in Section 1.1.11. This operation takes an integral homology sphere into an integral homology sphere.

Theorem 3.20. *Let Σ^τ be the result of mutation of Σ along a surface F. Then $\lambda(\Sigma^\tau) = \lambda(\Sigma)$.*

Proof. This is proved in [163] by first describing Σ as the result of surgery on a certain special link $\mathcal{L}$. Then the surgery formula for the Casson invariant is used to show that the surgeries on $\mathcal{L}$ and on the image of $\mathcal{L}$ under mutation produce homology spheres whose Casson invariants are the same. Another proof can be found in [264] and [265], see also Section 6.5.9.

3.4.3 Casson Invariant of Branched Coverings

Let k be a knot in an integral homology sphere Σ such that the n–fold cyclic branched covering Σ_n of Σ with branch set the knot k is an integral homology sphere. We wish to compute the Casson invariant of $\lambda(\Sigma_n)$ in terms of invariants of Σ and k. There exist several partial results in this direction.

Theorem 3.21. *(Mullins [233]) Let k be a knot in S^3 and Σ_2 a double branched covering of S^3 with branch set the knot k. If Σ_2 is an integral homology sphere then*

$$\lambda(\Sigma_2) = \frac{1}{8}\operatorname{sign} k - \frac{1}{12}\left.\frac{d}{dt}\right|_{t=-1} \ln V_k(t)$$

where sign k *is the knot signature and* V_k *is the Jones polynomial of* k.

Proof. The idea of the proof is to study how $\lambda(\Sigma)$ changes with the change of a crossing of k. The double branched cover of a 3-ball D^3 with branch set consisting of two untangled arcs limiting to ∂D^3 is a solid torus. Due to this fact, every crossing change of k induces a surgery of Σ_2, which can be studied with the help of the Casson surgery formula.

The Mullins formula holds more generally for the Walker invariant when Σ_2 is a $\mathbb{Z}/2$–homology sphere, see Section 4.4. For n–fold branched coverings with $n > 2$, the technique of crossing changes, which is used in the proof of Mullins' formula, induces a higher genus surface replacement. As a result, the corresponding change in the Casson invariant is much more difficult to trace, see discussion in Section 4.4. Recently, Garoufalidis and Kricker [122] managed to express the Casson invariant of an n–fold cyclic branched cover of S^3 with branch set a knot k in terms of equivariant knot signatures of k and residues of certain rational functions related to the Kontsevich integral of k, compare with Example 5.27.

Example 3.22. Brieskorn homology sphere $\Sigma(2,3,7)$ is a double branched covering of S^3 with branch set a Montesinos knot k of type $(2,3,7)$. In this instance, the Mullins formula can be verified directly. The manifold $\Sigma(2,3,7)$ is obtained by (-1)–surgery on the right handed trefoil so that $\lambda(\Sigma(2,3,7)) = -1$ by the Casson surgery formula (3.3). The usual algorithm for computing the knot signature gives sign $k = 8$, see [128]. Since k is in fact a pretzel knot, its Jones polynomial can be computed by using the `Maple` program of [180]. The answer is $V_k(t) = t^{-5} + t^{-7} - t^{-11} + t^{-12} - t^{-13}$ so that $V_k'(-1) = 24$, which is of course consistent with the Mullins formula.

Theorem 3.23. *(Hoste [148]) Let* k *be a knot in an integral homology sphere* Σ *and* $D_\varepsilon k$ *its untwisted Whitehead double with* ε*–clasp. If* Σ_n *is the* n*–fold cyclic branched cover of* Σ *with branch set* $D_\varepsilon k$ *then*

$$\lambda(\Sigma_n) = n \cdot \lambda(\Sigma) + \varepsilon \cdot n\, \Delta_k''(1).$$

Here, $\Delta_k(t)$ is the Alexander polynomial of k normalized so that $\Delta_k(1) = 1$ and $\Delta_k(t) = \Delta_k(t^{-1})$. The Alexander polynomial of $D_\varepsilon k$ is equal to 1 hence Σ_n is an integral homology sphere for any n. Recall that the *untwisted Whitehead double* of a knot k with ε–clasp is obtained by embedding the circle in a solid torus as shown in Figure 3.6 and then embedding that solid torus in Σ as a tubular neighborhood of k, with the longitude of the solid torus going to the canonical longitude of k preserving orientations.
Theorem 3.23 can be viewed as a special case of more general surgery formula for the Casson invariant, see Theorem 4.9. An extension of Theorem 3.23 to m–twisted Whitehead doubles of knots can be found in [64] and [151].

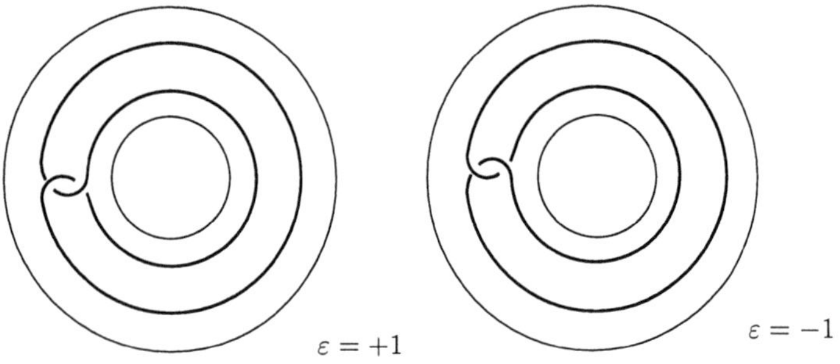

Fig. 3.6.

Theorem 3.24. *(Collin–Saveliev [57]) Let k be a graph knot in an integral homology sphere Σ such that the n-fold cyclic branched cover Σ_n over k is an integral homology sphere. Then*

$$\lambda(\Sigma_n) = n \cdot \lambda(\Sigma) + \frac{1}{8} \sum_{m=1}^{n-1} \operatorname{sign}^{m/n}(k), \tag{3.13}$$

where $\operatorname{sign}^{m/n}(k)$ *is the equivariant signature of the knot k, see (2.7).*

The definition of a graph knot can be found in Section 1.1.8. It is proved in [57] that in fact the formula (3.13) holds for any knot k in an integral homology sphere, not necessarily graph, if one replaces $\lambda(\Sigma_n)$ by the so called equivariant Casson invariant $\lambda^\tau(\Sigma_n)$, see Section 5.5 for details. Theorem 3.24 holds as stated because $\lambda(\Sigma_n) = \lambda^\tau(\Sigma_n)$ for all graph knots, see Theorem 5.23.

Theorem 3.25. *(Némethi [239]) Let k be a graph knot in an integral homology sphere Σ such that the n-fold cyclic branched cover Σ_n over k is an integral homology sphere. Then*

$$\lambda(\Sigma_n) = n \cdot \lambda(\Sigma) + \frac{1}{8}\,(\eta(\mathcal{I}, n) - n \cdot \eta(\mathcal{I}, 1)). \tag{3.14}$$

Here, the eta-invariant $\eta(\mathcal{I}, n)$ is defined as the following Dedekind sum

$$\eta(\mathcal{I}, n) = 4 \cdot \sum_{w \in \mathcal{R}} \sum_{v \in \mathcal{V}_w} \sum_{k=1}^{|m_w|-1} \left(\left(\frac{k m_v}{m_w}\right)\right) \cdot \left(\left(\frac{kn}{|m_w|}\right)\right),$$

where $\mathcal{R}$ is the set of the nodes in a plumbing diagram of the knot, $\mathcal{V}_w$ is the set of vertices adjacent to w including the arrowhead representing the knot, and m_w and m_v are their respective multiplicities, see [76]. We also use the following notation

$$((x)) = \begin{cases} x - [x] - \frac{1}{2}, & \text{if } x \notin \mathbb{Z}, \\ 0, & \text{if } x \in \mathbb{Z}, \end{cases} \tag{3.15}$$

where $[x]$ stands for the greatest integer less than or equal to x. This result is proved by methods different from those in [57]. A direct comparison of (3.13) and (3.14) shows that the correction term in Némethi's formula is nothing else but a sum of equivariant knot signatures (up to a factor of 1/8).

3.4.4 Casson Invariant of Fibered Knots

A knot k in an oriented closed 3–manifold M is called a *fibered knot* of genus g if its exterior K fibers over S^1 with fiber an oriented Riemann surface F (with boundary) of genus g and monodromy $f : F \to F$, so that

$$K = \frac{F \times [-1,1]}{\langle\, (x,1) \sim (f(x),-1)\,\rangle}.$$

Given a fibered knot k in an integral homology 3–sphere Σ, consider

$$\mathcal{R}_- = \{\, \alpha : \pi_1 F \to SU(2) \mid \alpha(\partial F) = -1\,\}/SO(3),$$

and let $f^* : \mathcal{R}_- \to \mathcal{R}_-$ be the map induced by the monodromy f. Akbulut and McCarthy show in [5] that $\lambda'(k \subset \Sigma)$ equals up to a sign the Lefschetz number of $f^* : \mathcal{R}_- \to \mathcal{R}_-$. If F has genus one, the space $\mathcal{R}_-$ consists of a single point. Therefore, $\lambda'(k \subset \Sigma) = \pm 1$ for any fibered knot k of genus one in an integral homology sphere Σ. In particular, $\lambda'(k \subset S^3) = \pm 1$ for both trefoil and figure–eight knots, compare with Section 3.2.8.

3.4.5 Finite Type Invariants

Finite type invariants for integral homology spheres were first introduced by Ohtsuki [250]. They are modeled after Vassiliev's invariants of knots, see [300] and also [22], [27] and [173]. A rational valued invariant ν of integral homology spheres is said to have *AS–type* (respectively, *B–type*) less than or equal to n if, for any algebraically split (respectively, boundary) link $\mathcal{L}$ of $(n+1)$ components framed by ± 1 in an integral homology sphere Σ, the following holds,

$$\sum_{I \subset \{\,1,2,\dots,n+1\,\}} (-1)^{|I|}\, \nu(\chi_\Sigma(\mathcal{L}_I)) = 0.$$

Here, $\mathcal{L}_I$ is the sublink of $\mathcal{L}$ consisting of the components numbered by the integers in I, $|I|$ is the cardinality of I, and $\chi_\Sigma(\mathcal{L}_I)$ is the result of surgery of Σ along $\mathcal{L}_I$. Recall that a link $\mathcal{L}$ is called *boundary* if all of its components bound disjoint Seifert surfaces. A link is *algebraically split* if every two of its components have zero linking number. All boundary links are algebraically split but not vice versa.

The Casson invariant λ is a finite type invariant of B–type 1, which follows from Property (2) of the Casson invariant, see Section 3.1. It is also a finite type invariant of AS–type 3, see [148]. For any integer n, a degree n polynomial in λ is a finite type invariant of B–type n and of AS–type $3n$. There exist however finite type invariants which cannot be expressed as polynomials in λ, see [181], and not all invariants of integral homology spheres are of finite type. Finite type invariants are closely related to quantum invariants of 3–manifolds, see [298].

3.4.6 Further Properties of the Casson Invariant

Casson's λ–invariant as it was defined in Section 3.1 is an invariant of orientation preserving homeomorphism. It would be interesting to know if $\lambda(\Sigma)$ is in fact an invariant with respect to weaker equivalences of integral homology spheres. Recall that two integral homology spheres Σ_0 and Σ_1 are called *h–cobordant* if there is a smooth compact oriented 4–manifold W such that $\partial W = -\Sigma_0 \cup \Sigma_1$ and the inclusions $\Sigma_k \to W$, $k = 0, 1$, are homotopy equivalences (in particular, $\pi_1(\Sigma_0) = \pi_1(\Sigma_1) = \pi_1(W)$). It is not difficult to show that $\lambda(\Sigma_0) = \lambda(\Sigma_1)$ for h–cobordant homology spheres Σ_0 and Σ_1, at least when $\mathcal{R}^*(\Sigma_0) = \mathcal{R}^*(\Sigma_1)$ are non-degenerate, see Corollary 6.18.

Any two h–cobordant homology spheres are homotopy equivalent. It is not known if the opposite is true. Preprint [196] addresses the invariance of the Casson invariant with respect to orientation preserving homotopy equivalence. It should be noted that this is a rather delicate problem because, for instance, the equality $\pi_1(\Sigma_0) = \pi_1(\Sigma_1)$ in general does not imply that $\lambda(\Sigma_0) = \lambda(\Sigma_1)$. For example, $\lambda(\Sigma(2,3,5)\,\#\,\Sigma(2,3,5)) = -2$ while $\lambda(\Sigma(2,3,5)\,\#\,-\Sigma(2,3,5)) = 0$. It should also be noted that for many integral homology spheres homotopy equivalence implies homeomorphism, see for example [118].

Let Σ_0 and Σ_1 be integral homology spheres obtained from each other by (± 1)–surgery along a knot with trivial Alexander polynomial. It is an immediate consequence of the surgery formula (3.3) that $\lambda(\Sigma_0) = \lambda(\Sigma_1)$. Lescop [183] provided a geometric converse, namely that if Σ_0 and Σ_1 are two integral homology spheres such that $\lambda(\Sigma_0) = \lambda(\Sigma_1)$ then Σ_1 can be obtained from Σ_0 by a sequence of (± 1)–surgeries on knots in integral homology spheres with trivial Alexander polynomials.

Let F_g be a closed oriented surface of genus g and let $\mathcal{T}_g$ be the Torelli group, see Section 1.1.13. Let us fix an embedding $F_g \to S^3$ such that the decomposition $S^3 = M_1 \cup M_2$ along F_g is a Heegaard splitting. Cutting S^3 open along F_g and regluing M_1 and M_2 by a map $f \in \mathcal{T}_g$ results in an integral homology sphere $\Sigma_f = M_1 \cup_f M_2$. Every integral homology sphere can be obtained by this construction. Morita [230] gives a formula for $\lambda(\Sigma_f)$ in terms of certain invariants of $f \in \mathcal{T}_g$, compare with Section 2.3.5. More precisely, he proves that $\lambda(\Sigma_f) = d(f)/24 + q(f)$ where $q(f)$ is defined in terms of Johnson's homomorphism [154] and $d(f)$ is defined in terms of secondary invariants associated with the signature of oriented surface bundles over surfaces. It should be noted that the definitions of d and q do not involve the Casson invariant $\lambda(\Sigma_f)$ and hence Morita's formula gives an alternative definition of $\lambda(\Sigma_f)$. Curiously enough, this definition does not use any $SU(2)$–representation spaces. An extension of Morita's results to rational homology spheres can be found in [184].

3.5 Casson Invariant of Seifert Fibered and Graph Homology Spheres

Seifert fibered and graph homology spheres form a rather large class of integral homology spheres whose Casson invariant can be given by an explicit closed form formula. In fact, the Casson invariant only needs to be calculated for Seifert fibered homology spheres as all other graph homology spheres are obtained from them by splicing, and the Casson invariant is splicing additive.

3.5.1 Casson Invariant of $\Sigma(p, q, r)$

In this section we describe $SU(2)$–representations of Brieskorn homology spheres $\Sigma(p, q, r)$ and compute $\lambda(\Sigma(p, q, r))$ following [83]. The fundamental group of $\Sigma(p, q, r)$ has presentation

$$\langle x_1, x_2, x_3, h \,|\, [h, x_k] = 1,\; x_1^p = h^{-b_1},\, x_2^q = h^{-b_2},\, x_3^r = h^{-b_3},\, x_1 x_2 x_3 = 1 \rangle,$$

where $b_1 qr + b_2 pr + b_3 pq = 1$, see (1.3). If α is an irreducible $SU(2)$–representation of $\pi_1(\Sigma(p, q, r))$ then $\alpha(h) = \pm 1$ and none of $\alpha(x_k)$ can be equal to ± 1. After conjugation, one may assume that $\alpha(x_1) \in \mathbb{C}$. The condition $\alpha(x_1)^p = \alpha(h)^{-b_1} = \pm 1$ implies that $\alpha(x_1)$ is a degree p root of 1 or -1, depending on the sign of $\alpha(h)$ and the parity of b_1. More precisely, $\alpha(x_1) = e^{\pi i \ell_1 / p}$, where $0 < \ell_1 < p$ and ℓ_1 is even if $\alpha(h)^{b_1} = 1$ and ℓ_1 is odd if $\alpha(h)^{b_1} = -1$. One can determine the conjugacy classes of $\alpha(x_2)$ or $\alpha(x_3)$ as follows: x_2 belongs to S_2, the conjugacy class of $e^{\pi i \ell_2 / q}$, where $0 < \ell_2 < q$ and ℓ_2 is even if and only if $\alpha(h)^{b_2} = 1$, and x_3 belongs to S_3, the conjugacy class of $e^{\pi i \ell_3 / r}$, where $0 < \ell_3 < r$ and ℓ_3 is even if and only if $\alpha(h)^{b_3} = 1$.

Therefore, one can associate with each irreducible representation α a triple (ℓ_1, ℓ_2, ℓ_3) of *rotation numbers* according to the rules stated above. Conversely, not every triple (ℓ_1, ℓ_2, ℓ_3) will determine a representation because the last relation $\alpha(x_1)\alpha(x_2)\alpha(x_3) = 1$ may not be satisfied. The product $\alpha(x_1)\alpha(x_2)$ belongs to $\alpha(x_1) \cdot S_2$, the image of S_2 under a rigid motion of S^3 taking 1 to $\alpha(x_1)$. The relation $x_1 x_2 x_3 = 1$ can be written in the form $x_1 x_2 = x_3^{-1}$ with $\alpha(x_3)^{-1}$ conjugated to $\alpha(x_3)$. Therefore, a triple (ℓ_1, ℓ_2, ℓ_3) defines a representation if and only if the intersection $(\alpha(x_1) \cdot S_2) \cap S_3$ is not empty. One can easily see that the latter translates into the following inequality:

$$\left| \frac{\ell_1}{p} - \frac{\ell_2}{q} \right| < \frac{\ell_3}{r} < 1 - \left| 1 - \left(\frac{\ell_1}{p} + \frac{\ell_2}{q} \right) \right|. \tag{3.16}$$

If the intersection $(\alpha(x_1) \cdot S_2) \cap S_3$ is not empty, it is a circle. All representations on this circle are conjugate to each other and thus belong to a single conjugacy class. This implies in particular that the representation space $\mathcal{R}^*(\Sigma(p,q,r))$ is finite for any $\Sigma(p,q,r)$.

Example 3.26. The fundamental group of $\Sigma(3,4,7)$ has presentation

$$\langle x_1,\dots,x_n,h \mid [h,x_k]=1,\ x_1^3=h^2,\ x_2^4=h^{-1},\ x_3^7=h^{-3},\ x_1x_2x_3=1\rangle.$$

If $\alpha(h)=1$ then the choices for (ℓ_1,ℓ_2,ℓ_3) are $(2,2,2)$, $(2,2,4)$, and $(2,2,6)$. The interval $[1/6,5/6]$ contains $2/7$ and $4/7$, but not $6/7$. If $\alpha(h)=-1$ then $\ell_1=2$, $\ell_2=1$ or 3, and $\ell_3=1,3$, or 5. The interval $[5/12,11/12]$ does not contain $1/7$, and the interval $[1/12,7/12]$ does not contain $5/7$. Therefore, the triples corresponding to representations are $(2,2,2)$, $(2,2,4)$, $(2,1,3)$, $(2,1,5)$, $(2,3,1)$, and $(2,3,3)$.

One can show by an explicit calculation with group cohomology that the representation space $\mathcal{R}^*(\Sigma(p,q,r))$ is non-degenerate, see [83]. Hence the Casson invariant of $\Sigma(p,q,r)$ is obtained by simply counting points in $\mathcal{R}^*(\Sigma(p,q,r))$ with signs. A key step in the calculation is therefore the following result.

Proposition 3.27. *All representations $\alpha \in \mathcal{R}^*(\Sigma(p,q,r))$ are counted with the same (negative) sign. Thus*

$$\lambda(\Sigma(p,q,r)) = -\frac{1}{2}\,\#\mathcal{R}^*(\Sigma(p,q,r)).$$

Proof. This result has two different proofs. The one due to Fintushel and Stern [83] uses the gauge-theoretical interpretation of the Casson invariant, see Chapter 5. Another proof is due to Lescop [185]. It is more straightforward in that it uses an explicit Heegaard splitting of $\Sigma(p,q,r)$ and calculates the intersection numbers in local charts.

There are many curious formulas for the cardinality of $\mathcal{R}^*(\Sigma(p,q,r))$ and hence for the Casson invariant of $\Sigma(p,q,r)$.

Proposition 3.28. *Denote by $\tau(p,q,r)$ the number of integral lattice points in the tetrahedron $\Delta(p,q,r)$ with vertices $(0,0,0)$, $(0,q,r)$, $(p,0,r)$, $(p,q,0)$. Then $\#\mathcal{R}^*(\Sigma(p,q,r)) = \tau(p,q,r)/4$ and hence $\lambda(\Sigma(p,q,r)) = -\tau(p,q,r)/8$.*

Proof. The inequality (3.16) is equivalent to the following system:

$$\begin{aligned} -\frac{\ell_1}{p} + \frac{\ell_2}{q} + \frac{\ell_3}{r} &> 0, \\ \frac{\ell_1}{p} - \frac{\ell_2}{q} + \frac{\ell_3}{r} &> 0, \\ \frac{\ell_1}{p} + \frac{\ell_2}{q} - \frac{\ell_3}{r} &> 0, \\ \frac{\ell_1}{p} + \frac{\ell_2}{q} + \frac{\ell_3}{r} &< 2. \end{aligned} \tag{3.17}$$

If we allow triples (ℓ_1, ℓ_2, ℓ_3) of arbitrary parities then this system will describe all integral lattice points inside the tetrahedron $\Delta(p,q,r)$. The group $\mathbb{Z}/2 \oplus \mathbb{Z}/2$ with generators s, t acts freely on these points by the rule $s(\ell_1, \ell_2, \ell_3) = (\ell_1, q - \ell_2, r - \ell_3)$, $t(\ell_1, \ell_2, \ell_3) = (p - \ell_1, q - \ell_2, \ell_3)$. Each orbit consists of four triples; exactly one triple in each orbit has the right parities.

Notice that the cardinality of $\mathcal{R}^*(\Sigma(p,q,r))$ also equals one quarter the number of integral lattice points in the tetrahedron with vertices $(p,0,0)$, $(0,q,0)$, $(0,0,r)$, and (p,q,r). These points are identified with the points in $\Delta(p,q,r)$ via the map $(\ell_1, \ell_2, \ell_3) \mapsto (p - \ell_1, q - \ell_2, r - \ell_3)$. Complete proofs of the following formulas for $\lambda(p,q,r)$ can be found in [83], [185], [243] and [275].

Let us think of $\Sigma(p,q,r)$ as the link of singularity at zero of the complex polynomial $f(x,y,z) = x^p + y^q + z^r$, see (1.1). The formula

$$\varphi(x,y,z) = \frac{f(x,y,z)}{|f(x,y,z)|}$$

defines a map $\varphi : S^5 \setminus \Sigma(p,q,r) \to S^1$ onto the circle of unit complex numbers. Milnor [219] showed that φ is the projection of a locally trivial bundle whose fiber F is a smooth simply-connected 4-manifold. A natural compactification of F is a smooth manifold $M(p,q,r)$ with boundary $\Sigma(p,q,r)$ called the *Milnor fiber*.

Theorem 3.29. *(Fintushel-Stern [83])*

$$\lambda(\Sigma(p,q,r)) = \operatorname{sign} M(p,q,r)/8.$$

Proof. Given an integer k, let τ_k be the number of integral lattice points (ℓ_1, ℓ_2, ℓ_3) such that $0 < \ell_1 < p$, $0 < \ell_2 < q$, $0 < \ell_3 < r$, and $k - 1 < \ell_1/p + \ell_2/q + \ell_3/r < k$. According to Brieskorn [40], $\operatorname{sign} M(p,q,r) = \tau_1 - \tau_2 + \tau_3$. It is an easy exercise in combinatorics to show that the latter quantity also equals $-\tau(p,q,r)$. For a more conceptual proof of this formula see Example 5.20.

Another formula for $\lambda(\Sigma(p,q,r))$ involves the so-called Dedekind sums. For any pair of relatively prime integers a, b ($a \neq 0$) the *Dedekind sum* $s(b,a)$ is given by

$$s(b,a) = \operatorname{sign}(a) \cdot \sum_{k=1}^{|a|-1} \left(\left(\frac{k}{a}\right)\right) \cdot \left(\left(\frac{kb}{a}\right)\right), \tag{3.18}$$

where the notation $((x))$ is explained in (3.15). Then

$$\lambda(\Sigma(p,q,r)) = -\frac{1}{2}\left[s(pq,r) + s(pr,q) + s(qr,p)\right] - \frac{pqr}{24}\left(1 - \frac{1}{p^2} - \frac{1}{q^2} - \frac{1}{r^2}\right) + \frac{1}{24pqr} - \frac{1}{8}.$$

It is proved in Rademacher–Grosswald [256], Section 2.C, formula (25), that

$$s(b,a) = \frac{1}{4a} \sum_{m=1}^{a-1} \frac{1+\xi^m}{1-\xi^m} \cdot \frac{1+\xi^{-bm}}{1-\xi^{-bm}},$$

where ξ is any primitive ath root of unity, $a > 0$. This implies the following formula for $\lambda(\Sigma(p,q,r))$, which can be found in [110] and [243],

$$\lambda(\Sigma(p,q,r)) = -\frac{1}{8} + \frac{1}{24pqr}\left(1 - p^2q^2r^2 + p^2q^2 + q^2r^2 + p^2r^2\right) + \frac{1}{8}\left(d(p,qr) + d(q,pr) + d(r,pq)\right). \tag{3.19}$$

Here we define, for relatively prime integers a, c with $a > 0$,

$$\begin{aligned} d(a,c) &= \frac{1}{a} \sum_{\xi^a = 1, \xi \neq 1} \frac{\xi + 1}{\xi - 1} \cdot \frac{\xi^c + 1}{\xi^c - 1}, \qquad (3.20)\\ &= -\frac{1}{a} \sum_{k=1}^{a-1} \cot\left(\frac{\pi k}{a}\right) \cot\left(\frac{\pi c k}{a}\right), \quad \text{by specifying } \xi = \exp(2\pi i k). \end{aligned}$$

Example 3.30. One can use these formulas to show that $\lambda(\Sigma(p,q,pqm \pm 1)) = -m(p^2-1)(q^2-1)/24$. An easier way to deduce this formula is to notice that $\Sigma(p,q,pqm \pm 1)$ is the result of $(-1/m)$–surgery on a (p,q)–torus knot in S^3, and apply the Casson surgery formula (3.3), compare with Example 2.12 and Example 3.33.

3.5.2 Casson Invariant of $\Sigma(a_1, \dots, a_n)$

The algorithm of the previous section describing $\mathcal{R}^*(\Sigma(p,q,r))$ in terms of rotation numbers (ℓ_1, ℓ_2, ℓ_3) can be generalized to Seifert fibered homology spheres with more than three singular fibers, see [83] and [168].

Let us consider a Seifert fibered homology sphere $\Sigma(a_1, a_2, a_3, a_4)$ with four singular fibers and choose Seifert invariants so that $b = 0$, see Section 1.1.4. The group $\pi_1 \Sigma(a_1, a_2, a_3, a_4)$ has presentation (1.3) with $b = 0$. It has two types of irreducible $SU(2)$–representations. In the first type, one of the generators x_k is mapped to ± 1, and in the second, the images of all generators differ from ± 1. Representations of the first type can be handled as in the previous section so we will concentrate on representations of the second type.

Let $\alpha : \pi_1 \Sigma(a_1, a_2, a_3, a_4) \to SU(2)$ be such that $\alpha(x_k) \neq \pm 1$. Then $\alpha(x_k)$ is conjugate to $e^{\pi i \ell_k / a_k}$ which uniquely defines a quadruple of *rotation numbers* $(\ell_1, \ell_2, \ell_3, \ell_4)$ such that $0 < \ell_k < a_k$ and ℓ_k is even if and only if $\alpha(h)^{b_k} = 1$. These two conditions are necessary for a quadruple $(\ell_1, \ell_2, \ell_3, \ell_4)$ to come from a representation, however, they may be not sufficient due to the presence of the extra relation $x_1 x_2 x_3 x_4 = 1$.

We may assume, perhaps after conjugation, that $\alpha(x_1) = e^{\pi i \ell_1 / a_1}$. Let S_k, $k = 2, 3, 4$, be the conjugacy classes of $\alpha(x_2)$, $\alpha(x_3)$ and $\alpha(x_4)$, respectively. The equation $\alpha(x_1)\alpha(x_2)\alpha(x_3)\alpha(x_4) = 1$ means that, in $SU(2) = S^3$, there exist radii r_2, r_3, and r_4 of the 2–spheres $\alpha(x_1) \cdot S_2$, $\alpha(x_1)\alpha(x_2) \cdot S_3$, and $\alpha(x_1)\alpha(x_2)\alpha(x_3) \cdot S_4$, respectively, which form a linkage spanning from $\alpha(x_1)$ to 1. A quadruple $(\ell_1, \ell_2, \ell_3, \ell_4)$ satisfying the necessary conditions corresponds to a representation α if and only if such a linkage can be formed. The connected component of α is then the configuration space of the linkage modulo rotations leaving the complex circle in $SU(2)$ invariant.

To see linkages, consider a stereographic projection $S^3 \to \mathbb{R}^3$ identifying S^3 with $\mathbb{R}^3 \cup \{\infty\}$ and mapping the complex circle in S^3 onto the z–axis. The 2–spheres in S^3 correspond to 2–spheres in $\mathbb{R}^3 \cup \{\infty\}$, but the concept of linkage is changed since spheres of the same radius in S^3 can project to spheres with different radii in $\mathbb{R}^3 \cup \{\infty\}$. A linkage with radii r_1, r_2, r_3, and r_4 will map to a linkage with radii r'_1, r'_2, r'_3, and r'_4, see Figure 3.7.

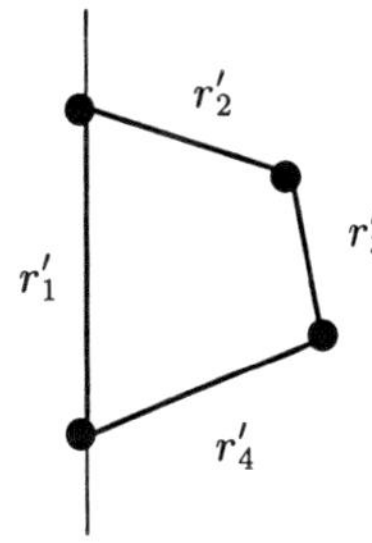

Fig. 3.7.

Suppose that $r'_1 \geq r'_2 \geq r'_3 \geq r'_4$. Then there are exactly two rigid configurations in the configuration space shown in Figure 3.8.

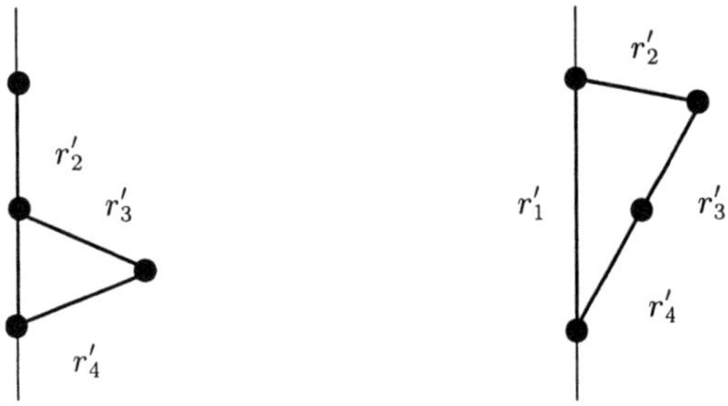

Fig. 3.8.

If we fix a generic r'_2, the configurations (divided out by rotations preserving the z–axis) form a circle, see Figure 3.9. For all possible choices of r'_2 we obtain a family of these circles, which together with the two points arising from the rigid configurations, form a 2–sphere.

A similar argument shows that the configuration spaces for all other choices of the radii are also 2–spheres. This proves that the representation space $\mathcal{R}^*(\Sigma(a_1, a_2, a_3, a_4))$ consists of a finite number of isolated points and a finite number of 2–spheres.

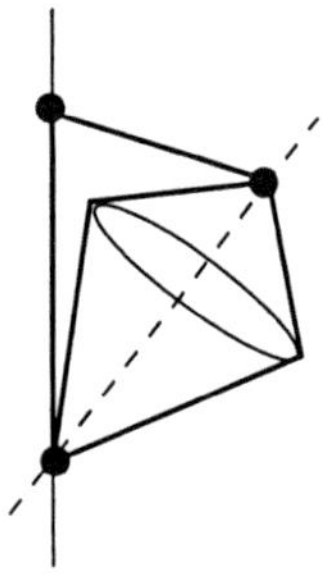

Fig. 3.9.

Example 3.31. The fundamental group of $\Sigma(2,3,5,7)$ has presentation

$$\langle x_1, \dots, x_n, h \mid [h, x_k] = 1,$$
$$x_1^2 = h^{-1},\ x_2^3 = h^2,\ x_3^5 = h^2,\ x_4^7 = h^{-4},\ x_1x_2x_3x_4 = 1\rangle.$$

Suppose that $\alpha(h) = 1$ then $\alpha(x_1) = \pm 1$. In the case $\alpha(x_1) = 1$, we have $\ell_1 = 0$ and $\alpha(x_2)\alpha(x_3)\alpha(x_4) = 1$, and we check if a quadruple $(0, \ell_2, \ell_3, \ell_4)$ corresponds to a representation by checking the condition

$$\left|\frac{\ell_2}{3} - \frac{\ell_3}{5}\right| < \frac{\ell_4}{7} < 1 - \left|1 - \left(\frac{\ell_2}{3} + \frac{\ell_3}{5}\right)\right|. \tag{3.21}$$

We end up with 4 representations $(0,2,2,2)$, $(0,2,2,4)$, $(0,2,2,6)$, and $(0,2,4,2)$. If $\alpha(x_1) = -1$ then $\ell_1 = 2$ and $\alpha(x_2)\alpha(x_3)\alpha(x_4) = -1$, so instead of (3.21), the condition

$$\left|\frac{\ell_2}{3} - \frac{\ell_3}{5}\right| < 1 - \frac{\ell_4}{7} < 1 - \left|1 - \left(\frac{\ell_2}{3} + \frac{\ell_3}{5}\right)\right| \tag{3.22}$$

should be checked. From this, we get 4 more representations, $(2,2,2,2)$, $(2,2,2,4)$, $(2,2,4,4)$, and $(2,2,4,6)$. Similar computations with $\alpha(h) = -1$ and at least one of the $\alpha(x_k)$ equal to ± 1 produce the representations $(1,0,2,2)$, $(1,0,2,4)$, $(1,0,2,6)$, $(1,0,4,4)$, $(1,2,0,2)$, $(1,2,0,4)$, $(1,2,2,0)$, and $(1,2,4,0)$.

Suppose that none of the $\alpha(x_k)$ equals ± 1. We have the following choices: $\ell_1 = 1$, $\ell_2 = 2$, $\ell_3 = 2$ or 4, and $\ell_4 = 2, 4$, or 6. To determine if any of them define representations, we need to check whether the intersection $(\alpha(x_1)\cdot S_2) \cap (\alpha(x_4)\cdot S_3)$ is non–empty. The property of this intersection to be non–empty does not change if we replace $\alpha(x_4)\cdot S_3$ by its conjugate. Therefore, we may assume that $\alpha(x_4) \in \mathbb{C}$, for the purposes of this computation. The computation then reduces to checking whether the intervals on the upper half of the complex circle corresponding to $\alpha(x_1)\cdot S_2$ and $\alpha(x_4)\cdot S_3$ intersect each other. It turns out that in our example all possible combinations of the ℓ_k can be realized, so we get six 2–dimensional spheres in $\mathcal{R}(\Sigma(2,3,5,7))$ (in addition to 16 isolated points we already have).

A similar algorithm describes $SU(2)$–representations of $\pi_1\Sigma(a_1,\dots,a_n)$ for all n via n–tuples $(\ell_1,\dots,\ell_n)$ of rotation numbers, see [83] and [168]. An n–tuple gives rise to at most one connected component of $\mathcal{R}^*(\Sigma(a_1,\dots,a_n))$. If m is the number of the ℓ_k different from 0 and a_k then the corresponding connected component in $\mathcal{R}^*(\Sigma(a_1,\dots,a_n))$ is a smooth closed orientable manifold of dimension $2m-6$ which is non-degenerate in the Morse–Bott sense, see [83]. This manifold admits a Morse function with only even indices of the critical points, see [83] and [168], and is in fact a non-singular rational projective variety, see [24] and [116].

For all $n \le 5$ the types of connected components of $\mathcal{R}^*(\Sigma(a_1,\dots,a_n))$ are completely determined. If $n = 3$, all components are points. If $n = 4$, they are either points or 2–spheres. If $n = 5$, in addition to points and 2–spheres there exist 4–dimensional components. They are homeomorphic to either $S^2 \times S^2$ or $\mathbb{C}P^2 \# n\cdot(-\mathbb{C}P^2)$, where $n = 0,1,2,3,4$, or 5. Moreover, there exist $\Sigma(a_1,\dots,a_5)$ for which each of these 4–manifolds actually occurs as a component, see [168]. An algebro–geometric description of $\mathcal{R}^*(\Sigma(a_1,\dots,a_n))$ for all n can be found in [23]. Poincaré polynomials of $\mathcal{R}^*(\Sigma(a_1,\dots,a_n))$ were computed for all n in [23] and [116].

Because of the complicated geometry of $\mathcal{R}^*(\Sigma(a_1,\dots,a_n))$, these spaces are of limited use in computing $\lambda(\Sigma(a_1,\dots,a_n))$. More efficient way is to use the splicing additivity of the Casson invariant and Example 1.10 which describes $\Sigma(a_1,\dots,a_n)$ as the splice of two Seifert fibered homology spheres, each with less than n singular fibers. The resulting formula is as follows, see [110] or [243],

$$\lambda(\Sigma(a_1,\dots,a_n)) =$$
$$-\frac{1}{8}+\frac{1}{24a}\left(1-(n-2)a^2+\sum_{k=1}^{n} c_k^2\right)+\frac{1}{8}\sum_{k=1}^{n} d(a_k,c_k), \quad (3.23)$$

where $a = a_1 \cdots a_n$ is the product of the a_k, $c_k = a/a_k$ and $d(a,c)$ are defined as in (3.20). Theorem 3.29 generalizes as follows. Let us think of $\Sigma(a_1,\dots,a_n)$ as the link of singularity (1.2), and denote by $M(a_1,\dots,a_n)$ its Milnor fiber viewed as a smooth 4-manifold with boundary $\Sigma(a_1,\dots,a_n)$.

Theorem 3.32. *(Neumann–Wahl [243])*

$$\lambda(\Sigma(a_1,\dots,a_n)) = \operatorname{sign} M(a_1,\dots,a_n)/8.$$

A geometric proof of this result can be found in Example 5.21. Theorem 3.32 allows in particular for easy proofs of some number–theoretic results, compare with [147].

Example 3.33. The result of (-1)–surgery of a homology sphere $\Sigma(a_1,\dots,a_n)$ along the fiber k of degree a_n is $\Sigma(a_1,\dots,a_{n-1},a_n+c_n)$ where $c_n = a/a_n = a_1\cdots a_{n-1}$, see Example 1.5. Therefore,

$$\lambda(\Sigma(a_1,\dots,a_{n-1},a_n+c_n)) = \lambda(\Sigma(a_1,\dots,a_n)) - \lambda'(k),$$

where $\lambda'(k)$ can be computed by using Theorem 3.32 as follows, see [243],

$$\lambda'(k) = \frac{1}{24}\left((n-2)c_n^2 - \sum_{k=1}^{n-1}\frac{c_n^2}{a_k^2}+1\right). \quad (3.24)$$

It should be noted that $\lambda'(k)$ is independent of a_n. For a Brieskorn homology sphere $\Sigma(p,q,r)$, the above formula specifies to $\lambda(\Sigma(p,q,r+pq)) = \lambda(\Sigma(p,q,r)) - (p^2-1)(q^2-1)/24$, compare with Example 3.30.

3.5.3 The Neumann–Wahl Conjecture

All links of singularities Σ with $H_*(\Sigma,\mathbb{Z}) = H_*(S^3,\mathbb{Z})$ are graph homology spheres. Therefore, every such link Σ can be obtained by splicing Seifert links, and $\lambda(\Sigma)$ can be effectively computed by using the splicing additivity of λ and the above formulas for $\lambda(\Sigma(a_1,\dots,a_n))$. We saw above that $\lambda(\Sigma(a_1,\dots,a_n))$ can be expressed in terms of other invariants of the singularity such as the signature of its Milnor fiber, see Theorem 3.29 and Theorem 3.32. A natural question arises if these results can be extended to other links of singularities.

More precisely, let Σ be the link of a normal complete intersection surface singularity and let M be the associated Milnor fiber viewed as a smooth compact manifold with boundary Σ. Assuming that Σ is an integral homology sphere, is it true that $\lambda(\Sigma) = \text{sign}(M)/8$? Neumann and Wahl conjectured in [243] that the answer to this question is positive. The importance of identities like $\lambda(\Sigma) = \text{sign}(M)/8$ in algebraic geometry is that they allow to express analytic invariants of a normal surface singularity in terms of the topology of its link.

The Neumann–Wahl conjecture has been confirmed for Seifert fibered homology spheres with any number of singular fibers, see Theorem 3.32, for the links of isolated singularities of hyper–surfaces $f(x,y) + z^n = 0$, and for a few other families of complete intersection singularities. The proofs in [243] rely crucially on the additivity of the Casson invariant under splicing and the original Fintushel–Stern formula, see Theorem 3.29. A different geometric approach can be found in [57]. In its full generality, the conjecture remains unresolved.

Remark 3.34. Neumann and Wahl asked in [243] if their conjecture can be generalized to the links of singularities which are rational homology spheres. Simply replacing the Casson invariant by its extension, the Walker invariant, will not do the job, see Examples 4.12 and 4.13. A possible extension of the Neumann–Wahl conjecture to this case was recently suggested by Némethi and Nicolaescu [240].

3.6 Applications of the Casson Invariant

Most topological applications of the Casson invariant λ stem from the fact that it is an integral lift of the Rokhlin invariant of integral homology spheres. It should be noted however that λ is not a lift of the *Rokhlin homomorphism* $\mu : \Theta^3_{\mathbb{Z}} \to \mathbb{Z}/2$ because λ is not an invariant of homology cobordism, see Remark 3.3.

3.6.1 Triangulating Topological 4–Manifolds

The concepts of topological, smooth, and PL manifolds coincide in dimensions 1, 2, and 3, see [26] and [222]. Relations between these three categories of manifolds are more complicated in higher dimensions. In dimension four, every PL–manifold has a unique smooth structure and vice versa, see [45] and [145]. However, topological manifolds in dimension four are utterly different from PL or smooth manifolds. On the one hand, there exist closed simply connected topological 4–manifolds which admit several non-diffeomorphic smooth structures, see Example 6.75. On the other hand, there exist topological manifolds without any smooth structures.

Example 3.35. According to Freedman [98], for any unimodular even integral form Q, there exists a unique (up to homeomorphism) simply connected closed topological 4–manifold whose intersection form is isomorphic to Q. Let W_8 be such a manifold with intersection form E_8 given by the graph in Figure 1.14. Suppose that W_8 admits a smooth structure, then its signature must be divisible by 16 by Theorem 2.1. However, $\operatorname{sign} E_8 = -8$. This contradiction shows that W_8 does not have any smooth (or PL) structure.

Recall that PL manifold of dimension n is a simplicial complex that admits a combinatorial triangulation. A triangulation is called *combinatorial* if the link of each of its vertices is homeomorphic to the sphere S^{n-1}. The closed topological 4–manifold W_8 constructed in Example 3.35 is not smooth, therefore, is not homeomorphic to a combinatorial simplicial complex.

Theorem 3.36. *(Casson) The manifold W_8 is not homeomorphic to a simplicial complex.*

Thus W_8 is not triangulable in the weakest possible sense – it is not homeomorphic to *any* simplicial complex, not just a combinatorial one. The key point in the proof of this result is the existence of the Casson invariant λ with the property that $\lambda(\Sigma) = 0$ for homotopy spheres Σ, see Theorem 3.1, Property (8).

Proof. Suppose that W_8 is homeomorphic to a simplicial complex. Passing to a star subdivision if necessary, we can assume that the triangulation of W_8 is combinatorial in the complement of the open star of a vertex v. This gives the link of the vertex v a PL structure Σ so that the star of v is homeomorphic to a cone $C(\Sigma)$ over Σ. Since W_8 is a topological manifold, Σ is simply-connected. Therefore its Casson and hence Rokhlin invariants vanish. On the other hand, Σ bounds a smooth simply connected compact 4–manifold $W_8 \setminus C'(\Sigma)$, where C' stands for an open cone, with the intersection form E_8. The Rokhlin invariant computed with the help of this manifold equals 1 modulo 2, which leads to a contradiction.

A similar construction with other even unimodular forms of signature 8 modulo 16 can be used to produce more examples of closed topological manifolds in dimension 4 which are not triangulable.

The relations between topological, smooth, and PL manifolds in dimensions five and higher are more complicated than in dimension 4, and will be discussed in Section 7.3.1. In particular, it is not known if every topological manifold of dimension greater than or equal to five is homeomorphic to a simplicial complex.

3.6.2 Amphicheiral Homology Spheres

An oriented integral homology sphere Σ is called *amphicheiral* if it is orientation preserving homeomorphic to $-\Sigma$, the manifold Σ with reversed orientation. Theorem 3.1, Property (6), implies that the Casson invariant and therefore the Rokhlin invariant of an amphicheiral homology sphere are zero. This solves a long-standing conjecture in the 3–manifold theory related to the Hauptvermutung, see [285], and prevents us from finding an easy solution to the triangulation problem for topological manifolds in dimensions five and higher, see Section 7.3.1.

It should be pointed out that the vanishing of the Rokhlin invariant for all amphicheiral $\mathbb{Z}/2$–homology spheres was actually known before the Casson invariant was introduced. This was proved independently by Galewski and Stern [120], Birman, and also Hsiang and Pao [149], and later generalized by several other people, most notably by Siebenmann [286].

3.6.3 Property P for Knots

A knot k in S^3 has *Property P* if the group $\pi_1(S^3 + (1/q)\cdot k)$ is trivial if and only if $q = 0$. Equivalently, the knot k has Property P if a counterexample to the Poincaré conjecture cannot be constructed by surgery on k, see Section 1.1.5. Many classes of knots have been shown to have Property P, see for instance [44] and [160] and references therein. However, it is still not known if *all* non–trivial knots have this property. The surgery formula (3.3) for the Casson invariant implies that

$$\lambda\left(S^3 + \frac{1}{q}\cdot k\right) = \frac{q}{2}\cdot \Delta_k''(1),$$

where $\Delta_k(t)$ is the Alexander polynomial of the knot k normalized so that $\Delta_k(1) = 1$ and $\Delta_k(t) = \Delta_k(t^{-1})$. Therefore, all knots k with $\Delta_k''(1) \neq 0$, which form a rather large class, have Property P. However, there exist many non–trivial knots k such that $\Delta_k''(1) = 0$. For example, all untwisted Whitehead doubles of knots have this property. The ongoing program of Kronheimer and Mrowka [175] aims at proving Property P for all knots (or at least for those of genus two and higher) by using an extension of the Casson invariant, the Floer homology.

4 Invariants of Walker and Lescop

The Casson invariant, originally defined for integral homology 3–spheres, was extended by Walker [302] to rational homology 3–spheres. Partial extensions were also proposed by Boyer and Lines [35] and Boyer and Nicas [36]. Walker defined his invariant by extending Casson's $SU(2)$ intersection theory to include reducible representations, which arise as long as the first integral homology of the rational homology sphere does not vanish. Most remarkably, Walker's invariant also admits a purely combinatorial definition in terms of surgery presentations. This definition was later generalized by Lescop [182], who defined an invariant for all oriented closed 3–manifolds thus extending both Casson's and Walker's invariants.

4.1 Definition of the Walker Invariant

Let $\mathcal{S}$ be the class of oriented rational homology 3–spheres modulo orientation preserving homeomorphism. A *Walker invariant* is a map $\lambda_W : \mathcal{S} \to \mathbb{Q}$ satisfying conditions **(0)** and **(1)** below:

(0) $\lambda_W(S^3) = 0$.

(1) *Let k be a knot in a rational homology sphere, K its exterior, and $\ell \in \partial K$ a longitude. Then*

$$\lambda_W(K_a) = \lambda_W(K_b) + \frac{1}{2}\,\tau(a,b;\ell) + \frac{< a,b >}{< a,\ell >< b,\ell >}\cdot\frac{1}{2}\,\Delta''_k(1) \qquad (4.1)$$

for primitive $a, b \in H_1(\partial K,\mathbb{Z})$ such that $< a,\ell > \neq 0$ and $< b,\ell > \neq 0$.

A homology class $c \in H_1(\partial K,\mathbb{Z})$ is called *primitive* if it is not a multiple of any other class in $H_1(\partial K,\mathbb{Z})$. A primitive class can be represented by a non–separating simple closed curve on ∂K. Associated with every primitive class c is a manifold K_c, see (1.13). Any simple closed curve representing a generator of $\ker\{\, H_1(\partial K,\mathbb{Z}) \to H_1(K,\mathbb{Z})\,\}$ is called a *longitude*; it is well–defined up to a sign. The brackets $<,>$ denote the intersection pairing $H_1(\partial K,\mathbb{Z}) \otimes H_1(\partial K,\mathbb{Z}) \to \mathbb{Z}$, and $\Delta_k(t)$ stands for the normalized Alexander polynomial of k, see [302]. Given a longitude ℓ, choose a basis x, y of $H_1(\partial K,\mathbb{Z})$ such that $< x,y >= 1$ and $\ell = dy$ for some $d \in \mathbb{Z}$. Then

$$\tau(a,b;\ell) = -s(<x,a>,<y,a>) \\ + s(<x,b>,<y,b>) + \frac{d^2-1}{12} \cdot \frac{<a,b>}{<a,\ell><b,\ell>}, \quad (4.2)$$

where $s(q,p)$ is the Dedekind sum (3.18). The formula (4.1) is referred to as the *surgery formula* for λ_W, and it is discussed in more detail in Section 4.2.3.

The existence and uniqueness of a Walker invariant together with some basic properties are given by the following theorem of Walker [302]. The invariant introduced in this theorem is referred to as the *Walker* or the *Casson–Walker invariant.*

Theorem 4.1. *There exists a unique Walker invariant λ_W. Moreover, it has the following properties :*

(2) *If Σ is an integral homology 3–sphere then $\lambda_W(\Sigma) = \lambda(\Sigma)$, the Casson invariant of Σ.*

(3) *The number $12 \cdot |H_1(\Sigma,\mathbb{Z})| \cdot \lambda_W(\Sigma)$ is an integer for any rational homology sphere.*

(4) *For any $\mathbb{Z}/2$–homology 3–sphere Σ, one has $|H_1(\Sigma,\mathbb{Z})|^2 \cdot \lambda_W(\Sigma) = \mu(\Sigma) \bmod 2$ where μ is the Rokhlin invariant.*

(5) *$\lambda_W(-\Sigma) = -\lambda_W(\Sigma)$ where $-\Sigma$ stands for Σ with reversed orientation.*

(6) $\lambda_W(\Sigma_0 \# \Sigma_1) = \lambda_W(\Sigma_0) + \lambda_W(\Sigma_1)$

Note that our definition of λ_W differs from that in [302] by a factor of 1/2 to match it better with the Casson invariant.

4.2 Construction of the Walker Invariant

The Casson invariant for integral homology 3–spheres was defined via the intersection theory of $SU(2)$–representation varieties. The main problem in generalizing this approach to a rational homology 3–sphere Σ is the presence of non-trivial reducible representations. Any reducible representation $\alpha : \pi_1(\Sigma) \to SU(2)$ factors through a representation $H_1(\Sigma,\mathbb{Z}) \to SU(2)$. If Σ is an integral homology sphere there is only one reducible representation, the trivial one. In general, $H_1(\Sigma,\mathbb{Z})$ is a finite abelian group, so there may exist (finitely many) non-trivial reducible representations. They must be included in the count together with the irreducible representations in order to obtain a well-defined invariant.

4.2.1 $SU(2)$–Representation Varieties

Let π be a finitely presented group and $\mathcal{R}(\pi)$ its $SU(2)$-character variety, see Section 3.2.1. The variety $\mathcal{R}(\pi)$ contains two singular strata : $\mathcal{S}$, consisting of (the conjugacy classes of) reducible representations or, equivalently, representations with abelian image, and $\mathcal{P}$, consisting of central representations, that is, representations of π into $\mathbb{Z}/2 = \{\pm 1\}$. We will use notations $\mathcal{R}^c(\pi) = \mathcal{P}$ for central representations, $\mathcal{R}^a(\pi) = \mathcal{S} \setminus \mathcal{P}$ for abelian (non-central) representations, and, as before, $\mathcal{R}^*(\pi)$ for irreducible representations.

The variety $\mathcal{R}^*(\pi)$ was described in detail in Section 3.2.1. The space $\mathcal{R}^c(\pi)$ is discrete, and the space $\mathcal{R}^a(\pi)$ can be identified as follows. Think of $SU(2)$ as unit quaternions and let $U(1) \subset SU(2)$ be the circle of unit complex numbers. Then every abelian representation can be conjugated to a $U(1)$–representation, two such representations being $SU(2)$–conjugate if and only if they are equal or conjugate by $j \in SU(2)$. Therefore, $\operatorname{Hom}(\pi, U(1)) \setminus \operatorname{Hom}(\pi, \mathbb{Z}/2)$ is a regular double cover of $\mathcal{R}^a(\pi)$ by the action $\alpha \mapsto j\alpha j^{-1}$. In particular, the Zariski tangent space of $\mathcal{R}^a(\pi)$ at a point α is equal to $H^1(\pi, \operatorname{ad}\alpha) = H^1(\pi, \mathbb{R})$.

4.2.2 The Intersection Theory

Let Σ be a rational homology 3–sphere, $\Sigma = M_1 \cup_F M_2$ a Heegaard splitting of Σ, and $F_0 = F \setminus D^2$. We have the following commutative diagram of fundamental groups,

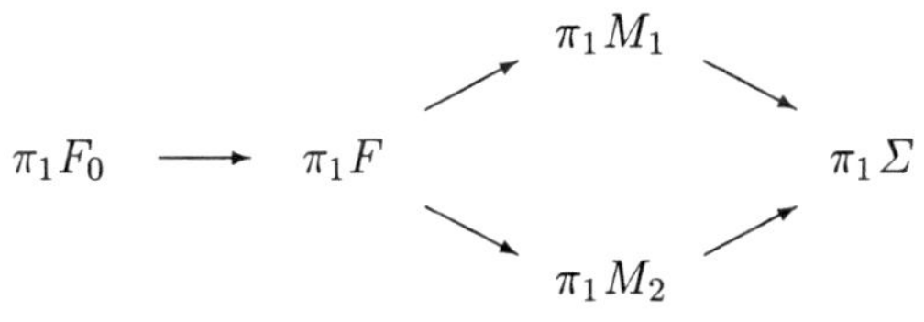

where all homomorphisms are surjective. This diagram induces the following two commutative diagrams, where $\mathcal{R}^*(\Sigma) = \mathcal{R}^*(\pi_1\Sigma)$ etc,

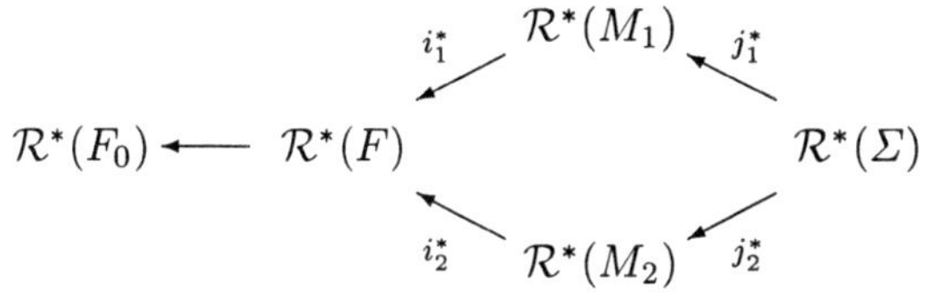

and

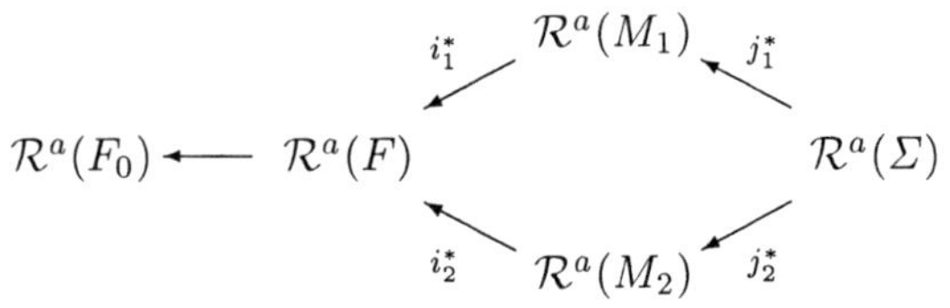

All arrows in the latter two diagrams are inclusions, therefore, we can view $\mathcal{R}^*(\Sigma)$ as the intersection of $\mathcal{R}^*(M_1)$ with $\mathcal{R}^*(M_2)$ inside $\mathcal{R}^*(F)$, and $\mathcal{R}^a(\Sigma)$ as the intersection of $\mathcal{R}^a(M_1)$ with $\mathcal{R}^a(M_2)$ inside $\mathcal{R}^a(F)$. Note that $\mathcal{R}^a(\Sigma)$ is finite.

Lemma 4.2. *The intersection $\mathcal{R}^a(M_1) \cap \mathcal{R}^a(M_2) = \mathcal{R}^a(\Sigma)$ in $\mathcal{R}^a(F)$ is transversal.*

Proof. Let g denote the genus of the Heegaard splitting $\Sigma = M_1 \cup M_2$. The intersection $\mathcal{R}^a(M_1) \cap \mathcal{R}^a(M_2)$ is transversal at α if and only if $i_1^* T_\alpha \mathcal{R}^a(M_1) \oplus i_2^* T_\alpha \mathcal{R}^a(M_2) = T_\alpha \mathcal{R}^a(F)$. With the identifications $T_\alpha \mathcal{R}^a(M_k) = H^1(M_k, \mathbb{R}) = \mathbb{R}^g$, $k = 1, 2$, and $T_\alpha \mathcal{R}^a(F) = H^1(F, \mathbb{R}) = \mathbb{R}^{2g}$, the space $i_1^* T_\alpha \mathcal{R}^a(M_1) \oplus i_2^* T_\alpha \mathcal{R}^a(M_2)$ is the image of the map $i_1^* \oplus i_2^*$ in the following Mayer-Vietoris exact sequence,

$$0 \to H^1(\Sigma, \mathbb{R}) \to H^1(M_1, \mathbb{R}) \oplus H^1(M_2, \mathbb{R}) \xrightarrow{i_1^* \oplus i_2^*} H^1(F, \mathbb{R}) \to \dots$$

Since Σ is a rational homology sphere, the map $i_1^* \oplus i_2^*$ is an isomorphism.

A similar argument shows that the intersection of $\mathcal{R}(M_1)$ and $\mathcal{R}(M_2)$ is transversal at all central representations, which proves the following result, compare with Lemma 3.6.

Corollary 4.3. *The intersection $\mathcal{R}^*(M_1) \cap \mathcal{R}^*(M_2) = \mathcal{R}^*(\Sigma)$ in $\mathcal{R}^*(F)$ is compact.*

The intersection $\mathcal{R}^*(M_1)$ and $\mathcal{R}^*(M_2)$ can be made transversal via a special isotopy as follows. We first choose a compactly supported isotopy of a tubular neighborhood of $\mathcal{R}^a(F)$ in $\mathcal{R}(F)$ which is identity on $\mathcal{R}^a(F)$ and which makes $\mathcal{R}^*(M_1)$ and $\mathcal{R}^*(M_2)$ transversal near abelian representations. Then we apply another compactly supported isotopy of $\mathcal{R}^*(F)$ to make $\mathcal{R}^*(M_1)$ and $\mathcal{R}^*(M_2)$ transversal at all intersection points. Throughout the procedure, the points in $\mathcal{R}^a(M_1) \cap \mathcal{R}^a(M_2)$ remain fixed.

Now both intersections $\mathcal{R}^*(M_1) \cap \mathcal{R}^*(M_2)$ and $\mathcal{R}^a(M_1) \cap \mathcal{R}^a(M_2)$ are transversal and consist of finitely many points. Following the analogy with the Casson invariant, we would like to assign a plus or minus one to each of these points and sum them up to get a well-defined invariant $\lambda_W(\Sigma)$ (the central representations in both instances are ignored). Unfortunately, this approach does not work because the resulting integer may depend on the choice of special isotopy.

Example 4.4. Let $g = 2$ and consider an isotopy of a tubular neighborhood of $\mathcal{R}^a(F)$ in $\mathcal{R}(F)$ which we may have to apply to make $\mathcal{R}^*(M_1)$ and $\mathcal{R}^*(M_2)$ transversal. View this tubular neighborhood as the total space of the singular normal bundle ξ. Let $\alpha \in \mathcal{R}^a(M_1) \cap \mathcal{R}^a(M_2)$ and let ξ_α be the fiber of ξ over α. Calculations with normal bundle in [302] show that the triple $(\mathcal{R}(F), \mathcal{R}(M_1), \mathcal{R}(M_2)) \cap \xi_\alpha$ is diffeomorphic to $(C(S^1), C(a_1), C(a_2))$ where $a_1, a_2 \in S^1$ and C stands for the cone, see Figure 4.1.

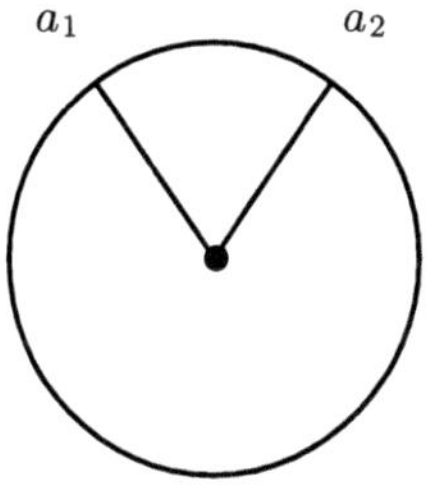

Fig. 4.1.

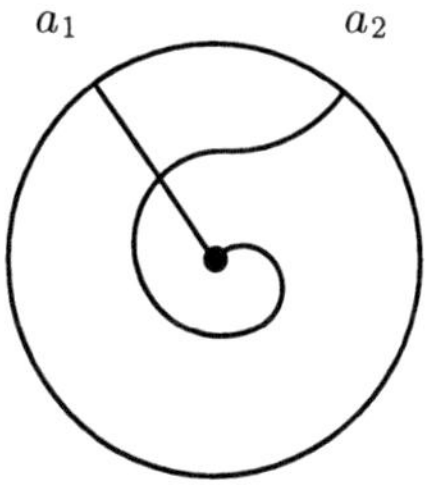

Fig. 4.2.

There is an isotopy of $C(S^1)$ supported away from $\partial C(S^1)$ which introduces an extra irreducible representation, see Figure 4.2.

The above example makes it clear that if each of the points of $\mathcal{R}^*(M_1) \cap \mathcal{R}^*(M_2)$ is to contribute $+1$ or -1 to the invariant λ_W, compare with (3.8), the contribution of points of $\mathcal{R}^a(M_1) \cap \mathcal{R}^a(M_2)$ must be the global twisting in the normal bundle of $\mathcal{R}^a(F)$. Thus, after applying a special isotopy to make $\mathcal{R}^*(M_1) \cap \mathcal{R}^*(M_2)$ transverse, we define

$$\#(\mathcal{R}(M_1) \cap \mathcal{R}(M_2)) = \sum_{\alpha \in \mathcal{R}^*(M_1) \cap \mathcal{R}^*(M_2)} \delta_\alpha + \sum_{\alpha \in \mathcal{R}^a(M_1) \cap \mathcal{R}^a(M_2)} I_\alpha,$$

where $\delta_\alpha = \pm 1$ and I_α is the twisting number, both defined precisely in [302]. Walker shows that $\#(\mathcal{R}(M_1) \cap \mathcal{R}(M_2))$ does not depend on the choice of isotopy used to make $\mathcal{R}^*(M_1)$ and $\mathcal{R}^*(M_2)$ transverse. The Walker invariant is now defined by the formula

$$\lambda_W(\Sigma) = \frac{1}{2} \cdot \frac{\#(\mathcal{R}(M_1) \cap \mathcal{R}(M_2))}{|H_1(\Sigma, \mathbb{Z})|}. \tag{4.3}$$

Note that definition (4.3) differs from that in [302] by a factor of 1/2. We introduced this factor to match λ_W with the Casson invariant λ in the case of integral homology spheres.

Walker [302] proves that λ_W is independent of the choice of Heegaard splitting and checks the properties of λ_W stated in Theorem 4.1. Most proofs are generalization of Casson's work for integral homology spheres, and therefore are omitted here.

4.2.3 The Surgery Formula

The surgery formula (4.1) simplifies a lot when applied to null-homologous knots in rational homology spheres.

Corollary 4.5. *Let k be a null-homologous knot in a rational homology 3–sphere Σ, and let $L(p,q)$ be the lens space obtained by $(-p/q)$–surgery on the unknot in S^3. Then*

$$\lambda_W\left(\Sigma + \frac{p}{q}\cdot k\right) = \lambda_W(\Sigma) - \lambda_W(L(p,q)) + \frac{q}{2p}\cdot \Delta''_k(1).$$

Corollary 4.6. *Let $L(p,q)$ be the lens space obtained by $(-p/q)$–surgery on the unknot in S^3 then $\lambda_W(L(p,q)) = s(q,p)/2$ where $s(q,p)$ is the Dedekind sum (3.18).*

Example 4.7. For any integer $n \geq 2$ we have $\lambda_W(L(n,1)) = s(1,n)/2 = (n-1)(n-2)/24n$. The latter identity follows from the reciprocity formula for Dedekind sums, see for example [256]. One can easily check that, for odd n, this calculation is consistent with property (4) of Theorem 4.1 and Example 2.9.

If k is a null-homologous knot in a rational homology sphere Σ then

$$\lambda'_W(k\subset\Sigma) = \lambda_W\left(\Sigma + \frac{1}{m+1}\cdot k\right) - \lambda_W\left(\Sigma + \frac{1}{m}\cdot k\right)$$

is independent of m which generalizes Property (1) of the Casson invariant, see Section 3.1. Given a pair of null-homologous knots k, ℓ in a rational homology sphere Σ with $\mathrm{lk}_\Sigma(k,\ell)=0$, define $\lambda''_W(k,\ell\subset\Sigma)$ to be

$$\lambda_W\left(\Sigma + \frac{1}{m+1}\cdot k + \frac{1}{n+1}\cdot \ell\right) - \lambda_W\left(\Sigma + \frac{1}{m}\cdot k + \frac{1}{n+1}\cdot \ell\right)$$
$$- \lambda_W\left(\Sigma + \frac{1}{m+1}\cdot k + \frac{1}{n}\cdot \ell\right) + \lambda_W\left(\Sigma + \frac{1}{m}\cdot k + \frac{1}{n}\cdot \ell\right),$$

compare with (3.2). The following result, which is proved in [302], is then a generalization of property (2) of the Casson invariant, see Section 3.1.

Proposition 4.8. *Let k and ℓ be null-homologous knots in a rational homology sphere Σ bounding disjoint Seifert surfaces. Then $\lambda''_W(k,\ell\subset\Sigma)=0$.*

4.2.4 Combinatorial Definition of the Walker Invariant

The surgery formula (4.1) not only uniquely determines the Walker invariant λ_W but also can be used to give an alternative definition of it. Roughly, the idea is to present a rational homology sphere Σ as a surgery on a link in

S^3, compute $\lambda_W(\Sigma)$ by surgering one component at a time and applying (4.1), and then show that the result is independent of the choice of surgery presentation. The latter is accomplished in [302] by introducing a complete set of moves on surgery presentations of Σ so that any two presentations are related by a sequence of the moves and their inverses, and then showing that $\lambda_W(\Sigma)$ does not change under the moves.

Note that this approach also shows that the Casson invariant λ is well defined by its surgery formula (3.3). This would be difficult to prove without first passing to λ_W : the above moves relate any two integral homology spheres through a sequence of rational, but perhaps not integral, homology spheres.

4.3 The Lescop Invariant

The surgery formula (4.1) was extended by Lescop [182] to framed links with arbitrarily many components. Her formula relates the Walker invariants of any two rational homology spheres obtained from each other by surgery on a link. It also extends naturally to all framed links in S^3 in such a way that the result is the same for any two framed links representing the same manifold. Hence it defines an invariant for all closed oriented 3–manifolds. We denote this invariant by λ_L and call it the *Lescop invariant.* The definition of λ_L is given below.

Let $\mathcal{L} = k_1 \cup \ldots \cup k_n$ be a framed link in a rational homology sphere M and let $A(\mathcal{L}) = (\ell_{ij})$, $i, j = 1, \ldots, n$, be its linking matrix, see Section 1.2.2. We introduce the following notations.

The *eight-linking number* of $\mathcal{L}$ is defined as

$$\mathbb{L}_8(\mathcal{L}) = \sum_{(j,\sigma)\in N\times S_n} \ell_{j\sigma(1)}\ell_{\sigma(1)\sigma(2)}\ell_{\sigma(2)\sigma(3)}\cdots\ell_{\sigma(n-1)\sigma(n)}\ell_{\sigma(n)j}$$

where S_n denotes the set of bijections from $N = \{1, \ldots, n\}$ to itself. The eight-linking number of $\mathcal{L}$ originates in the following combinatorial construction. Consider a graph Γ with n vertices whose underlying space is the figure eight, see Figure 4.3.

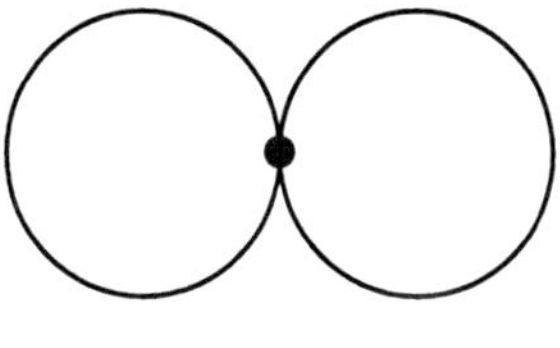

Fig. 4.3.

We assign the rational number ℓ_{ij} to every edge in the graph Γ connecting vertices v_i and v_j, and define the *linking number of $\mathcal{L}$ with respect to Γ* as the product of ℓ_{ij} over all the edges. The eight-linking number of $\mathcal{L}$ is then the sum of the linking numbers of $\mathcal{L}$ with respect to all possible figure–eight graphs Γ. The *modified eight-linking number* of $\mathcal{L}$ is defined as

$$\mathbb{M}_8(\mathcal{L}) = \begin{cases} \mathbb{L}_8(\mathcal{L}), & \text{if } n > 1, \\ \mathbb{L}_8(k_1) + 1/q_1^2 = (p_1^2+1)/q_1^2, & \text{if } n = 1. \end{cases}$$

We define the ζ-coefficient of $\mathcal{L}$ as follows. If $n = 1$ then

$$\zeta(\mathcal{L}) = \mathcal{O}_M(k_1) \cdot \frac{1}{2} \Delta''_{k_1}(1) - \frac{|H_1(M,\mathbb{Z})|}{24\, \mathcal{O}_M(k_1)^2},$$

where $\Delta_{k_1}(t)$ is the normalized Alexander polynomial of k_1 and $\mathcal{O}_M(k_1) = |H_1(M,\mathbb{Z})|/|\operatorname{tor}(H_1(M \setminus k_1, \mathbb{Z}))|$. If $n > 1$ we define

$$\zeta(\mathcal{L}) = \frac{\partial^n}{\partial t_1 \ldots \partial t_n} \Delta_{\mathcal{L}}(1, \ldots, 1),$$

where $\Delta_{\mathcal{L}}$ is the normalized multivariable Alexander polynomial of $\mathcal{L}$, see [182].

If I is a subset of $N = \{1, 2, \ldots, n\}$ then $\mathcal{L}_I$ denotes the framed link consisting of the components k_j of $\mathcal{L}$ numbered by $j \in I$. The cardinality of I is denoted by $|I|$.

Suppose that the components of the link $\mathcal{L} = k_1 \cup \ldots \cup k_n$ are framed by (p_i, q_i), $i = 1, \ldots, n$. Define the function

$$\begin{aligned} &\mathbb{F}_M(\mathcal{L}) = \\ &|H_1(\chi_M(\mathcal{L}), \mathbb{Z})| \sum_{I \neq \emptyset, I \subset N} \frac{\det A(\mathcal{L}_{N \setminus I})}{\det A(\mathcal{L})} \left(\frac{\zeta(\mathcal{L}_I)}{|H_1(M,\mathbb{Z})|} + \frac{(-1)^{|I|}}{24} \mathbb{M}_8(\mathcal{L}_I) \right) \\ &\qquad + |H_1(\chi_M(\mathcal{L}), \mathbb{Z})| \left(\frac{\operatorname{sign} A(\mathcal{L})}{8} + \frac{1}{2} \sum_{i=1}^{n} s(p_i - q_i \operatorname{lk}_M(k_i, k_i), q_i) \right), \end{aligned}$$

where $s(a, b)$ is the Dedekind sum (3.18). Note that $p_i = q_i \operatorname{lk}_M(k_i, k_i) \bmod \mathbb{Z}$ hence $p_i - q_i \operatorname{lk}_M(k_i, k_i)$ is an integer. The determinant of an empty matrix is one. Other equivalent definitions of $\mathbb{F}_M(\mathcal{L})$ can be found in [182], including one in terms of the one-variable Alexander polynomial.

The following theorem is proved in [182]. The invariant λ_L that it introduces is called the *Lescop invariant* or the *Casson–Walker–Lescop invariant.*

Theorem 4.9. *There exists a unique function λ_L from the class of closed oriented 3–manifolds to $\mathbb{Q}$ such that $\lambda_L(\chi_{S^3}(\mathcal{L})) = \mathbb{F}_{S^3}(\mathcal{L})$ for any framed link $\mathcal{L}$ in S^3. The function λ_L is an invariant of orientation preserving homeomorphism, and in addition it has the following properties :*

(1) $12 \cdot \lambda_L(M) \in \mathbb{Z}$ *for any closed oriented manifold M. Moreover, $4 \cdot \lambda_L(M) \in \mathbb{Z}$ if and only if $H_1(M, \mathbb{Z}/3) \neq \mathbb{Z}/3$.*

(2) *For any rational homology sphere M and for any framed link $\mathcal{L}$ in M, the following* surgery formula *holds*:

$$\lambda_L(\chi_M(\mathcal{L})) = \frac{|H_1(\chi_M(\mathcal{L}),\mathbb{Z})|}{|H_1(M,\mathbb{Z})|} \cdot \lambda_L(M) + \mathbb{F}_M(\mathcal{L}).$$

(3) $\lambda_L(-M) = (-1)^{b_1(M)+1}\, \lambda_L(M)$ *where $-M$ stands for M with reversed orientation, and $b_1(M) = \dim H_1(M,\mathbb{Q})$ is the first Betti number of M.*

(4) $\lambda_L(M_1 \# M_2) = |H_1(M_2,\mathbb{Z})|\, \lambda_L(M_1) + |H_1(M_1,\mathbb{Z})|\, \lambda_L(M_2)$ *with the agreement that $|A| = 0$ for an infinite abelian group A.*

(5) *If $b_1(M) = 0$, that is, if M is a rational homology sphere, then $\lambda_L(M) = |H_1(M,\mathbb{Z})| \cdot \lambda_W(M)$, where λ_W is the Walker invariant. In particular, if M is an integral homology sphere then $\lambda_L(M) = \lambda(M)$, the Casson invariant of M.*

(6) *If $b_1(M) = 1$ then*

$$\lambda_L(M) = \frac{1}{2}\, \Delta''_M(1) - \frac{1}{12}\, |\operatorname{tor}(H_1(M,\mathbb{Z}))|,$$

where $\Delta_M(t)$ is the normalized Alexander polynomial of M, see [182]. Note that if $M = \Sigma + 0 \cdot k$ where k is a knot in an integral homology sphere Σ then $\Delta_M(t) = \Delta_k(t)$, the Alexander polynomial of k normalized so that $\Delta_k(1) = 1$ and $\Delta_k(t^{-1}) = \Delta_k(t)$.

(7) *If $b_1(M) = 2$ then $\lambda_L(M) = -|\operatorname{tor}(H_1(M,\mathbb{Z}))| \cdot \operatorname{lk}_M(\gamma,\gamma')$ where γ and γ' are rationally null homologous curves defined as follows. Let S_1 and S_2 be oriented embedded surfaces in general position in M such that their homology classes generate $H_2(M,\mathbb{Z})$. Then γ is their oriented intersection, and γ' is the curve parallel to γ and inducing the trivialization of the tubular neighborhood of γ given by the surfaces.*

(8) *If $b_1(M) = 3$ then $\lambda_L(M) = |\operatorname{tor}(H_1(M,\mathbb{Z}))| \cdot ((a \cup b \cup c)[M])^2$ where a, b, c is a basis in $H^1(M,\mathbb{Z})$ and $[M]$ is the fundamental class of M.*

(9) *If $b_1(M) \geq 4$ then $\lambda_L(M) = 0$.*

Remark 4.10. Let M be obtained by 0–framed surgery on an algebraically split link $\mathcal{L} = k_1 \cup \ldots \cup k_n$ where $n = b_1(M)$. Then the Lescop invariants for $n = 1, 2$, and 3 are respectively the invariants λ', λ'', and λ''' of the link $\mathcal{L}$, see Section 3.3.5. Both sets of invariants vanish for $n \geq 4$.

4.4 Properties of the Walker and Lescop Invariants

Being extensions of the Casson invariant, both Walker and Lescop invariants inherit a good deal of its properties.

4.4.1 The Gluing Formula

We follow Lescop [184] and call an oriented 3–manifold A a $\mathbb{Q}$*–handlebody* if A has the same rational homology as the standard handlebody. For a 3–manifold A with boundary, the kernel $\mathcal{L}_A$ of the map $H_1(\partial A,\mathbb{Q}) \to H_1(A,\mathbb{Q})$ induced by the inclusion will be called the *Lagrangian* of A.

Theorem 4.11. *(Lescop [184]) Let A, A', B and B' be $\mathbb{Q}$–handlebodies such that ∂A, $\partial A'$, $-\partial B$ and $-\partial B'$ are identified via orientation preserving homeomorphisms with a genus g surface F. Assume that $\mathcal{L}_A = \mathcal{L}_{A'}$ and $\mathcal{L}_B = \mathcal{L}_{B'}$ and that $\mathcal{L}_A \cap \mathcal{L}_B = 0$ in $H_1(F,\mathbb{Q})$. Then the manifolds $A\cup B$, $A\cup B'$, $A'\cup B$ and $A'\cup B'$ are rational homology spheres, and*

$$\lambda_W(A\cup B) - \lambda_W(A\cup B') - \lambda_W(A'\cup B) + \lambda_W(A'\cup B') = R(A,A',B,B'),$$

where the function R is described below. The function R vanishes whenever $g \le 2$.

The function R is defined as follows. Consider the closed manifold $D_A = A\cup(-A')$ and the intersection form $\Lambda^3 H^1(D_A,\mathbb{Q}) \to \mathbb{Q}$ given by the formula $a\wedge b\wedge c \to (a\cup b\cup c)[D_A]$. Via Poincaré duality, it defines a form $\Lambda^3 H_2(D_A,\mathbb{Q}) \to \mathbb{Q}$. Now, since $\mathcal{L}_A = \mathcal{L}_{A'}$, the Mayer–Vietoris argument can be used to show that $H_2(D_A,\mathbb{Q}) = \mathcal{L}_A$ and thus carry the intersection form on $\Lambda^3 H_2(D_A,\mathbb{Q})$ to the intersection form $\mathcal{I}_{AA'}: \Lambda^3\mathcal{L}_A \to \mathbb{Q}$, compare with [291]. Define $\mathcal{I}_{BB'}$ similarly. Let $\alpha_1,\dots,\alpha_g$ and $\beta_1,\dots,\beta_g$ be bases in $\mathcal{L}_A$ and $\mathcal{L}_B$ such that $\langle\alpha_i,\beta_j\rangle = \delta_{ij}$ with respect to the intersection form $\langle\,,\,\rangle : H_1(F,\mathbb{Q})\otimes H_1(F,\mathbb{Q}) \to \mathbb{Q}$. Then

$$R(A,A',B,B') = -4 \sum_{1\le i,j,k\le g} \mathcal{I}_{AA'}(\alpha_i,\alpha_j,\alpha_k)\cdot\mathcal{I}_{BB'}(\beta_i,\beta_j,\beta_k).$$

Note that if $g=0$ and $A'=B'=D^3$ then Theorem 4.11 claims nothing else but the additivity of the Walker invariant λ_W with respect to connected sums. If $g=1$ and A' and B' are solid tori then Theorem 4.11 is an extension to the Walker invariant of the splicing additivity of Section 3.4.1, see also [106]. If $g=2$ and $A=A'$ then, for a proper choice of B and B', Theorem 4.11 is an extension of the mutation invariance of the Casson invariant, see Section 3.4.2.

An extension of Theorem 4.11 to the Lescop invariant was proved in [184]. Theorem 4.11 was also used in [184] to extend Morita's results to rational homology spheres, see Section 3.4.6.

4.4.2 Branched Covers

The following generalization of *Mullins' formula*, see Theorem 3.21, holds for the Walker invariant. Let Σ_2 be a double branched cover of S^3 with branch set a knot k. If Σ_2 is a $\mathbb{Z}/2$–homology sphere then

$$\lambda_W(\Sigma_2) = \frac{1}{8}\,\operatorname{sign} k - \frac{1}{12}\,\frac{d}{dt}\Big|_{t=-1} \ln V_k(t),$$

where V_k is the Jones polynomial of k. Theorem 3.23 holds without any change because all the branched covers Σ_p over an untwisted Whitehead double are in fact integral homology spheres. Extensions of Theorem 3.23 to twisted Whitehead doubles given in [64] and [151] hold for the Walker invariant as well for the Casson invariant. It is not known if Theorem 3.24 and Theorem 3.25 extend to rational homology spheres and the Walker invariant.

Example 4.12. Let p and q be positive relatively prime integers. The link $\Sigma(2,p,q)$ of singularity at zero of $x^2 + y^p + z^q = 0$ admits a $\mathbb{Z}/2$–action defined by the formula $(x,y,z) \mapsto (-x,y,z)$, which makes it into a double branched cover of S^3 with branch set the right-handed (p,q)–torus knot k. Mullins' formula then implies that

$$\lambda_W(\Sigma(2,p,q)) = -\frac{1}{12}\,\frac{d}{dt}\Big|_{t=-1} \ln V_k(t) + \frac{1}{8}\,\operatorname{sign} M(2,p,q)$$

where $M(2,p,q)$ is the Milnor fiber of the singularity and $V_k(t)$ is the Jones polynomial of k. Thus the logarithmic derivative of the Jones polynomial can be viewed as an obstruction for the Neumann–Wahl conjecture to hold in this situation, compare with Remark 3.34. This was first noticed in [56]. If both p and q are odd this obstruction vanishes. This is true for example because the Neumann–Wahl conjecture holds for the integral homology spheres $\Sigma(2,p,q)$. The vanishing of the obstruction can also be checked directly by using the explicit formula for $V_k(t)$, see [155],

$$V_k(t) = t^{(p-1)(q-1)/2}\,\frac{1 - t^{p+1} - t^{q+1} + t^{p+q}}{1 - t^2}.$$

On the other hand, if $p = 2$ and $q = n$ is odd, the obstruction is not zero because, according to the above formula,

$$\frac{d}{dt}\Big|_{t=-1} \ln V_k(t) = \frac{n^2 - 1}{n}.$$

Example 4.13. Example 4.12 demonstrates that substituting the Walker invariant for the Casson invariant in the Neumann–Wahl conjecture makes it fail. Another way to see this is to consider a lens space $L(n, n-1) = -L(n,1)$ with an odd $n \geq 3$. It arises as the link of singularity at 0 of the polynomial $x^2 + y^2 + z^n$, see Example 1.31. As such, it has a well defined Milnor fiber $M(2,2,n)$, whose signature can be computed as $\operatorname{sign} M(2,2,n) = 1 - n$ using the formula of Brieskorn [40]. On the other hand, the Walker invariant of $L(n, n-1)$ equals $(1-n)(n-2)/24n$, see Example 4.7. An extension of the Neumann–Wahl conjecture to the links of singularities which are rational homology spheres was recently proposed in [240].

4.4.3 Seifert Fibered Manifolds

The invariant λ_L was computed in [182] for all Seifert fibered manifolds. Here, we restrict ourselves to Seifert fibered manifolds which are rational homology spheres. For such manifolds, the formula of [182] also provides a formula for the Walker invariant λ_W since $\lambda_L(M) = |H_1(M,\mathbb{Z})| \cdot \lambda_W(M)$.

Let M be a Seifert fibered manifold and a rational homology sphere, then the quotient of M by the S^1–action is either S^2 or $\mathbb{R}P^2$, see Section 1.2.3.

If the quotient is S^2 then $M = M(b;(a_1,b_1),\dots,(a_n,b_n))$ is the manifold obtained by surgery on the link in Figure 1.1. Let $e = -b + \sum_{k=1}^n b_k/a_k$ then

$$\lambda_L(M) =$$

$$\frac{a_1\cdots a_k}{8}\left(\frac{\operatorname{sign}(e)}{3}\left(2-n+\sum_{k=1}^{n}\frac{1}{a_k^2}\right)+\frac{e\,|e|}{3}-e-4|e|\sum_{k=1}^{n}s(b_k,a_k)\right)$$

where $s(b,a)$ is the Dedekind sum (3.18). In the case of an integral homology sphere $\Sigma(a_1,\dots,a_n)$ this formula agrees with the formula (3.23).

If the quotient of M is $\mathbb{R}P^2$ then M is the manifold shown in Figure 1.32. Let $e = -b + \sum_{k=1}^n b_k/a_k$ as before then

$$\lambda_L(M) = 2\,a_1\cdots a_n\left(\frac{e}{8}-\sum_{k=1}^{n}s(b_k,a_k)\right).$$

4.5 Casson Type Invariants from Other Lie Groups

One can use an approach similar to that of Walker to define Casson type invariants of rational homology spheres starting with Lie groups other than $SU(2)$. Such invariants for the low rank Lie groups $SO(3)$, $SU(2)\times S^1$, $U(2)$, $\operatorname{Spin}(4)$ and $SO(4)$ were defined by Curtis [61]. It turns out that all these invariants can be expressed in terms of the Walker invariant λ_W. Namely, the $SO(3)$ invariant is identical to λ_W, the $SU(2)\times S^1$ and $U(2)$ invariants are proportional to λ_W, and the $\operatorname{Spin}(4)$ and $SO(4)$ invariants are proportional to the square of λ_W.

Apparently, the $SU(3)$ Casson invariant of Boden and Herald [30] defined by using gauge theory, see Section 5.6, is the first in the series of Casson type invariants for low rank Lie groups which cannot be expressed in terms of the $SU(2)$ Casson invariant. An approach to defining $SU(n)$ Casson invariants for all n was suggested in [49].

An $SU(n)$ Casson invariant of knots was defined in [102] and [103] as an extension of the Casson invariant $\lambda'(k\subset\Sigma)$, see Section 3.1. It can be viewed as an algebraic–topological count of the number of characters of $SU(n)$ representations of the knot group which take a longitude into a given conjugacy class. Explicit formulas which compute these invariants for all n are obtained in [33].

5 Casson Invariant and Gauge Theory

In this chapter, we give an account of $SU(2)$–gauge theory in dimension three. We discuss C. Taubes' gauge-theoretical definition of the Casson invariant as (roughly) the Euler number of the gradient field of the Chern–Simons function. The Chern–Simons function plays a central role in modern understanding of homology 3–spheres, so we discuss it in some detail. An infinite dimensional analogue of Morse theory applied to the Chern-Simons function produces the instanton Floer homology which will be discussed in the next chapter. This gauge-theoretical approach to the Casson invariant leads to several extensions in a direction different from that of Walker and Lescop. One of the extensions we discuss is the $SU(3)$ Casson invariant of H. Boden and C. Herald. Another one is the Casson-type invariant for knots in integral homology spheres introduced by X.-S. Lin and C. Herald, and finally, the equivariant Casson invariant of integral homology spheres with a finite cyclic group action by O. Collin and the author.

5.1 Gauge Theory in Dimension 3

Let M be a closed oriented 3–manifold and $E \to M$ a principal $SU(2)$–bundle over M. By topological reasons, E is necessarily trivial, and we will fix a trivialization $E = M \times SU(2)$. The (affine) space of $SU(2)$–connections on E will be denoted by $\mathcal{A}$. The above trivialization identifies $\mathcal{A}$ with $\Omega^1(M, \mathfrak{su}(2))$, the linear space of differential 1–forms with matrix $\mathfrak{su}(2)$–coefficients.

A connection $A \in \mathcal{A}$ allows to lift any curve $\gamma : [0,1] \to M$ to a horizontal curve $\tilde{\gamma} : [0,1] \to E$ which is uniquely determined by a lift $\tilde{\gamma}(0)$ of $\gamma(0)$. If $\gamma(0) = \gamma(1)$ then $\tilde{\gamma}(0)$ and $\tilde{\gamma}(1)$ differ by an element of $SU(2)$, which is called the *holonomy* of A along γ and is denoted by $\operatorname{hol}_A(\gamma)$. Holonomy defines a map from the monoid of based loops in M to $SU(2)$.

The group of automorphisms of E is isomorphic to $\mathcal{G} = \operatorname{Map}(M, SU(2))$ and is called a *gauge group*. Its elements are called *gauge transformations*. They act on $\mathcal{A}$ by the rule $g^*A = g^{-1}dg + g^{-1}Ag$, where $g \in \mathcal{G}$ and $A \in \mathcal{A}$. The action of $\mathcal{G}$ on $\mathcal{A}$ is not free. The stabilizer of a connection $A \in \mathcal{A}$ equals the stabilizer of its holonomy group in $SU(2)$. A connection is called *irreducible* if its stabilizer is $\{\pm 1\}$, and it is called *reducible* otherwise. More specifically, connections with stabilizer $U(1)$ are called *abelian*, and those with stabilizer

$SU(2)$ are called *central*. The latter include the product connection θ, which is also called the *trivial* connection.

We let $\mathcal{B} = \mathcal{A}/\mathcal{G}$ and denote by $\mathcal{B}^*$ the subset of $\mathcal{B}$ consisting of $\mathcal{G}$–orbits of irreducible connections. Completed in appropriate Sobolev norms, $\mathcal{B}^*$ is a smooth infinite dimensional Banach manifold. The group $\mathcal{G}$ is not connected. In fact, $\pi_0\,\mathcal{G} = [M, SU(2)] = \mathbb{Z}$, the latter isomorphism given by the degree of $g : M \to SU(2)$. The homotopy exact sequence now implies that $\pi_1\mathcal{B}^* = \mathbb{Z}$.

Associated with each connection $A \in \mathcal{A}$ is its *curvature*, $F_A = dA + A \wedge A$, which is an $\mathfrak{su}(2)$–valued differential 2–form on M (the wedge here stands for the matrix product on $\mathfrak{su}(2)$–coefficients and the wedge product on differential forms). A connection A is called *flat* if $F_A = 0$. The holonomy map for a flat connection A defines a homomorphism $A : \pi_1 M \to SU(2)$. This correspondence establishes a well-known identification

$$\operatorname{Hom}(\pi_1 M, SU(2))/SO(3) = \{\,\text{Flat } SU(2) \text{ connections on } M\,\}/\mathcal{G}.$$

Irreducible representations on the left correspond to irreducible flat connections on the right, so that the representation space $\mathcal{R}^*(M)$ can be viewed as a subset of $\mathcal{B}^*$. Abelian (resp. central) connections correspond to abelian (resp. central) representations, see Section 3.2.1.

A choice of Riemann metric on M provides us with an L^2–inner product on the spaces $\Omega^p(M)$ of differential p–forms for each $p \geq 0$. Together with the positive-definite pairing $(a,b) \mapsto -\operatorname{tr}(ab)$ on $\mathfrak{su}(2)$ it gives an L^2–inner product $(\ ,\ )$ on $\Omega^p(M, \mathfrak{su}(2))$. Given orientation on M, this defines the *Hodge star operator* $* : \Omega^p(M, \mathfrak{su}(2)) \to \Omega^{3-p}(M, \mathfrak{su}(2))$ by the rule

$$-\int_M \operatorname{tr}(a \wedge *b) = (a, b).$$

The tangent space to $\mathcal{A}$ at a point $A \in \mathcal{A}$ is of course $\Omega^1(M, \mathfrak{su}(2))$, and the normal space to the $\mathcal{G}$–orbit of A may be identified with the kernel of the differential operator

$$d_A^* = - * d_A * \,:\, \Omega^1(M, \mathfrak{su}(2)) \to \Omega^0(M, \mathfrak{su}(2))$$

which is formally adjoint to the operator d_A. Here, $d_A u = du + [A, u]\,(= du + Au - uA)$ is the covariant derivative of u with respect to A.

The operator d_A extends by the Leibniz rule to all spaces $\Omega^p(M, \mathfrak{su}(2))$. If A is flat, we obtain an elliptic complex

$$\Omega^0(M, \mathfrak{su}(2)) \xrightarrow{d_A} \Omega^1(M, \mathfrak{su}(2)) \xrightarrow{d_A} \Omega^2(M, \mathfrak{su}(2)) \xrightarrow{d_A} \Omega^3(M, \mathfrak{su}(2))$$

whose cohomology $H^*(M, \operatorname{ad} A)$ is the cohomology of M with coefficients in the twisted flat bundle $E \times_{SU(2)} \mathfrak{su}(2)$. In particular, $H^0(M, \operatorname{ad} A)$ can be identified with the Lie algebra of the stabilizer of A, and $H^1(M, \operatorname{ad} A)$ with the group cohomology of $\pi_1 M$ with coefficients in representation (3.5).

5.2 Chern-Simons Function

Let M be a closed oriented 3-dimensional manifold. For any connection $\alpha \in \mathcal{A}$, we define the *Chern-Simons function* of α by the formula

$$\mathbf{cs}(\alpha) = \frac{1}{8\pi^2} \int_M \operatorname{tr}(\alpha \wedge d\alpha + \frac{2}{3} \alpha \wedge \alpha \wedge \alpha). \tag{5.1}$$

The Stokes Theorem implies that, for any smooth compact oriented 4-manifold X with $\partial X = M$ and an $SU(2)$–connection A on X which restricts to α over M,

$$\mathbf{cs}(\alpha) = \frac{1}{8\pi^2} \int_X \operatorname{tr}(F_A \wedge F_A). \tag{5.2}$$

According to Chern-Weyl theory, the latter integral over a closed manifold represents a second Chern number and, in particular, is an integer. This implies that $\mathbf{cs}(g^*\alpha) = \mathbf{cs}(\alpha) + \deg(g)$, where $g \in \mathcal{G}$ so that Chern-Simons function is well-defined as a map $\mathbf{cs} : \mathcal{B} \to \mathbb{R}/\mathbb{Z}$.

Another equivalent definition of the Chern-Simons function is as follows. Given a connection α over M, we can take a path $[0,1] \to \mathcal{A}$ from the trivial connection to α in the trivial $SU(2)$–bundle. This path determines a connection A in the trivial $SU(2)$-bundle over $[0,1] \times M$, and we let

$$\mathbf{cs}(\alpha) = \frac{1}{8\pi^2} \int_{[0,1]\times M} \operatorname{tr}(F_A \wedge F_A). \tag{5.3}$$

The set of values of $\mathbf{cs}$ on its critical points is an invariant of a 3–manifold called the *Chern–Simons invariant.* Methods for computing this invariant and examples of computations can be found in Auckly [18], Fintushel-Stern [83] and Kirk-Klassen [165]. In all known cases, the values of $\mathbf{cs}$ on its critical points are rational numbers. Whether the same is true in general is an open problem.

Example 5.1. The Chern–Simons invariant for a lens space $L(p,q)$ is given by $-n^2r/p$ for $n = 0, \dots, [p/2]$ where r is any integer satisfying $qr = -1 \bmod p$, see [165]. Note that the Chern–Simons invariant distinguishes homotopy inequivalent lens spaces, compare with Section 1.2.1, but it cannot distinguish homotopy equivalent non-homeomorphic lens spaces.

Example 5.2. The Chern-Simons invariants for Seifert fibered homology spheres $\Sigma(a_1,\dots,a_n)$ are calculated in [83] and [165]. Suppose that $\Sigma(a_1,\dots,a_n)$ has Seifert invariants b, $(a_1,b_1),\dots,(a_n,b_n)$ with b even, see Section 1.1.4. Moreover, if one of the a_i is even, assume it is a_1 and arrange the Seifert invariants so that the b_i with $i \neq 1$ are even. If $\alpha : \pi_1\Sigma(a_1,\dots,a_n) \to SU(2)$ is a representation with rotation numbers $(\ell_1,\dots,\ell_n)$, see Section 3.5.2, then

$$\mathbf{cs}(\alpha) = -\frac{e^2}{4a_1\cdots a_n} \bmod \mathbb{Z} \quad \text{where} \quad e = a_1\cdots a_n \cdot \sum_{i=1}^n \frac{\ell_i}{a_i}. \tag{5.4}$$

Example 5.3. The results of Examples 5.1 and 5.2 are instances of the following more general theorem, see [165]. Let $k \subset M$ be a knot in a closed oriented 3–manifold M with exterior K, and let curves m and ℓ form a basis in $\pi_1(\partial K) = \mathbb{Z}^2$. Given two representations $\alpha_0, \alpha_1 : \pi_1 M \to SU(2)$, suppose that there is a path $\alpha_t : \pi_1 K \to SU(2)$, $0 \le t \le 1$, from α_0 to α_1 (both restricted to $\pi_1 K$). Let $(\mu(t), \lambda(t))$ be a path in $\mathbb{R}^2$ such that

$$\alpha_t(m) = e^{2\pi i \mu(t)}, \quad \alpha_t(\ell) = e^{2\pi i \lambda(t)}.$$

Then

$$\mathbf{cs}(\alpha_1) - \mathbf{cs}(\alpha_0) = -2 \int_0^1 \lambda(t)\mu'(t)\, dt \in \mathbb{R}/\mathbb{Z}. \tag{5.5}$$

This means that if we know the image of the path $\alpha_t : \pi_1 K \to SU(2)$, $0 \le t \le 1$, in the pillowcase, see Section 3.3.1, then we can compute the difference $\mathbf{cs}(\alpha_1) - \mathbf{cs}(\alpha_0)$. Thus the difference is determined by the pillowcase image of $\mathcal{R}(K)$. Application of this result to surgeries on the unknot gives results of Example 5.1, and its application to the surgeries on a regular fiber in a Seifert fibered homology sphere gives results of Example 5.2.

This method does not work in the situations when the representation variety $\mathcal{R}(K)$ is not connected. This happens for instance for the figure-eight knot, see Figure 3.3. This difficulty can sometimes be overcome by passing to the complex character variety $\mathrm{Hom}(\pi_1 K, SL(2,\mathbb{C}))/SL(2,\mathbb{C})$. In particular, the Chern–Simons invariants of surgeries on the figure-eight knot can be computed by this method, see [165].

An affirmative answer to the following question would make this theory more useful. Given a closed manifold M and representations $\alpha_0, \alpha_1 : \pi_1 M \to SU(2)$, does there exist a knot (or link) $k \subset M$ so that the restrictions of α_0 and α_1 onto the exterior K of k lie on a path component of $\mathcal{R}(K)$?

Example 5.4. The formula (5.5) was used to compute the Chern-Simons invariants for torus bundles over a circle in [165] and for higher genus surface bundles in [15]. The formula (5.5) was generalized to include general decompositions of 3–manifolds along tori in [166] and [170].

Example 5.5. Two representations, $\alpha_M : \pi_1 M \to SU(2)$ and $\alpha_N : \pi_1 N \to SU(2)$ are called ***flat cobordant*** if there exists an oriented cobordism W between M and N and a representation $\alpha : \pi_1 W \to SU(2)$ restricting to α_M and α_N over M and N, respectively. It follows immediately from (5.2) that $\mathbf{cs}(\alpha_M) = \mathbf{cs}(\alpha_N)$. Using this observation and the formula (5.5) David Auckly [18] developed a method which allows to compute the Chern–Simons invariants for all graph manifolds and for some hyperbolic manifolds.

Remark 5.6. The definition of $\mathbf{cs}(A) \in \mathbb{R}/\mathbb{Z}$ for $SU(2)$–connections A can be easily adapted to define $\mathbf{cs}(A') \in \mathbb{R}/4\mathbb{Z}$ for $SO(3)$–connections A'. The two Chern-Simons functions are related by the formula $4 \cdot \mathbf{cs}(A) = \mathbf{cs}(\mathrm{ad}_A)$ where ad_A is the induced $SO(3)$–connection in the adjoint bundle. If A is a flat connection viewed as an $SU(2)$ representation then the representation corresponding to ad_A is obtained by composing A with the adjoint representation $\mathrm{ad} : SU(2) \to SO(3)$.

The Chern-Simons function was first defined as a secondary characteristic class in [54]. It was shown later in [12], [13] and [14] that the ρ_α-invariants, which are defined as real numbers, are congruent to $\mathbf{cs}(\alpha)$ modulo $\mathbb{Z}$. Thus, $\mathbf{cs}(\alpha)$ can be viewed as the non-integral part of ρ_α, see Remark 6.17.

5.3 The Casson Invariant from Gauge Theory

A gauge–theoretical definition of the Casson invariant was given by Taubes [292]. His construction was motivated by an analogy with finite dimensional Morse theory, namely, the derivation of the Euler characteristic from critical points of a Morse function. Another description of the Casson invariant motivated by Seiberg–Witten gauge theory was recently suggested by Ozsváth and Szabó [254].

5.3.1 Morse Theory and Euler Characteristic

Let $h : X \to \mathbb{R}$ be a smooth function on a closed smooth manifold X. The *Hessian* of h at a point $p \in X$ is the bilinear form on the tangent space T_pX given in local coordinates by the matrix

$$\mathrm{Hess}_p\, h = \left(\frac{\partial^2 h}{\partial x_i \partial x_j} \right) (p). \tag{5.6}$$

If p is a critical point of h, we define its *Morse index* $\mu(p)$ as the number of negative eigenvalues of $\mathrm{Hess}_p(h)$ counted with multiplicities. We call $h : X \to \mathbb{R}$ a *Morse function* if its Hessian is non-degenerate at all of its critical points. The condition of being Morse ensures that the critical points of h are isolated and hence there are only finitely many of them. Then, according to the Poincaré–Hopf theorem,

$$e(X) = \sum_p \, (-1)^{\mu(p)},$$

where $e(X)$ is the Euler characteristic of X and the summation on the right extends over all critical points of h.

5.3.2 Critical Points of cs and Spectral Flow

Let Σ be an integral homology sphere endowed with a Riemannian metric and let $\mathbf{cs} : \mathcal{B}^* \to \mathbb{R}/\mathbb{Z}$ be the Chern–Simons function (5.1). The L^2*–gradient* of $\mathbf{cs}$ can be easily computed as

$$(\nabla\,\mathbf{cs})(A) = -\frac{1}{4\pi^2} * F_A, \qquad (5.7)$$

which can be viewed as a vector field on the (infinite dimensional) manifold $\mathcal{B}^*$. The critical points of **cs** are (the gauge equivalence classes of) irreducible flat connections, which we identified earlier with the points in $\mathcal{R}^*(\Sigma)$.

The *Hessian* of **cs** at a connection A is the first order differential operator

$$\left(-\frac{1}{4\pi^2}\right) * d_A : \ker d_A^* \to \ker d_A^*, \qquad (5.8)$$

which is essentially self–adjoint. Let us postpone discussion on the non-degeneracy of this operator until next section and see instead how it can be used to define indices of the critical points of **cs**. The problem is of course that the spectrum of (5.8) is unbounded in both directions, hence we cannot define the index of a critical point as the number of negative (or positive) eigenvalues, as we did it in the finite–dimensional Morse theory. One way around this problem is to compare the Hessians at two critical points by seeing how many eigenvalues change sign along a path connecting them. This will give a relative index, and the problem then will be to pick an overall normalization.

The natural choice of a flat connection with which to compare all other flat connections is the trivial connection θ. Unfortunately, θ does not belong to $\mathcal{B}^*$. Because of this, it is useful to replace the operator (5.8) by a Fredholm operator which gives the same relative index between irreducible flat connections but which also makes sense at reducibles. We define such an operator by

$$K_A = \begin{pmatrix} 0 & d_A^* \\ d_A & -*\,d_A \end{pmatrix} : (\Omega^0 \oplus \Omega^1)(\Sigma, \mathfrak{su}(2)) \to (\Omega^0 \oplus \Omega^1)(\Sigma, \mathfrak{su}(2)). \qquad (5.9)$$

For any connection A, the operator K_A is an elliptic operator on Σ. Its L^2–completion is a self–adjoint Fredholm operator. It has pure point real spectrum without accumulation points, which is unbounded in both directions.

Let A_0 and A_1 be flat connections and consider a continuously differentiable path of operators $K_{A(t)}$ with $A(0) = A_0$ and $A(1) = A_1$. The one–parameter family of spectra of operators $K_{A(t)}$ can be viewed as a collection of spectral curves in the plane connecting the spectrum of K_{A_0} to the spectrum of K_{A_1} see Figure 5.1. These curves are continuously differentiable functions of t, at least near zero. Consider the straight line connecting the points $(0, -\delta)$ and $(1, \delta)$ where $\delta > 0$ is chosen smaller than the absolute value of any non–zero eigenvalue of K_{A_0} and K_{A_1}. The number of eigenvalues, counted with multiplicities, which cross from negative to positive side of this line minus the number which cross from positive to negative is well–defined and finite along a generic path. This number is called the *spectral flow* of the family along the path. Note that we choose a $\delta > 0$ to help us deal with flat connections A whose associated operator K_A has non-trivial kernel; away from such connections we could choose $\delta = 0$.

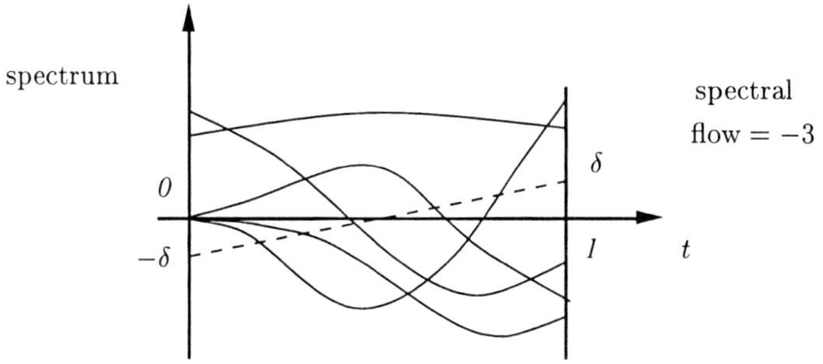

Fig. 5.1.

The spectral flow only depends on the homotopy class rel $\{0,1\}$ of the path $t \to A(t)$. Therefore, it defines a locally constant function on the space of continuous paths between A_0 and A_1. We denote this function by $\operatorname{sf}(A_0, A_1)$. Choosing different A_0 or A_1 within their gauge equivalence classes changes $\operatorname{sf}(A_0, A_1)$ to $\operatorname{sf}(A_0, A_1) + 8k$ for some integer k, see Corollary 6.3. Therefore, $\operatorname{sf}(A_0, A_1) \bmod 8$ is well defined on gauge equivalence classes of A_0 and A_1.

Applications of the spectral flow in this chapter will only use its modulo 2 reduction. The full strength of the modulo 8 spectral flow will be explored in Chapter 6, together with various spectral flow formulas.

5.3.3 Non–degenerate Case

Let Σ be an integral homology sphere such that $\mathcal{R}^*(\Sigma)$ is non-degenerate, that is, $H^1(\pi_1\Sigma, \operatorname{ad}\alpha) = 0$ for every representation $\alpha \in \mathcal{R}^*(\Sigma)$, see Lemma 3.16.

Lemma 5.7. *For every $\alpha : \pi_1\Sigma \to SU(2)$, the kernel of the operator K_α is isomorphic to $H^0(\pi_1\Sigma, \operatorname{ad}\alpha) \oplus H^1(\pi_1\Sigma, \operatorname{ad}\alpha)$.*

Proof. Since K_α is a self–adjoint operator and α is flat, we have $\ker K_\alpha = \ker K_\alpha^2$ where

$$K_\alpha^2 = \begin{pmatrix} d_\alpha^*\, d_\alpha & 0 \\ 0 & d_\alpha^*\, d_\alpha + d_\alpha\, d_\alpha^* \end{pmatrix}.$$

Therefore, the kernel of K_α is represented by harmonic 0– and 1–forms with coefficients in the flat bundle ad_α. Such forms are in one-to-one correspondence with vectors in $H^0(\pi_1\Sigma, \operatorname{ad}\alpha) \oplus H^1(\pi_1\Sigma, \operatorname{ad}\alpha)$, by Hodge theorem.

The cohomology groups $H^0(\pi_1\Sigma, \operatorname{ad}\alpha)$ vanish for irreducible representations α, therefore, $\mathcal{R}^*(\Sigma)$ is non-degenerate if and only if all irreducible flat connections are non-degenerate as critical points of the Chern–Simons function.

Suppose that $\mathcal{R}^*(\Sigma)$ is non-degenerate, then it is finite. For any $\alpha \in \mathcal{R}^*(\Sigma)$ define its *Floer index* by the formula

$$\mu(\alpha) = \operatorname{sf}(\theta, \alpha) \bmod 8, \tag{5.10}$$

compare with Section 6.2.2. Taubes showed in [292] that the quantity

$$\frac{1}{2} \sum_{\alpha \in \mathcal{R}^*(\Sigma)} (-1)^{\mu(\alpha)} \tag{5.11}$$

is independent of the metric on Σ and equals (up to an overall sign) the Casson invariant of Σ, compare with Theorem 5.9. Thus in the non-degenerate case, both (3.9) and (5.11) count the same points in the space $\mathcal{R}^*(\Sigma)$, and the essence of Taubes' theorem is that the two counts are consistent, one coming from intersection theory and the other from gauge theory. A formula similar to (5.11) will hold for all integral homology spheres Σ after we take care in the next section of (possible) degeneracies in $\mathcal{R}^*(\Sigma)$.

5.3.4 Perturbations

Let Σ be an integral homology 3–sphere and let $\gamma_i : S^1 \times D^2 \to \Sigma$, $i = 1, \dots, m$, be a collection of embeddings of solid tori in Σ with disjoint images. We will think of the γ_i as loops in Σ and refer to $\{\gamma_i\}$ as a link. Let $\eta(x)$ be a smooth rotationally symmetric bump function on the unit disc D^2 with the support away from the boundary of D^2 and with integral one. Define an *admissible perturbation function* $h : \mathcal{B} \to \mathbb{R}$ by the formula

$$h(A) = \sum_{i=1}^{m} \int_{D^2} h_i(\operatorname{tr} \operatorname{hol}_A(\gamma_i(S^1 \times \{x\}))) \cdot \eta(x)\, d^2x, \tag{5.12}$$

where h_i, $i = 1, \dots, m$, are smooth functions (one averages over the tubular neighborhood by analytical reasons). The function h is gauge invariant because a gauge transformation only changes holonomy by conjugation. Given an admissible perturbation function h, the *perturbed flat moduli space*

$$\mathcal{R}_h(\Sigma) = \{\, A \mid *F_A - 4\pi^2\, (\nabla h)(A) = 0 \,\}/\mathcal{G},$$

where ∇h is the L^2–gradient of h, is the critical point set of $\mathbf{cs} + h : \mathcal{B} \to \mathbb{R}/\mathbb{Z}$. Extend the definition (5.9) to obtain the operator

$$K_{A,h} = \begin{pmatrix} 0 & d_A^* \\ d_A & -*\,d_A + 4\pi^2 \operatorname{Hess}_A h \end{pmatrix} \tag{5.13}$$

where $\operatorname{Hess}_A h$ is the Hessian of h at A. A critical point A of the perturbed Chern-Simons function $\mathbf{cs} + h$ is *non-degenerate* if and only if $\ker K_{A,h} = 0$. We call $\mathcal{R}_h(\Sigma)$ *non-degenerate* if all of its points are non-degenerate. The following result is proved in [137] and [292], see also [30].

Theorem 5.8. *For any admissible perturbation function h, the space $\mathcal{R}_h(\Sigma)$ is compact. Moreover, there exists a link $\{\gamma_i\}$ with sufficiently many components such that, for a generic choice of the functions $\{h_i\}$, the critical point set $\mathcal{R}_h(\Sigma)$ is non-degenerate and contains the trivial connection.*

Proof. Uhlenbeck's compactness theorem [299] can be adapted to prove the compactness of $\mathcal{R}_h(\Sigma)$, see [138]. Recall that the compactness of the non-perturbed space $\mathcal{R}(\Sigma)$ is immediate from the fact that $\mathcal{R}(\Sigma)$ is a real algebraic variety, see Section 3.2.1. To prove the second part of the proposition, one needs to find an abundant collection of admissible perturbation functions such that, in any direction tangent to $\mathcal{R}(\Sigma)$, there is a function whose derivative is non-zero. Given such a collection, an argument with the implicit function theorem completes the proof.

Let h be a generic admissible perturbation function as in Theorem 5.8 so that $\mathcal{R}_h(\Sigma)$ is non-degenerate and hence finite. Denote by $\mathcal{R}^*_h(\Sigma)$ the subset of $\mathcal{R}_h(\Sigma)$ consisting of the orbits of irreducible perturbed flat connections. Given $A \in \mathcal{R}^*_h(\Sigma)$, define its *Floer index* $\mu(A)$ as the spectral flow mod 8 of the family of operators $K_{A(t),h}$ where $A(t)$ is a path of connections from θ to A.

Theorem 5.9. *(Taubes [292]) The quantity*

$$\frac{1}{2} \sum_{A \in \mathcal{R}^*_h(\Sigma)} (-1)^{\mu(A)} \tag{5.14}$$

is independent of h and the metric on Σ and equals up to an overall sign the Casson invariant of Σ.

Remark 5.10. This is a rather technical result and we refer the reader to Taubes' paper [292] for the proof. We only mention that the requirement that Σ be an integral homology sphere simplifies things a lot in that it prevents the existence of non-trivial reducible flat connections before and after a small perturbation. This allows one to use the standard cobordism argument to show that the quantity (5.14) is independent of the choice of perturbation h, since the parametrized moduli space

$$\bigcup_{t \in [0,1]} \mathcal{R}^*_{h(t)}(\Sigma) \times \{t\} \tag{5.15}$$

gives a cobordism between $\mathcal{R}^*_{h(0)}(\Sigma)$ and $\mathcal{R}^*_{h(1)}(\Sigma)$. If the assumption that $H_1(\Sigma, \mathbb{Z}) = 0$ is dropped, there might exist non-trivial reducible flat connections which will prevent (5.15) from being compact. In the case of rational homology spheres, this problem is addressed in preprint [187]. This leads to a gauge theoretical interpretation of the Walker invariant.

5.3.5 Morse–type Perturbations

In practice, it is common that a degenerate representation variety $\mathcal{R}^*(\Sigma)$ turns out to be non-degenerate in the following weaker sense: at every point A in $\mathcal{R}^*(\Sigma)$, the kernel of the Hessian is equal to $T_A\mathcal{R}^*(\Sigma)$ and the Hessian is non-degenerate in the direction of the normal bundle of $\mathcal{R}^*(\Sigma)$ in $\mathcal{B}^*$. In particular, $\mathcal{R}^*(\Sigma)$ is smooth. If this is the case we say that $\mathcal{R}^*(\Sigma)$ is non-degenerate in *the Morse–Bott sense.* Each connected component of the manifold $\mathcal{R}^*(\Sigma)$ admits a Morse function, and these functions combined together can be used instead of the admissible perturbations (5.12) to perturb $\mathcal{R}^*(\Sigma)$ into a finite collection of non-degenerate points. Of course, one needs to show that the invariants resulting from the two different sets of perturbations are the same. This is widely believed to be true although we are not aware of any proof of this fact that has ever appeared in the literature, at least in such a generality. The following is based on Christopher Herald's notes he kindly supplied us with.

Theorem 5.11. *Suppose that $\mathcal{R}^*(\Sigma)$ is non-degenerate in the Morse–Bott sense. Then there exists an admissible perturbation function $h : \mathcal{B}^* \to \mathbb{R}$ such that its restriction to each connected component $\mathcal{R}_k$ of $\mathcal{R}^*(\Sigma)$ is a Morse function $f_k : \mathcal{R}_k \to \mathbb{R}$. Moreover, for any sufficiently small $\varepsilon > 0$ there is an open neighborhood U_k of $\mathcal{R}_k$ disjoint from other components of $\mathcal{R}^*(\Sigma)$ such that all the critical points of $\mathbf{cs} + \varepsilon \cdot h$ in U_k are non-degenerate and can naturally be identified with the critical points of f_k.*

This theorem is proved in [29]. Its downside is that we do not have control of the Morse functions f_k, and they may turn out to be unnecessarily complicated. Before we go on, we need to slightly enlarge the class of admissible perturbation functions. Define $g_i : \mathcal{B}^* \to \mathbb{R}$ by the formula

$$g_i(A) = \int_{D^2} h_i(\operatorname{tr}\operatorname{hol}_A(\gamma_i(S^1 \times \{x\}))) \cdot \eta(x)\, d^2x,$$

and view an m–tuple $(g_1, \dots, g_m)$ as a function from $\mathcal{B}^*$ to $\mathbb{R}^m$. We extend the class of perturbations by allowing to compose $(g_1, \dots, g_m)$ with arbitrary smooth functions $\psi : \mathbb{R}^m \to \mathbb{R}$, not just linear functions $\psi(t_1, \dots, t_m) = t_1 + \dots + t_m$ as in (5.12). This extended class of admissible perturbations leads to the same invariants.

Theorem 5.12. *Let $\mathcal{R}^*(\Sigma)$ be non-degenerate in the Morse–Bott sense, and let $\mathcal{R}_1, \dots, \mathcal{R}_n$ be its connected components. For any choice of Morse functions $f_k : \mathcal{R}_k \to \mathbb{R}$, there is an admissible perturbation function h such that the restriction of h onto $\mathcal{R}_k$ is equal to f_k for each $k = 1, \dots, n$.*

Proof. We start with an abundant collection of admissible perturbation functions as in the proof of Theorem 5.8. We enlarge this collection if necessary so that there are admissible perturbation functions $g_1, \dots, g_m$ which separate points in the flat moduli space $\mathcal{R}(\Sigma)$. The proof that this is possible is essentially due to Taubes [292], and we only recast it briefly. Abundance guarantees that finitely many admissible perturbation functions are needed to separate points $(A, A') \in \mathcal{R}(\Sigma) \times \mathcal{R}(\Sigma)$ in an open neighborhood W of the diagonal. For any pair (A, A') not on the diagonal, there is a curve γ in Σ such that the trace of the holonomy along γ separates points in an open neighborhood of (A, A'). Compactness of $\mathcal{R}(\Sigma) \times \mathcal{R}(\Sigma) \setminus W$ allows us to choose a finite collection of additional curves to assure the point separation property.

The point separation property for $g_1, \dots, g_m$ is not nearly as strong as saying that any smooth function on the $\mathcal{R}_k$ can be obtained as a linear combination of $g_1, \dots, g_m$. This is the reason why we enlarged the class of admissible perturbations to include functions $\psi(g_1, \dots, g_m)$ for any smooth $\psi : \mathbb{R}^m \to \mathbb{R}$. The theorem now follows by composing the function $(g_1, \dots, g_m) : \mathcal{A} \to \mathbb{R}^m$ with the f_k and suitable bump functions on $\mathbb{R}^m$.

5.3.6 Casson Invariant and Seiberg–Witten Equations

Let Σ be an oriented integral homology sphere equipped with a Riemannian metric g and consider the bundle $\mathrm{Cl}(T\Sigma) \otimes \mathbb{C}$ of complexified Clifford algebras over Σ. There exists a unique $SU(2)$ vector bundle W_0 over Σ acted upon by $\mathrm{Cl}(T\Sigma) \otimes \mathbb{C}$ such that the action of the volume form of Σ is the identity on W_0. Let $W = W_0 \otimes L$ where L is a trivial complex line bundle over Σ, then W is a $U(2)$–bundle. The Levi–Civita connection on $T\Sigma$ lifts to a connection on W_0, and for any choice of a $U(1)$–connection on L we have the corresponding Dirac operator $\eth_A : \Gamma(W) \to \Gamma(W)$ on the sections of W.

Let $\mathcal{A} = \mathcal{C} \times \Gamma(W)$ where $\mathcal{C}$ is the space of smooth $U(1)$–connections in L. The gauge group $\mathcal{G} = \mathrm{Map}(\Sigma, S^1)$ acts on $\mathcal{A}$ by the rule $g^*(A, \psi) = (A + 2g^{-1}dg, g^{-1}\psi)$. This action is free on the subset $\mathcal{A}^*$ away from the points $(A, 0)$. In proper Sobolev completions, $\mathcal{B}^* = \mathcal{A}^*/\mathcal{G}$ is an infinite dimensional Banach manifold. Define the *Chern–Simons–Dirac function* $\mathbf{cs}_D : \mathcal{B}^* \to \mathbb{R}$ by the formula

$$\mathbf{cs}_D(A, \psi) = \int_\Sigma A \wedge F_A + \int_\Sigma \langle \psi, \eth_A \psi \rangle_{\mathrm{Re}}\, dV,$$

compare with (5.1). The set of critical points of $\mathbf{cs}_D$ is identified with the moduli space $\mathcal{M}$ of solutions of Seiberg–Witten equations, see [177] and [228]. After a generic perturbation, the moduli space $\mathcal{M}$ is finite and non–degenerate and its points are canonically oriented. A count of these points results in an invariant $\chi_{SW}(\Sigma, g) = \#\mathcal{M}$, which is independent of all arbitrary choices made in its definition but one: it depends on the Riemannian metric g. However, according to [53], the quantity

$$\alpha(\Sigma) = \chi_{SW}(\Sigma, g) - \left(\operatorname{ind}\eth + \frac{1}{8}\operatorname{sign} W\right)$$

is a topological invariant of Σ, where W is any compact spin 4–manifold with $\partial W = \Sigma$ and $\eth$ is the Dirac operator on W satisfying the boundary conditions of [12]. The invariant $\alpha(\Sigma)$ can also be expressed in terms of $\chi_{SW}(\Sigma, g)$ and certain η–invariants of Σ, see for example [198].

Theorem 5.13. *For any integral homology sphere Σ, the quantity $\alpha(\Sigma)$ is equal up to an overall sign to the Casson invariant $\lambda(\Sigma)$.*

This result was conjectured to be true by Kronheimer and Mrowka. It was first proved modulo 2 in [53] then checked by an explicit calculation for all Brieskorn homology spheres $\Sigma(p, q, r)$ in [248]. A complete proof of this conjecture can be found in [198] where it is shown that the invariant α satisfies the Casson's surgery formula and hence coincides with λ up to an overall sign by Theorem 3.1. The proof in [198] crucially relies on the paper of Meng and Taubes [213]. Another approach to proving Theorem 5.13 can be found in a series of preprints [205].

The above invariant arising from the Seiberg–Witten gauge theory bears close resemblance to the invariant defined by Ozsváth and Szabó [254] using Heegaard splittings and theta divisors on Riemann surfaces. Their invariant is defined for all rational homology spheres and, if properly normalized, coincides with the Walker invariant (and the Casson invariant for integral homology spheres).

5.4 Casson–type Invariants of Knots

By Casson–type invariants of knots we mean the invariants introduced by X.-S. Lin [199] and C. Herald [137]. They are obtained by (creatively) counting irreducible $SU(2)$ representations of the knot group which have a fixed trace along the knot meridian. It turns out that these invariants can be expressed in terms of the (regular) Casson invariant and certain equivariant knot signatures. Ramifications of these ideas lead to the equivariant Casson invariant, see Section 5.5, and the Floer homology of knots, see Section 6.7.5.

5.4.1 Representation Varieties of Knot Groups

Let $k \subset \Sigma$ be a knot in an oriented integral homology 3–sphere Σ and let K be its exterior. Then K is a compact oriented 3–manifold with boundary $\partial K = T^2$, a 2–dimensional torus. The inclusion $\partial K \to K$ induces a map of representation varieties $r : \mathcal{R}(K) \to \mathcal{R}(T^2)$, where $\mathcal{R}(T^2)$ is the pillowcase variety, see Section 3.3.1. In a generic situation, the variety $\mathcal{R}(K)$ and its image in $\mathcal{R}(T^2)$ are described by the following theorem of Herald [138].

Theorem 5.14. *For a generic admissible perturbation h, the perturbed character variety $\mathcal{R}_h(K)$ consists of two central orbits, a smooth open arc of abelian orbits with one non-compact end limiting to each central orbit, and a smooth 1-dimensional manifold of irreducible orbits with finitely many ends, each limiting to a different point on the abelian arc. Suppose that the abelian arc is parametrized by ϕ, $0 < \phi < \pi$. If a point on the abelian arc with coordinate ϕ is the limiting points of an irreducible orbit then $e^{2i\phi}$ is a root of the Alexander polynomial $\Delta_k(t)$.*

The map $r : \mathcal{R}_h(K) \to \mathcal{R}(T^2)$ is an immersion taking the 1–dimensional strata of $\mathcal{R}_h(K)$ into the smooth part of $\mathcal{R}(T^2)$.

The points on the abelian arc which are the limiting points of irreducible components are called *bifurcation points.*

Proof. To put the variety $\mathcal{R}(K)$ into general position one can use essentially the same admissible perturbations as in Section 5.3.4. We only need to make sure that the image of the $\gamma_i : S^1 \times D^2 \to K$ is disjoint from ∂K. This way all perturbed flat connections restrict to flat connections over ∂K and we still have the the restriction map $r : \mathcal{R}_h(K) \to \mathcal{R}(T^2)$ from the perturbed flat moduli space $\mathcal{R}_h(K)$ to the pillowcase. The complete proof is rather technical and we refer the reader to [138] for all the details.

On the other hand, the statement about the roots of the Alexander polynomial is relatively easy to see, compare with [6]. Assume for the sake of simplicity that k is a knot in S^3 and that no perturbations are needed. Every reducible representation $\alpha : \pi_1(K) \to SU(2)$ factors through $U(1) \to SU(2)$. Therefore, Zariski tangent space $H^1(K, \operatorname{ad}\alpha)$ at α can be identified with $H^1_\alpha(K, \mathfrak{u}(1)) \oplus H^1_\alpha(K, \mathbb{C})$ according to the splitting $\mathfrak{su}(2) = \mathfrak{u}(1) \oplus \mathbb{C}$, with $U(1)$ acting on $\mathbb{C}$ with weight 2. Here, H^1_α stands for the first cohomology with coefficients in the corresponding reduction of α. An easy calculation shows that $H^1_\alpha(K, \mathfrak{u}(1)) = \mathbb{R}$ for all abelian α. At a bifurcation point the dimension of the Zariski tangent space jumps from one to at least two, which is due to the non-vanishing of $H^1_\alpha(K, \mathbb{C})$ at such a point.

Let us compute $H^1_\alpha(K, \mathbb{C})$. The knot exterior K is a $K(\pi, 1)$–space for $\pi = \pi_1 K$. The group π has Wirtinger presentation with generators $a_1, \dots, a_n$ and relations $r_1, \dots, r_n$. Let $\mathbb{Z}[\pi]$ be its group ring then the $\mathbb{Z}[\pi]$–chain complex of its universal cover in degrees 0, 1, and 2 is of the form

$$\mathbb{Z}[\pi] \longleftarrow \mathbb{Z}[\pi]^n \xleftarrow{\;A\;} \mathbb{Z}[\pi]^n \longleftarrow \cdots$$

where $A = (\partial r_i / \partial a_j)$. Every representation $\alpha : \pi_1 K \to U(1)$ factors through $H_1(K, \mathbb{Z}) = \mathbb{Z}$, which is generated by a meridian m of the knot k. Let $\alpha(m) = e^{i\phi}$ then the induced representation $H_1(K, \mathbb{Z}) \to \mathbb{C}^*$ sends m to $e^{2i\phi}$. The cohomology $H^*_\alpha(K, \mathbb{C})$ is the cohomology of the complex

$$\mathbb{C}[t, t^{-1}] \longleftarrow \mathbb{C}[t, t^{-1}]^n \xleftarrow{\;A(t,t^{-1})\;} \mathbb{C}[t, t^{-1}]^n \longleftarrow \cdots$$

where $A(t,t^{-1})$ is the classical Alexander matrix, and the formal variable t should be replaced by $e^{2i\phi}$. The Alexander polynomial is the greatest common divisor of the minors of $A(t,t^{-1})$ of order $n-1$. Since $H^0_\alpha(K,\mathbb{C}) = 0$, the non-vanishing of $H^1_\alpha(K,\mathbb{C})$ at a bifurcation point ϕ implies that $\Delta_k(e^{2i\phi}) = 0$.

Remark 5.15. Conversely, one may expect that every point ϕ on the abelian arc with $\Delta_k(e^{2i\phi}) = 0$ will be a bifurcation point. This was shown to be the case by Frohman and Klassen in [101] under the assumption that the point corresponds to a simple root of the Alexander polynomial. This result continues to hold under the much weaker hypothesis that the Tristram–Levine signature function of the knot has a jump at the point, see [136] and [141]. This implies, in particular, that under either assumption, the space $\mathcal{R}^*(K)$ is non-empty.

Example 5.16. The Alexander polynomial of a trefoil knot is $t + t^{-1} - 1$. It has two roots on the unit circle, $\exp(\pi i/3)$ and $\exp(5\pi i/3)$. These correspond to two bifurcation points $\phi = \pi/6$ and $\phi = 5\pi/6$ on the arc of abelian representations, see Figure 3.2. The Alexander polynomial of the figure-eight knot equals $3-t-t^{-1}$, so it does not have roots on the unit circle. Respectively, there are no bifurcation points in its representation variety, see Figure 3.3.

5.4.2 The Invariants

Let m be the meridian of a knot $k \subset \Sigma$ and fix an $\alpha \in [0,\pi]$. Let $\mathcal{S}_\varphi$ consist of representations $\alpha \in \mathcal{R}(T^2)$ such that $\operatorname{tr}\alpha(m) = 2\cos\varphi$. For a generic perturbation h, the intersection $r(\mathcal{R}_h(K)) \cap \mathcal{S}_\varphi$ is transversal and hence consists of finitely many points. By counting these points with proper signs as defined in [137], one obtains an integer $\#(r(\mathcal{R}_h(K)) \cap \mathcal{S}_\varphi)$. The following theorem is due to Herald [137].

Theorem 5.17. *For every $0 \le \varphi \le \pi$ such that $\Delta_k(e^{2i\varphi}) \ne 0$, the integer $\#(r(\mathcal{R}_h(K)) \cap \mathcal{S}_\varphi)$ is independent of generic admissible perturbation h and defines an integral valued invariant $h_\varphi(\Sigma,k)$ of the knot $k \subset \Sigma$. Moreover,*

$$h_\varphi(\Sigma,k) = 4\lambda(\Sigma) + \frac{1}{2}\operatorname{sign} B_k(e^{2i\varphi}),$$

where $B_k(t) = (1-t)S + (1-t^{-1})S^t$ and S is a Seifert matrix of $k \subset \Sigma$, compare with (2.7).

Proof. One starts by checking that $h_0(\Sigma,k) = 4\lambda(\Sigma)$. As φ increases from 0 to π, the integer h_φ only changes when φ passes through a bifurcation point on the arc of reducibles. By studying the local picture at a bifurcation points, one shows that $h_\varphi(\Sigma,k)$ changes by the same rule as half the signature of the matrix $B_k(e^{2i\varphi})$.

The invariant h_φ defined in Theorem 5.17 is called the *Herald–Lin invariant.* Originally, a special case of Theorem 5.17 appeared in [199]. That paper used topological methods and only covered the case of $\varphi = \pi/2$ and $\Sigma = S^3$. The invariant introduced in [199] is sometimes referred to as the *Casson–Lin invariant.* Heusener and Kroll [141] and Austin (unpublished) have found a topological proof of Theorem 5.17 in the case of $\Sigma = S^3$ and arbitrary φ.

5.5 Equivariant Casson Invariant

Let Σ be an integral homology 3–sphere with a finite cyclic group action making it into a branched cover of another integral homology sphere with branch set a knot k. Equivariant Casson invariants of Σ are defined in [57] by counting $SU(2)$-representations of $\pi_1\Sigma$ equivariant with respect to the induced action. These invariants can be identified with certain equivariant knot signatures of k. Application of this theory to various natural cyclic group actions on the links of singularities leads to geometric proofs of the Fintushel-Stern and Neumann-Wahl formulas and their extensions, and gives a closed form formula for the Floer homology of Brieskorn homology spheres. The definition of equivariant Casson invariants in [57] is gauge theoretical. Cappell, Lee and Miller in an unpublished manuscript utilized an intersection theory to define an equivariant Casson invariant for integral homology spheres acted upon by a cyclic group of odd prime order.

5.5.1 Equivariant Gauge Theory

Let Σ be an integral homology 3-sphere and $\tau : \Sigma \to \Sigma$ an orientation preserving diffeomorphism of order n. It induces a map $\tau^* : \mathcal{R}^*(\Sigma) \to \mathcal{R}^*(\Sigma)$ on irreducible $SU(2)$–representations of $\pi_1\Sigma$ with the fixed point set $\mathcal{R}^\tau(\Sigma) = \operatorname{Fix}(\tau^*)$. For any conjugacy class $[\alpha] \in \mathcal{R}^\tau(\Sigma)$ there exists a matrix $u_\alpha \in SU(2)$ such that $\tau^*\alpha = u_\alpha \alpha u_\alpha^{-1}$. The irreducibility of α implies that u_α is defined uniquely up to sign. Since $u_{g\alpha g^{-1}} = g u_\alpha g^{-1}$, there is a well-defined correspondence $[\alpha] \mapsto |\operatorname{tr} u_\alpha|$. Observe that u_α^n belongs to the stabilizer of α, so that $u_\alpha^n = \pm 1$ and hence $|\operatorname{tr} u_\alpha| = 2\cos(\pi m/n)$, for some uniquely defined integer m such that $0 \le m \le [n/2]$. Thus, we have a splitting

$$\mathcal{R}^\tau(\Sigma) = \bigsqcup_{m=0}^{[n/2]} \mathcal{R}_m(\Sigma),$$

where $\mathcal{R}_m(\Sigma)$ consists of the conjugacy classes of irreducible representations $\alpha : \pi_1\Sigma \to SU(2)$ such that $\sigma^*\alpha = u\alpha u^{-1}$ for some $u \in SU(2)$ with $|\operatorname{tr} u| = 2\cos(\pi m/n)$, $m = 0, \dots, [n/2]$.

Let E be a trivialized $SU(2)$-bundle over Σ. Any endomorphism of E which lifts $\tau : \Sigma \to \Sigma$ induces an action on connections on E by pull-back.

As any two such lifts of τ differ by a gauge transformation, the induced action $\tau^* : \mathcal{B}^*(\Sigma) \to \mathcal{B}^*(\Sigma)$ is well-defined. Let $\mathcal{B}^\tau(\Sigma) \subset \mathcal{B}^*(\Sigma)$ be the fixed point set of τ^*. It can be broken into a disjoint union as follows.

For any connection α with $[\alpha] \in \mathcal{B}^\tau(\Sigma)$ there exists a lift ν_α of τ such that $\nu_\alpha^*\alpha = \alpha$. As the bundle E is trivialized, ν_α can be given in the base–fiber coordinates by the formula $\nu_\alpha(x,\xi) = (\tau(x), \rho_\alpha(x)\xi)$ where $\rho_\alpha : \Sigma \to SU(2)$ is a gauge transformation. Observe that ρ_α^n belongs to the stabilizer of α and therefore $\rho_\alpha^n = \pm 1$ by the irreducibility of α. The solution set of the equation $\rho^n = \pm 1$ in $SU(2)$ is a collection of conjugacy classes described by the equations $|\operatorname{tr}\rho| = 2\cos(\pi m/n)$ with $m = 0, \ldots, [n/2]$. The space $\mathcal{B}^\tau(\Sigma)$ splits into disjoint components according to which conjugacy class the image of ρ_α belongs,

$$\mathcal{B}^\tau(\Sigma) = \bigsqcup_{m=0}^{[n/2]} \mathcal{B}_m(\Sigma).$$

In particular, $\mathcal{R}_m(\Sigma)$ is contained in $\mathcal{B}_m(\Sigma)$, a refinement of the usual holonomy correspondence between flat connections over Σ and representations of the fundamental group of Σ. We will regard $\mathcal{R}_m(\Sigma)$ as the critical point set of the Chern-Simons function **cs** restricted to $\mathcal{B}_m(\Sigma)$. The Zariski tangent space of $\mathcal{R}_m(\Sigma)$ at a point α can be identified with the equivariant group cohomology

$$H^1_\tau(\pi_1\Sigma, \operatorname{ad}\alpha) = \operatorname{Fix}\{\,\tau^* : H^1(\pi_1\Sigma, \operatorname{ad}\alpha) \to H^1(\pi_1\Sigma, \operatorname{ad}\alpha)\,\}.$$

The representation space $\mathcal{R}_m(\Sigma)$ is called *non-degenerate* if the cohomology groups $H^1_\tau(\pi_1\Sigma, \operatorname{ad}\alpha)$ vanish for all $\alpha \in \mathcal{R}_m(\Sigma)$.

5.5.2 Definition of the Invariants

Let Σ be an integral homology 3–sphere with an orientation preserving diffeomorphism $\tau : \Sigma \to \Sigma$ of order n. The diffeomorphism τ induces a $\mathbb{Z}/n$–action on Σ whose quotient is always homeomorphic to a rational homology sphere, which we denote by Σ'. From this point on we will assume that Σ' is an integral homology sphere, which is equivalent to saying that τ has a fixed point on Σ. This condition ensures that Σ' has a distinguished branch set, a knot k, corresponding to the non–empty fixed point set of τ.

Refining the construction of the Floer index outlined in Section 5.3.2, we associate with any non-degenerate representation $\alpha \in \mathcal{R}^\tau(\Sigma)$ an *equivariant Floer index* as follows. Let us fix a Riemannian metric on Σ which is invariant with respect to τ, and a lift $\nu : E \to E$ of τ such that $\nu^*\alpha = \alpha$. Choose a path $\alpha(t)$, $t \in \mathbb{R}$, of connections such that $\alpha(t) = \theta$ near $-\infty$, $\alpha(t) = \alpha$ near $+\infty$, and $\nu^*\alpha(t) = \alpha(t)$ for all t (the latter can be achieved by averaging). Consider the path of self-adjoint Fredholm operators

$$K^\nu_{\alpha(t)} : (\Omega^0 \oplus \Omega^1)_\nu(\Sigma, \mathfrak{su}(2)) \to (\Omega^0 \oplus \Omega^1)_\nu(\Sigma, \mathfrak{su}(2))$$

obtained by restricting the operators (5.9) onto the differential forms invariant with respect to the induced action of ν. The Floer index $\mu^\tau(\alpha)$ is then defined to be the spectral flow of $K^\nu_{\alpha(t)}$ reduced mod 4.

Note that $\mu^\tau(\alpha) = \mu(\alpha) \bmod 2$ whenever n is odd, see [59]. In general, this is not true if n is even, see Example 5.24.

If $\mathcal{R}^\tau(\Sigma)$ is non-degenerate, the *equivariant Casson invariant* is defined by the formula

$$\lambda^\tau(\Sigma) = 1/2 \sum_{\alpha \in \mathcal{R}^\tau(\Sigma)} (-1)^{\mu^\tau(\alpha)}. \tag{5.16}$$

In addition, *refined equivariant Casson invariants* are defined as

$$\lambda^\tau_m(\Sigma) = \kappa/2 \sum_{\alpha \in \mathcal{R}_m(\Sigma)} (-1)^{\mu^\tau(\alpha)}, \tag{5.17}$$

where $\kappa = 1/2$ if $m/n \neq 1/2$ and $\kappa = 1$ if $m/n = 1/2$. The extra factor of κ is explained by the fact that the correspondence $\alpha \mapsto \operatorname{tr} u_\alpha$ is only well-defined up to a sign. Because of that, the count of representations with fixed $|\operatorname{tr} u_\alpha|$ is roughly twice the count of those with fixed $\operatorname{tr} u_\alpha$, unless $\operatorname{tr} u_\alpha = 0$.

If $\mathcal{R}^\tau(\Sigma)$ fails to be non-degenerate, an approach similar to that of Section 5.3.4 and Section 5.4.1 provides a family of generic admissible perturbation functions $h : \mathcal{B}^\tau(\Sigma) \to \mathbb{R}$ which yield a non-degenerate equivariant representation space, see [57] for details. The equivariant Casson invariant is then defined by counting points in this space.

5.5.3 Equivariant Casson and Knot Signatures

The equivariant Casson invariants of a homology sphere with a cyclic group action are expressed in [57] in terms of equivariant knot signatures of the branch set of the action. The equivariant knot signatures $\operatorname{sign}^{m/n}(k)$ are defined as in (2.7).

Theorem 5.18. *Let Σ be an integral homology 3–sphere and $\tau : \Sigma \to \Sigma$ an orientation preserving diffeomorphism of order n. Suppose that the quotient of Σ by the induced $\mathbb{Z}/n$–action is homeomorphic to an integral homology sphere Σ' with branch set a knot $k \subset \Sigma'$. Then, for any $m = 0, \dots, [n/2]$,*

$$\lambda^\tau_m(\Sigma) = \lambda(\Sigma') + \frac{1}{8} \cdot \operatorname{sign}^{m/n}(k),$$

and therefore, for the (global) equivariant Casson invariant,

$$\lambda^\tau(\Sigma) = n \cdot \lambda(\Sigma') + \frac{1}{8} \cdot \sum_{m=0}^{n-1} \operatorname{sign}^{m/n}(k).$$

Proof. A non-degenerate equivariant representation $\alpha : \pi_1 \Sigma \to SU(2)$ such that $\tau^* \alpha = u \alpha u^{-1}$ can be pushed down to the quotient $\Sigma/\tau = \Sigma'$, away from the branch set k, to give a representation $\alpha' : \pi_1(\Sigma' \setminus k) \to SU(2)$ with $\alpha'(m)$ conjugate to u (where m is the meridian of k). According to Theorem 5.17, a certain count of representations α' with a fixed trace of the meridians essentially gives the sum of an equivariant signature of k and the Casson invariant of Σ'. Therefore, to prove the theorem, one compares the count of representations α given by (5.16) with that of Theorem 5.17. This is done with the help of the orbifold gauge theory developed in [59], see also Section 6.7.5. In the degenerate case, special attention should be paid to matching perturbations in the two theories. For a complete proof, see [57].

5.5.4 Applications

Let k be a knot in an integral homology 3–sphere Σ, which is the boundary of a smooth oriented 4–manifold M with $H_1(M, \mathbb{Z}) = 0$. Let Σ_n be the n–fold cyclic branched cover of Σ with branch set the knot k. Given a Seifert surface F of the knot k, let M_n be the n–fold cyclic branched cover of M whose branch set is the surface F with its interior pushed slightly into M. Then M_n is a smooth 4–dimensional manifold with boundary Σ_n and with a natural $\mathbb{Z}/n$–action. Fix a generator $\tau : M_n \to M_n$ of $\mathbb{Z}/n$.

Although the manifold M_n might depend on the choice of F, this will not be the case for its signature. More precisely, the vector space $H_2(M_n, \mathbb{C})$ can be split as $H_2(M_n, \mathbb{C}) = \mathcal{H}_0 \oplus \mathcal{H}_1 \oplus \ldots \mathcal{H}_{n-1}$, where $\mathcal{H}_m$ is the eigenspace of the operator $\tau_* : H_2(M_n, \mathbb{C}) \to H_2(M_n, \mathbb{C})$ with eigenvalue $e^{2\pi i m/n}$. The signature of the intersection form of M_n restricted to $\mathcal{H}_m$ is denoted by $\operatorname{sign}^m(M_n)$ and is called the mth *equivariant signature* of M_n. Obviously,

$$\operatorname{sign}(M_n) = \sum_{m=0}^{n-1} \operatorname{sign}^m(M_n),$$

and it is a classical result of Viro (see Theorem 4.4 and Section 4.8 in [301]) that $\operatorname{sign}^m(M_n) = \operatorname{sign}(M) + \operatorname{sign}^{m/n}(k)$. This fact together with Theorem 5.18 proves the following result.

Corollary 5.19. *Suppose that the 3–manifold $\Sigma_n = \partial M_n$ constructed above from a knot $k \subset \Sigma = \partial M$ is an integral homology sphere. Then*

$$\lambda^\tau(\Sigma_n) - n \cdot \lambda(\Sigma) = \frac{1}{8}\left(\operatorname{sign}(M_n) - n \cdot \operatorname{sign}(M)\right).$$

Similarly, for the refined equivariant Casson invariants (5.17)

$$\lambda^\tau_m(\Sigma_n) - \lambda(\Sigma) = \frac{1}{8}\left(\operatorname{sign}^m(M_n) - \operatorname{sign}(M)\right).$$

Example 5.20. Let $p, q, r \geq 2$ be pairwise relatively prime integers then the link $\Sigma(p,q,r)$ of singularity (1.1) is an integral homology sphere. The map $\tau(x,y,z) = (x, y, e^{2\pi i/r} z)$ is an orientation-preserving diffeomorphism of order r, which makes $\Sigma(p,q,r)$ into an r–fold cyclic branched cover of S^3 with branch set a (p,q)–torus knot. According to Corollary 5.19,

$$\lambda^\tau(\Sigma(p,q,r)) = \frac{1}{8} \operatorname{sign} M(p,q,r),$$

where $M(p,q,r)$ is the Milnor fiber (for a proper choice of the branching surface). Theorem 5.23 below implies that in fact $\lambda^\tau(\Sigma(p,q,r)) = \lambda(\Sigma(p,q,r))$, so that we arrive at the Fintushel–Stern formula, see Theorem 3.29. The second part of Corollary 5.19 refines this formula in that it relates the equivariant signatures of $M(p,q,r)$ to the invariants $\lambda^\tau_m(\Sigma(p,q,r))$.

Example 5.21. Let $a_1, \ldots, a_n$ be pairwise relatively prime integers, and let $\Sigma(a_1, \ldots, a_n)$ be the link of singularity (1.2). Then $\Sigma(a_1, \ldots, a_n)$ is an integral homology sphere which can be represented as an a_n–fold branched cover of integral homology sphere $\Sigma(a_1, \ldots, a_{n-1})$ via the action $\tau(z_1, \ldots, z_{n-1}, z_n) = (z_1, \ldots, z_{n-1}, e^{2\pi i/a_n} z_n)$. The branch set of this covering is a regular fiber of the Seifert fibration of $\Sigma(a_1, \ldots, a_{n-1})$. According to Theorem 5.23 below, invariants λ^τ and λ coincide in this case, and an induction by n then proves that

$$\lambda(\Sigma(a_1, \ldots, a_n)) = \frac{1}{8} \operatorname{sign} M(a_1, \ldots, a_n),$$

where $M(a_1, \ldots, a_n)$ is the Milnor fiber, compare with Theorem 3.32.

Example 5.22. Let Σ be an integral homology sphere obtained as the link of singularity at zero of $f(x,y) + z^n = 0$, see Example 1.19. The map $\tau(x,y,z) = (x, y, e^{2\pi i/n} z)$ is an orientation preserving diffeomorphism of Σ of order n, which makes Σ into a cyclic branched cover of S^3 with branch set a graph knot, compare with Example 1.23. The invariants $\lambda^\tau(\Sigma)$ and $\lambda(\Sigma)$ again coincide, see Theorem 5.23, therefore, $\lambda(\Sigma)$ equals one eighth of the Milnor fiber signature.

The last three examples provide geometric proofs of the Neumann-Wahl conjecture, see Section 3.5.3, for several families of algebraic links. The conjecture was originally verified for all these families in [83] and [243] by purely combinatorial methods. The crucial fact used in the above examples is of course the equality $\lambda = \lambda^\tau$. This equality is a consequence of the following theorem proved in [57] (for the definition of graph knot see Section 1.1.8).

Theorem 5.23. *Let $k \subset \Sigma$ be a graph knot in an integral homology sphere and let Σ_n be the n–fold cyclic branched cover of Σ with branch set the knot k. If Σ_n is an integral homology sphere then $\lambda^\tau(\Sigma_n) = \lambda(\Sigma_n)$.*

The following examples demonstrate that Theorem 5.23 is specific to graph knots and that in general the equivariant Casson invariant need not be equal to the Casson invariant.

Example 5.24. Let $\Sigma(p,q,r)$ be a Brieskorn homology sphere (1.1) and consider the complex conjugation involution $\tau(x,y,z) = (\bar{x},\bar{y},\bar{z})$. The quotient of $\Sigma(p,q,r)$ by this $\mathbb{Z}/2$–action is S^3, with branch set a Montesinos knot of type (p,q,r), see Example 1.22. The induced action of τ on the representation space $\mathcal{R}^*(\Sigma(p,q,r))$ is identity. More precisely, for every representation $\alpha : \pi_1(\Sigma(p,q,r)) \to SU(2)$, there exists $u \in SU(2)$ such that $\operatorname{tr} u = 0$ and $\tau^*\alpha = u\alpha u^{-1}$. According to [272], the Floer indices are related by the formula $\mu(\alpha) = 2\mu^\tau(\alpha) + 1 \bmod 4$. Therefore, the equivariant Casson invariant $\lambda^\tau(\Sigma(p,q,r))$ is obtained by counting all the representations in $\mathcal{R}^*(\Sigma(p,q,r))$ but with signs different from those giving $\lambda(\Sigma(p,q,r))$. This observation was used in [272] to give a closed form formula for the Floer homology of $\Sigma(p,q,r)$, see also Section 6.4.2. These results were extended to all Seifert fibered homology spheres $\Sigma(a_1,\dots,a_n)$ in [278].

Example 5.25. Every plumbed integral homology sphere Σ admits a preferred involution $\tau : \Sigma \to \Sigma$ generalizing the complex conjugation involution on links of singularities, see Example 1.26. Let $\lambda^\tau(\Sigma)$ be the corresponding equivariant Casson invariant then $\lambda^\tau(\Sigma) = \bar{\mu}(\Sigma)$, the $\bar{\mu}$–invariant of Neumann and Siebenmann, see Section 7.2.3.

Example 5.26. Let k be a knot in S^3 and denote by $D_\varepsilon k$ its untwisted Whitehead double with ε–clasp, see Figure 3.6. Let Σ be the n-fold cyclic branched cover of S^3 with branch set $D_\varepsilon k$. Then Σ is an integral homology sphere and $\lambda^\tau(\Sigma) = 0$ for any knot k because all the equivariant signatures of $D_\varepsilon k$ are equal to zero. On the other hand, $\lambda(\Sigma) = n\varepsilon \cdot \Delta''_k(1)$ according to Theorem 3.23. Therefore, in general, $\lambda(\Sigma) \neq \lambda^\tau(\Sigma)$.

Example 5.27. Let Σ_n be an integral homology sphere represented as an n–fold cyclic branched cover of S^3 with branch set a knot k. If $n = 2$, the difference between the Casson and equivariant Casson invariants can be measured in terms of the Jones polynomial of k, see Theorem 3.21 (where $(1/8) \cdot \operatorname{sign} k = \lambda^\tau(\Sigma_2)$). For arbitrary n, this difference can be expressed in terms of residues of certain rational functions related to the Kontsevich integral of k, see [122].

5.6 The $SU(3)$ Casson Invariant

This section describes the $SU(3)$ Casson invariant defined by H. Boden and C. Herald in [30]. Their definition extends the gauge theoretical approach to the Casson invariant outlined in Section 5.3. Unlike the regular Casson invariant, the invariant of Boden and Herald takes real values. Recently, two integer valued versions of the $SU(3)$ Casson invariant were announced, one by Cappell, Lee and Miller [46] and the other by Boden, Herald and Kirk [31]. An approach to defining $SU(n)$ Casson invariants for all n was proposed in [49].

5.6.1 Some $SU(3)$–Gauge Theory

Let Σ be an oriented integral homology 3–sphere and let $\mathcal{A}$ be the space of $SU(3)$–connections on a trivialized bundle $\Sigma \times SU(3)$. By fixing a trivial connection θ, we identify $\mathcal{A}$ with the linear space $\Omega^1(\Sigma, \mathfrak{su}(3))$ of $\mathfrak{su}(3)$–valued differential 1–forms on Σ. The gauge group $\mathcal{G} = \operatorname{Map}(\Sigma, SU(3))$ consisting of the automorphisms of the bundle $\Sigma \times SU(3)$ acts on $\mathcal{A}$ by the rule $g^*A = g^{-1}dg + g^{-1}Ag$, where $g \in \mathcal{G}$ and $A \in \mathcal{A}$, compare with Section 5.1. This action is not free, and according to the size of the stabilizer, there is a natural stratification of $\mathcal{A}$ and also of the quotient space $\mathcal{B} = \mathcal{A}/\mathcal{G}$. An $SU(3)$–connection A is called *irreducible* if its stabilizer consists of constant maps into $\mathbb{Z}/3$, the center of $SU(3)$. Irreducible connections form the top stratum $\mathcal{A}^*$ and its quotient $\mathcal{B}^* = \mathcal{A}^*/\mathcal{G}$ has the structure of a Banach manifold (in proper Sobolev completions). Below the top stratum, there are strata whose stabilizers are respectively $U(1)$, $S(U(1)\times U(1)\times U(1))$, $S(U(1)\times U(2))$, and $SU(3)$.

Let us consider the space $\mathcal{R}(\Sigma)$ of the gauge equivalence classes of flat $SU(3)$ connections. As usual, it can be identified with the $SU(3)$ character variety

$$\mathcal{R}(\Sigma) = \operatorname{Hom}(\pi_1(\Sigma), SU(3))/\,SU(3)$$

consisting of $SU(3)$ representations of $\pi_1(\Sigma)$ modulo conjugation. The only strata relevant to $\mathcal{R}(\Sigma)$ are those with stabilizers $\mathbb{Z}/3, U(1)$, and $SU(3)$. The reason is that, for the integral homology sphere Σ, there exist no non-trivial representations $\pi_1(\Sigma) \to U(1)$. According to the three types of stabilizers, $\mathbb{Z}/3$, $U(1)$, and $SU(3)$, the space $\mathcal{R}(\Sigma)$ splits into a disjoint union $\mathcal{R}(\Sigma) = \mathcal{R}^*(\Sigma) \cup \mathcal{R}^a(\Sigma) \cup \{\theta\}$.

5.6.2 Definition of the Invariant

As in the $SU(2)$ case, the space $\mathcal{R}(\Sigma)$ can be viewed as the critical point set of the Chern–Simons function $\mathbf{cs} : \mathcal{B} \to \mathbb{R}/\mathbb{Z}$, see Section 5.3.2. The Chern–Simons function may have to be perturbed, which is done with the help of admissible perturbation functions $h : \mathcal{B} \to \mathbb{R}$. These functions are modeled after admissible perturbation functions (5.12); their precise definition can be found in [29] and [30]. Let

$$\mathcal{R}_h(\Sigma) = \{\, A \mid\ *F_A - 4\pi^2\,(\nabla h)(A) = 0 \,\}/\mathcal{G}$$

be the perturbed flat moduli space. Boden and Herald [30] show that there exists an $\varepsilon > 0$ such that, for all admissible perturbation functions h with $||h|| \le \varepsilon$, the space $\mathcal{R}_h(\Sigma)$ splits as

$$\mathcal{R}_h(\Sigma) = \mathcal{R}^*_h(\Sigma) \cup \mathcal{R}^a_h(\Sigma) \cup \{\,\theta\,\},$$

where $\mathcal{R}^*_h(\Sigma)$ and $\mathcal{R}^a_h(\Sigma)$ are compact 0–dimensional manifolds consisting of gauge orbits that satisfy certain cohomological non-degeneracy condition, see [29] or [30]. Moreover, for any perturbed flat connection A there exists a unique component $\hat{A}$ of flat connections which is within ε–distance of A.

Given an $SU(3)$–connection A, define a self–adjoint elliptic operator K_A as in (5.9) with only difference that now it will act on the space $\Omega^0(\Sigma, \mathfrak{su}(3)) \oplus \Omega^1(\Sigma, \mathfrak{su}(3))$. Let A be a flat connection, and choose a path $A_t \in \mathcal{A}$ with $A_0 = \theta$ and $A_1 = A$. Define $\mathrm{sf}\,(\theta, A)$ to be the spectral flow of the path K_{A_t}, see Section 5.3.2. If $[A] \in \mathcal{R}^a(\Sigma)$ then we can choose the path so that each A_t has stabilizer $U(1)$ for all $0 < t \leq 1$. Adjusting by a path of gauge transformations, we can assume that, for all $0 \leq t \leq 1$,

$$A_t \in \mathcal{A}_{S(U(2)\times U(1))},$$

the space of connections on $\Sigma \times S(U(2) \times U(1))$. Let $\mathfrak{h} = \mathfrak{s}(\mathfrak{u}(2) \times \mathfrak{u}(1))$ be the Lie algebra of $S(U(2) \times U(1))$ viewed as a Lie subalgebra of $\mathfrak{su}(3)$. Then $\mathfrak{su}(3) = \mathfrak{h} \oplus \mathbb{C}^2$, and the spectral flow decomposes as

$$\mathrm{sf}\,(\theta, A) = \mathrm{sf}\,_{\mathfrak{h}}(\theta, A) + \mathrm{sf}\,_{\mathbb{C}^2}(\theta, A).$$

All the above can be perturbed if necessary by using an admissible perturbation function h. The operator K_A will then be replaced by the operator $K_{A,h}$, see (5.13), and the above spectral flows will be the spectral flows of families of the $K_{A,h}$. For a sufficiently small admissible perturbation h define

$$\lambda'_{BH}(\Sigma) = \sum_{[A]\in\mathcal{R}^*_h(\Sigma)} (-1)^{\mathrm{sf}\,(\theta,A)}$$

and

$$\lambda''_{BH}(\Sigma) = \sum_{[A]\in\mathcal{R}^a_h(\Sigma)} (-1)^{\mathrm{sf}\,(\theta,A)} \cdot \left(\mathrm{sf}\,_{\mathbb{C}^2}(\theta, A) - 4\,\mathbf{cs}(\hat{A}) + 2\right),$$

where $\hat{A}$ is the unique component of flat connections within $||h||$–distance of A, see above. The following theorem is proved in [29].

Theorem 5.28. *For a generic sufficiently small admissible perturbation h, the quantity $\lambda_{BH}(\Sigma) = \lambda'_{BH}(\Sigma) - 1/2 \cdot \lambda''_{BH}(\Sigma)$ is an invariant of integral homology 3–sphere Σ.*

The invariant $\lambda_{BH}(\Sigma)$ is called the $SU(3)$ *Casson invariant* of Σ. The number $\lambda'_{BH}(\Sigma)$ may look like the natural generalization of the Casson invariant, compare with (5.11). Unfortunately, it depends on the choice of perturbation h, which is due to the fact that reducible connections can perturb into irreducible ones. A similar situation occurs in Walker's generalization of the Casson invariant, see Example 4.4.

Whenever an irreducible flat connection sinks into or emerges from reducible connections, an integer jump occurs in the spectral flow $\mathrm{sf}\,_{\mathbb{C}^2}(\theta, A)$. Thus the discrepancy in $\lambda'_{BH}(\Sigma)$ is compensated by the second term,

$\lambda''_{BH}(\Sigma)$. It should be noted that only the difference $\mathrm{sf}_{\mathbb{C}^2}(\theta, A) - 4\,\mathbf{cs}(\hat{A})$ is well defined on the gauge orbit; each term individually depends on the choice of representative for $[A]$. Adding in the constant part of the second term only adds a multiple of $SU(2)$ Casson invariant. It is needed to assure certain properties of the resulting invariant.

5.6.3 Properties and Computations

The $SU(3)$ Casson invariant λ_{BH} defined in Theorem 5.28 takes real values. The conjectured rationality of the Chern–Simons function $\mathbf{cs}$ on flat connections, see Section 5.2, would imply that λ_{BH} is rational. The invariant λ_{BH} is insensitive to the change of orientation,

$$\lambda_{BH}(-\Sigma) = \lambda_{BH}(\Sigma).$$

It satisfies the following connected sum formula, see [29],

$$\lambda_{BH}(\Sigma_1 \# \Sigma_2) = \lambda_{BH}(\Sigma_1) + \lambda_{BH}(\Sigma_2) + 4\,\lambda(\Sigma_1)\,\lambda(\Sigma_2),$$

where λ is the original Casson invariant. Although λ_{BH} is not additive under the connected sum operation, the difference $\lambda_{BH} - 2\,\lambda^2$ is. If $\lambda_{BH}(\Sigma) \neq 0$ then $\pi_1(\Sigma)$ admits a non-trivial representation into $SU(3)$ or $SU(2)$.

The invariant λ_{BH} is rather difficult to compute because one lacks algebraic tools such as the surgery formula, which are so useful in computing the $SU(2)$ Casson invariant. The following calculations can be found in [32].

Example 5.29. Let $\Sigma_k(p,q)$ be the integral homology sphere obtained by $(1/k)$–surgery on the right handed (p,q)–torus knot so that $\Sigma_{-k}(p,q) = \Sigma(p,q,pqk+1)$ and $\Sigma_k(p,q) = -\Sigma(p,q,pqk-1)$ for $k \geq 0$. Then

$$\lambda_{BH}(\Sigma_k(2,3)) = \frac{84k^3 - 138k^2 + 19k}{6(6k-1)},$$

$$\lambda_{BH}(\Sigma_k(2,5)) = \frac{3100k^3 - 2850k^2 + 241k}{10(10k-1)},$$

$$\lambda_{BH}(\Sigma_k(2,7)) = \frac{26264k^3 - 16156k^2 + 970k}{14(14k-1)},$$

$$\lambda_{BH}(\Sigma_k(2,9)) = \frac{124200k^3 - 59004k^2 + 2758k}{18(18k-1)}.$$

Note that the $SU(3)$ Casson invariants of the manifolds obtained by $(1/k)$–surgery and $(-1/k)$–surgery on the same torus knot have different absolute values while their $SU(2)$ Casson invariants only differ in sign.

6 Instanton Floer Homology

Andreas Floer [93] associated with every integral homology 3–sphere eight finitely generated abelian groups $I_n(\Sigma)$, $n = 0, \dots, 7$, which are now referred to as the (instanton) Floer homology. The Floer homology is an invariant of orientation preserving diffeomorphism. It is a refinement of the Casson invariant $\lambda(\Sigma)$ in that $\lambda(\Sigma)$ is half the Euler characteristic of $I_*(\Sigma)$. The definition of $I_*(\Sigma)$ relies heavily on gauge theory in dimensions three and four. In this chapter, we define $I_*(\Sigma)$, review its properties, and give examples of calculations. We also explain relations between $I_*(\Sigma)$ and the symplectic Floer homology, between $I_*(\Sigma)$ and Donaldson polynomials, and describe various extensions of $I_*(\Sigma)$.

6.1 Gauge Theory in Dimension 4

This section reviews $SU(2)$ gauge theory, first on closed 4–manifolds and then on open 4–manifolds with cylindrical ends. The latter is the set up needed for the Floer theory. This section mirrors to certain extent Section 5.1 which discussed gauge theory in dimension three.

6.1.1 Gauge Theory on Closed 4–Manifolds

Let $E \to X$ be a principal $SU(2)$–bundle over a closed smooth oriented connected 4–manifold X. Such a bundle is classified by its second Chern number $c_2(E)[X] \in \mathbb{Z}$. Choose a Riemannian metric on X and let $\mathcal{A}(E)$ be the infinite dimensional Banach space of connections on E completed with respect to a proper Sobolev L^p_k–norm. The space $\mathcal{A}(E)$ is acted upon by the *gauge group* $\mathcal{G}(E) = \operatorname{Aut}(E)$, in proper completion. In local charts, this action is given by the formula $g^*A = g^{-1}dg + g^{-1}A\,g$, where $g \in \mathcal{G}(E)$ and $A \in \mathcal{A}(E)$. The quotient space $\mathcal{B}(E) = \mathcal{A}(E)/\mathcal{G}(E)$ is singular at the orbits of reducible connections. However, its subset $\mathcal{B}^*(E)$ consisting of the orbits of irreducible connections is open and dense in $\mathcal{B}(E)$ and has the structure of an infinite dimensional smooth Banach manifold. The tangent space to $\mathcal{B}^*(E)$ at an orbit $[A]$ can be identified with the kernel of the operator

$$d_A^* : \Omega^1(X, \operatorname{ad} E) \to \Omega^0(X, \operatorname{ad} E)$$

which is adjoint to the covariant derivative d_A. Here, $\Omega^*(X, \operatorname{ad} E)$ stands for differential forms on X with coefficients in the vector bundle $\operatorname{ad} E = E \times_{\operatorname{ad}} \mathfrak{su}(2)$. Given a connection A, its *curvature* F_A is a 2–form with coefficients in $\operatorname{ad} E$ defined in local charts by the formula $F_A = dA + A \wedge A$. One can easily check that $F_{g^*A} = g^{-1} F_A\, g$.

A choice of Riemannian metric on X induces the *Hodge operator* $* : \Omega^k(X, V) \to \Omega^{4-k}(X, V)$ such that $*^2 = (-1)^{k(4-k)}$, for any vector bundle V. Restricted to 2–forms, the Hodge operator is an involution so that it determines a decomposition

$$\Omega^2(X, V) = \Omega^2_+(X, V) \oplus \Omega^2_-(X, V)$$

of the 2–forms into the $(+1)$ and (-1) eigenspaces for $*$. These subspaces are respectively the subspaces of *self-dual* (SD) and *anti-self-dual* (ASD) forms. A reverse of orientation on X changes SD forms into ASD forms, and vice versa, so the two concepts are completely equivalent. From now on we choose to work with ASD forms. A connection A is called ASD if its curvature F_A is an ASD form, that is, if

$$*F_A + F_A = 0. \tag{6.1}$$

Equation (6.1) is called the ASD *equation*. Its solutions, ASD connections, are also referred to as *instantons*. The subset of all ASD connections is invariant under the action of the gauge group. For a fixed bundle E, we denote the quotient by $\mathcal{M}(E)$ and call it *the moduli space* of ASD connections. Naturally, $\mathcal{M}(E)$ is a subset of $\mathcal{B}(E)$, and we call the intersection $\mathcal{M}^*(E) = \mathcal{M}(E) \cap \mathcal{B}^*(E)$ the *moduli space* of irreducible ASD connections.

The moduli spaces $\mathcal{M}(E)$ and $\mathcal{M}^*(E)$ depend on the choice of Riemannian metric on X. If $c_2(E)$ is non-zero then for almost all metrics on X the moduli space $\mathcal{M}^*(E)$ is a smooth orientable submanifold of $\mathcal{B}^*(E)$ whose dimension equals the index of the elliptic operator

$$\mathcal{D}_A = d_A^* + d_A^+ : \Omega^1(X, \operatorname{ad} E) \to \Omega^0(X, \operatorname{ad} E) \oplus \Omega^2_+(X, \operatorname{ad} E), \tag{6.2}$$

where d_A^+ is the operator $d_A : \Omega^1(X, \operatorname{ad} E) \to \Omega^2(X, \operatorname{ad} E)$ followed by the projection onto the subspace $\Omega^2_+(X, \operatorname{ad} E)$. The operator (6.2) is called the ASD *operator*.

Proposition 6.1. *Let $b_1 = \operatorname{rank} H_1(X, \mathbb{Q})$ be the first Betti number of X and b_2^+ the number of positive eigenvalues of the intersection form Q of X, see (2.1). Then*

$$\operatorname{ind} \mathcal{D}_A = 8\, c_2(E)[X] - 3(1 - b_1 + b_2^+)(X). \tag{6.3}$$

Proof. Application of the Atiyah–Singer index theorem to the ASD operator $\mathcal{D}_A$ provides us with the formula

$$\operatorname{ind} \mathcal{D}_A = -\int_X \hat{A}(X)\,\mathrm{ch}(V_+)\,\mathrm{ch}(\mathrm{ad}\, E),$$

where $\hat{A}$ is the $\hat{A}$–polynomial and ch is the Chern character, see [11]. Since $\mathrm{ch}(\mathrm{ad}\, E) = 3 - 8\,c_2(E)$ up to degree four and $\hat{A}(X)\,\mathrm{ch}(V_+) = (\mathcal{L} + \mathcal{E})/2$, where $\mathcal{L}$ is the $\mathcal{L}$–polynomial and $\mathcal{E}$ is the Euler class, we further obtain

$$\begin{aligned} \operatorname{ind} \mathcal{D}_A &= 8 \int_X c_2(E) - \frac{3}{2} \int_X (\mathcal{L} + \mathcal{E}) \\ &= 8\,c_2(E)[X] - \frac{3}{2}\,(\operatorname{sign} X + e(X)) = 8\,c_2(E)[X] - 3(1 - b_1 + b_2^+)(X). \end{aligned}$$

Thus we have established that, for $c_2(E) \neq 0$ and a generic metric, the moduli space $\mathcal{M}^*(E)$ is a smooth manifold of dimension given by (6.3). Another fact vital in the applications is that $\mathcal{M}^*(E)$ can be canonically oriented. The orientation of $\mathcal{M}^*(E)$ is determined by choosing an orientation of $H^1(X, \mathbb{R}) \oplus H^2_+(X, \mathbb{R})$, called a *homology orientation*, see [71] and [227].

The difficult part of the theory is non-linear analysis of the moduli spaces $\mathcal{M}(E)$. The moduli spaces are usually not compact although because of theorems of Uhlenbeck and Taubes one has good control on the nature of neighborhoods at infinity, see [71].

More precisely, if $b_2^+(X) > 0$ then, for almost all Riemannian metrics on X, the moduli space $\mathcal{M}(E)$ is completely contained in $\mathcal{B}^*(E)$ so that $\mathcal{M}(E) = \mathcal{M}^*(E)$ is a smooth orientable manifold of dimension given by (6.3), see [71]. If $b_2^+(X) = 0$ then $\mathcal{M}(E)$ has singular points each of which has a neighborhood homeomorphic to a cone over $\mathbb{C}P^2$ with either orientation, and the number of singular points equals half the number of solutions of the equation $Q(a, a) = -1$, see [67]. The latter is a main ingredient in the proof of Donaldson's theorem [67] about smooth 4-manifolds with definite intersection forms, see Theorem 7.13.

In all of the above, one can replace $SU(2)$ by $SO(3)$ with minimal changes. The principal $SO(3)$–bundles $E \to X$ are classified by the first Pontryagin class $p_1(E) \in H^4(X, \mathbb{Z})$ and the second Stiefel–Whitney class $w_2(E) \in H^2(X, \mathbb{Z}/2)$ related by the equation $w_2(E)^2 = p_1(E) \bmod 4$ if $w_2(E)$ lifts to the integers. The moduli spaces of ASD connections in E are defined as above, and they have dimension

$$-2\,p_1(E)[X] - 3(1 - b_1 + b_2^+)(X). \tag{6.4}$$

6.1.2 Gauge Theory on Open 4–Manifolds

All open manifolds we consider in this section are obtained by the following construction. Let X be a smooth compact connected oriented 4–manifold with boundary ∂X. The boundary will be oriented by the "outside normal goes first" rule (this is precisely the orientation that makes the Stokes Theorem come out right). Note that with this orientation, $\partial([0,1]\times M) = M\times\{1\} - M\times\{0\}$ for any oriented manifold M. We make X into an open manifold with cylindrical end(s) by attaching the infinite cylinder $\mathbb{R}_+ \times \partial X$ to its boundary, where $\mathbb{R}_+ = [0,\infty)$. The number of ends equals the number of connected components of ∂X. If X is a Riemannian manifold, the manifold $X \cup (\mathbb{R}_+ \times \partial X)$ will be a complete Riemannian manifold with the product metric on its open ends. We will abuse notations by using the same symbol X to denote the open manifold $X \cup (\mathbb{R}_+ \times \partial X)$.

Given a manifold X with cylindrical ends and $\delta \geq 0$, let $e_\delta : X \to \mathbb{R}$ be a smooth positive function such that $e_\delta(t,x) = \exp(\delta t)$ at each of the ends, $t \in \mathbb{R}_+$. We define the weighted Sobolev $L^p_{k,\delta}$–norm by the formula

$$\|\xi\|_{L^p_{k,\delta}} = \|e_\delta \cdot \xi\|_{L^p_k}.$$

It will become clear later on why it may be necessary to deal with non-zero δ. Let E be a principal $SU(2)$–bundle over X translationally invariant over the ends. Denote by $\mathcal{A}(E)$ the infinite dimensional Banach space of connections on E completed in a proper $L^p_{k,\delta}$–norm. The space $\mathcal{A}(E)$ is acted upon by the gauge group $\mathcal{G}(E) = \operatorname{Aut}(E)$, in proper completion, with quotient space $\mathcal{B}(E) = \mathcal{A}(E)/\mathcal{G}(E)$. As in the closed case, the subset $\mathcal{B}^*(E)$ of irreducible connections is open and dense in $\mathcal{B}(E)$ and has the structure of an infinite dimensional Banach manifold.

A connection $A \in \mathcal{A}(E)$ is called ASD if its curvature F_A satisfies the equation (6.1). A connection A is said to have *finite Yang–Mills action* if the L^2–norm of its curvature F_A is bounded on X. The subset of all ASD connections in $\mathcal{A}(E)$ having finite Yang–Mills action is invariant under the action of the gauge group. We denote the quotient by $\mathcal{M}(E)$ and call it the *moduli space* of ASD connections, so that

$$\mathcal{M}(E) = \{\, A \in \mathcal{A}(E) \mid *F_A + F_A = 0,\ \|F_A\|^2_{L^2} < \infty \,\}/\,\mathcal{G}(E). \tag{6.5}$$

We denote the intersection $\mathcal{M}(E) \cap \mathcal{B}^*(E)$ by $\mathcal{M}^*(E)$ and call it the *moduli space* of irreducible ASD connections. Note that the finite–action condition forces each ASD connection approach asymptotically a flat connection at each of the ends, see [229]. In the next two sections we will show that generically, perhaps after a perturbation, $\mathcal{M}^*(E)$ is a smooth orientable manifold, and will compute its dimension.

6.1.3 Linear Analysis

The (formal) dimension of $\mathcal{M}^*(E)$ equals the index of the ASD operator $\mathcal{D}_A$, see (6.2), where by $\Omega^*(X, \operatorname{ad} E)$ we mean the space of differential forms on X with coefficients in the bundle $\operatorname{ad} E$ completed in a proper $L^p_{k,\delta}$–norm. For index calculations, connection A does not have to be ASD; any connection having finite Yang–Mills action will do. For the sake of simplicity, we choose a connection A which is equal to a flat connection near infinity at each of the ends. The operator $\mathcal{D}_A$ is elliptic but not necessarily Fredholm since the manifold X is not compact. In fact, according to [202], the operator $\mathcal{D}_A$ is Fredholm for all but countably many choices of δ, and when it is, its index depends on δ. This has to do with the behavior of $\mathcal{D}_A$ at the ends of X which we will now discuss.

Suppose for simplicity that X has only one end of the form $\mathbb{R}_+ \times M$, and let A be a connection over X equal to a flat connection near infinity. Applying a gauge transformation if necessary, we may assume that, over the end, the component of A in the $\mathbb{R}_+$–direction vanishes (this gauge is called *temporal*). Moreover, the following identifications can be made,

$$\begin{aligned} \Omega^1(M \times \mathbb{R}_+) &= \Omega^0(M) \oplus \Omega^1(M), \qquad && dt \wedge \omega_0 + \omega_1 \mapsto (\omega_0, \omega_1), \\ \Omega^2_+(M \times \mathbb{R}_+) &= \Omega^1(M), && dt \wedge \omega_1 + *\omega_1 \mapsto \omega_1, \end{aligned}$$

where $*$ stands for the Hodge operator on M. An easy calculation shows that, over the end,

$$\mathcal{D}_A = \frac{\partial}{\partial t} + B_{A(t)},$$

where $A(t)$ is the restriction of A onto the slice $\{\, t \,\} \times M$ and

$$B_A = \begin{pmatrix} 0 & -d_A^* \\ -d_A & *d_A \end{pmatrix}. \tag{6.6}$$

Note that $B_A = -K_A$ where K_A is the operator (5.9). In particular, B_A is an elliptic operator with pure point real spectrum without accumulation points.

Let α be the limiting flat connection of A as $t \to +\infty$, then for all δ not in the spectrum of B_α, the operator $\mathcal{D}_A$ is Fredholm and hence has a well defined index $\operatorname{ind}_\delta \mathcal{D}_A$, see [202]. In particular, if B_α has trivial kernel, then δ can be equal to zero. For any δ_1 and δ_2 not in the spectrum of B_α, the indices $\operatorname{ind}_{\delta_1} \mathcal{D}_A$ and $\operatorname{ind}_{\delta_2} \mathcal{D}_A$ differ by the number of eigenvalues of B_α between δ_1 and δ_2 counted with multiplicities.

If X has more than one end, the operator $\mathcal{D}_A$ will be Fredholm and will have a well defined index $\operatorname{ind}_\delta \mathcal{D}_A$ as long as δ is not in the spectrum of any of the operators B_α corresponding to the ends of X. We define

$$\operatorname{ind} \mathcal{D}_A = \operatorname{ind}_\delta \mathcal{D}_A \tag{6.7}$$

where δ is a positive number less than the smallest positive eigenvalue of all the B_α.

An important special case of the above discussion is as follows. Let M be a closed oriented 3–manifold and $X = \mathbb{R} \times M$ the infinite cylinder over M. Let E be an $SU(2)$–bundle over M and $\{\, A(t) \,\}$, $t \in \mathbb{R}$, a path of connections in E such that $A(t) = \alpha$ near $-\infty$ and β near $+\infty$. This path of connections defines a connection A in the pull-back bundle over X. The index of the corresponding ASD operator $\mathcal{D}_A$ will be denoted by $\operatorname{ind} \mathcal{D}_A(\alpha, \beta)$.

Proposition 6.2. *The index* $\operatorname{ind}\mathcal{D}_A(\alpha,\beta)$ *equals modulo 8 the spectral flow* $\operatorname{sf}(\beta,\alpha)$ *of a path of operators* K_A *connecting* K_β *to* K_α*, as defined in Section 5.3.2.*

Proof. The spectral flow of a family K_A connecting K_β to K_α is the same as the spectral flow of the family $B_A = -K_A$ connecting B_α to B_β. Therefore, we only need to compare the latter with the index of $\mathcal{D}_A = \partial/\partial t + B_{A(t)}$.

The cylinder X has two ends, $\mathbb{R}_+ \times M$ and $\mathbb{R}_+ \times (-M) = \mathbb{R}_- \times M$, where $\mathbb{R}_- = (-\infty, 0]$. Therefore, we can work with the $L^p_{k,\delta}$–spaces where $\delta > 0$ is less then the smallest absolute value of the eigenvalues of both B_α and B_β.

For the sake of concreteness, suppose that $A(t) = \alpha$ for all $t \le 0$ and $A(t) = \beta$ for all $t \ge 1$. Every $\xi \in \ker\mathcal{D}_A$ can be expanded for $t \ge 1$ in terms of an orthonormal basis $\{\,\varphi_\lambda\,\}$ of eigenforms of the operator B_β as follows,

$$\xi(x,t) = \sum_\lambda \xi^+_\lambda(t)\,\varphi_\lambda(x).$$

Since $d\xi/dt + B_\beta\xi = 0$ we have that $\xi^+_\lambda(t) = e^{-\lambda t}\xi^+_\lambda(1)$. Thus if $\xi \in L^p_{k,\delta}(\mathbb{R}\times M)$ then $\delta < \lambda$ and

$$\xi(x,t) = \sum_{\delta<\lambda} e^{-\lambda t}\,\xi^+_\lambda(1)\,\varphi_\lambda(x).$$

The same argument applies at the other end. If $\{\,\psi_\lambda\,\}$ is an orthonormal basis of eigenforms of B_α we have

$$\xi(x,t) = \sum_{\lambda<-\delta} e^{-\lambda t}\,\xi^-_\lambda(0)\,\psi_\lambda(x).$$

Hence the dimension of $\ker\mathcal{D}_A$ equals the number of eigenvalues of B_α smaller than $-\delta$ which change to eigenvalues of B_β greater than δ as t changes from 0 to 1. Since $\operatorname{coker}\mathcal{D}_A = \ker\mathcal{D}^*_A$ and $\mathcal{D}^*_A = \partial/\partial t - B_{A(t)}$, we similarly conclude that the dimension of $\operatorname{coker}\mathcal{D}_A$ equals the number of eigenvalues of B_α greater than $-\delta$ which change to eigenvalues of B_β smaller than δ as t changes from 0 to 1. Therefore, $\operatorname{ind}\mathcal{D}_A = \dim\ker\mathcal{D}_A - \dim\operatorname{coker}\mathcal{D}_A$ equals the spectral flow of $B_{A(t)}$.

Corollary 6.3. *For any flat connections* α *and* β*, the spectral flow* $\operatorname{sf}(\alpha,\beta)$ *is well defined modulo 8.*

Proof. We need to show that $\operatorname{sf}(\alpha,\beta)$ is independent modulo 8 of the choice of a path $A(t)$ connecting α to β. Given two such paths, A and B, the difference of the respective spectral flows equals the index of the ASD operator $\mathcal{D}_C$ over the 4–manifold $S^1 \times M$ obtained by gluing the two cylinders $\mathbb{R} \times M$ together at the ends. Here, $C = A \cup (-B)$. According to (6.3), we have $\operatorname{ind}\mathcal{D}_C = 8\,c_2(E) - 3(1 - b_1 + b_2^+)(S^1 \times M) = 0 \bmod 8$.

Corollary 6.4. *Let* $\mu(\alpha) = \operatorname{sf}(\theta,\alpha)$ *be the Floer index defined in (5.10) then* $\mu(\alpha) = \operatorname{ind}\mathcal{D}_A(\alpha,\theta) \bmod 8$*, where* $\operatorname{ind}\mathcal{D}_A(\alpha,\theta)$ *is defined as in (6.7).*

6.1.4 Non–linear Analysis

Let $\mathbb{R}_+ \times M$ be an end of X, and $A = \{ A(t) \}$ an asymptotically flat connection over the end in temporal gauge. In the product metric at the end, the ASD equation $* F_A + F_A = 0$ takes the form

$$\frac{\partial A(t)}{\partial t} + * F_{A(t)} = 0. \tag{6.8}$$

In the last equation, $*$ refers to the Hodge operator on M, and $F_{A(t)}$ is the curvature 2–form on M.

Example 6.5. We will be particularly interested in the case when $X = \mathbb{R} \times M$ is a cylinder. The equation (6.8) can then be viewed as the gradient flow equation for the Chern–Simons function, up to a constant,

$$\frac{\partial A(t)}{\partial t} = 4\pi^2\, (\nabla\, \mathbf{cs})(A(t)), \tag{6.9}$$

see (5.7). Therefore, ASD connections on $\mathbb{R} \times M$ can be identified with the trajectories of the gradient flow of $\mathbf{cs}$. Moreover,

$$\begin{aligned}\frac{d}{dt}\, \mathbf{cs}(A(t)) &= (\nabla\, \mathbf{cs})(A(t)) \cdot \frac{\partial A}{\partial t} \\ &= \frac{1}{4\pi^2} \int_M \operatorname{tr}\left(\frac{\partial A}{\partial t} \wedge F_{A(t)} \right) = \frac{1}{4\pi^2} \left\| \frac{\partial A}{\partial t} \right\|^2_{L^2(M)}.\end{aligned}$$

Therefore, $\mathbf{cs}$ is non-decreasing along gradient flow trajectories, and these trajectories have finite length as long as the Yang–Mills action of A over $\mathbb{R} \times M$ is finite.

In the case of a closed 4–manifold X, see Section 6.1.1, we perturbed the Riemannian metric on X to put the ASD moduli spaces $\mathcal{M}(E)$ into a general position. In the case of a manifold X with cylindrical ends, we are only allowed to perturb the metric so that it remains a product metric at the ends. In the case of $X = \mathbb{R} \times M$, for instance, this means that we are only allowed to perturb the metric on the 3-manifold M. We may end up with not having enough perturbations to put $\mathcal{M}(E)$ into a general position. A way around this problem is to perturb the ASD equation rather than the metric. Perturbations we have in mind are described below, compare with [68] and [93]. They are analogous to the perturbations described in Section 5.3.4.

Let $\gamma : S^1 \to X$ be an embedded loop and "thicken" it to an embedding $\gamma : S^1 \times B^3 \to X$ where B^3 is an open oriented 3–manifold. Thus, for each $z \in B^3$, we have a parallel copy of γ which we call γ_z. Given $A \in \mathcal{A}(E)$, denote by $\operatorname{hol}_A(\gamma, x)$ the holonomy of A starting at a point $x \in \gamma(S^1 \times B^3)$ around the loop γ_z passing through x. Let $\Pi : SU(2) \to \mathfrak{su}(2)$ be given by

$$\Pi(u) = u - \frac{1}{2}\,\mathrm{tr}(u)\cdot \mathrm{Id}\,.$$

It is equivariant with respect to the adjoint action on both $SU(2)$ and $\mathfrak{su}(2)$, hence $\Pi\,\mathrm{hol}_A(\gamma, x)$ defines a section of $\mathrm{ad}\,E$ over $\gamma(S^1 \times B^3)$. If ν is a differential 2–form supported in $\gamma(S^1 \times B^3)$, define a section

$$\tau(\nu, \gamma, A) \in \Omega^2(X, \mathrm{ad}\,E)$$

by tensoring $\Pi\,\mathrm{hol}_A(\gamma, x)$ with ν over $\gamma(S^1 \times B^3)$ and letting it be zero otherwise. For fixed ν and γ, this gives an equivariant map from $\mathcal{A}(E)$ to $\Omega^2(X, \mathrm{ad}\,E)$.

Given a collection $\gamma_i : S^1 \times B^3_i \to X$, $i = 1, \dots, m$, of disjoint embeddings, repeat the above construction for each γ_i. Let

$$\pi(A) = \sum_{i=1}^{m} \varepsilon_i \cdot \tau(\nu_i, \gamma_i, A)$$

be a linear combination of the forms $\tau(\nu_i, \gamma_i, A)$ and let $\pi_+(A)$ be its orthogonal projection onto the space $\Omega^2_+(X, \mathrm{ad}\,E)$. The perturbed ASD moduli space is the space of solution of the perturbed ASD equation

$$F_A + *F_A = \pi_+(A).$$

When $\pi_+ = 0$ this is the usual moduli space of ASD connections. Note also that these perturbations match the perturbations at the ends of X described in Section 5.3.4. For generic choices of a Riemannian metric on X and a perturbation, the perturbed ASD moduli space is a smooth manifold of dimension $\mathrm{ind}\,\mathcal{D}_A$, see [70] and [93]. We say in this case that the ASD moduli space is in *general position.*

As in the closed case, the moduli spaces $\mathcal{M}^*(E)$ over a manifold X with cylindrical end $\mathbb{R}_+ \times \partial X$ are orientable. An orientation for $\mathcal{M}^*(E)$ is determined by choosing an orientation of $H^1(X, \mathbb{R}) \oplus H^2_+(X, \mathbb{R}) \oplus H^0(\partial X, \mathbb{R}) \oplus H^1(\partial X, \mathbb{R})$, called a *homology orientation*, see [229]. Results of this section are also valid for $SO(3)$ ASD moduli spaces.

6.2 Definition of the Floer Homology

The Floer homology is roughly a Morse theory of the Chern–Simons function $\mathbf{cs} : \mathcal{B}^*(\Sigma) \to \mathbb{R}/\mathbb{Z}$. In this section, after reviewing the finite dimensional Morse theory, we will show how to extend it to infinite dimensions, and define Floer homology groups. Next we will show that the Floer homology is functorial with respect to cobordisms of integral homology spheres, and use this to prove that the Floer homology is well-defined as an invariant of orientation preserving diffeomorphism.

6.2.1 Review of the Morse Theory

It has been known for a long time [216], see also [34] and [281], that the homology of a smooth manifold M (satisfying certain technical conditions) can be defined by using a Morse function. A smooth function $h : M \to \mathbb{R}$ is called a *Morse function* if all its critical points are non-degenerate. A critical point is called *non-degenerate* if the Hessian of h is a non-degenerate symmetric bilinear form, see (5.6). Being Morse is a generic condition. Given a Morse function, one defines *Morse index* $\mu(p)$ of a critical point p as the number of negative eigenvalues of the Hessian at p, compare with Section 5.3.1. The critical points p of index $\mu(p) = n$ generate a free abelian group $C_n(M)$. The groups $C_n(M)$ are organized into a chain complex by defining a boundary operator $\partial : C_n(M) \to C_{n-1}(M)$ with the property that $\partial^2 = 0$. The operator ∂ can be defined by counting the gradient flow lines of h connecting critical points. To assure that this count makes sense, the function h may have to be perturbed to make it Morse–Smale. The homology of the chain complex $(C_*(M), \partial)$ is the usual homology $H_*(M, \mathbb{Z})$ of the manifold M, and as such, is independent of the choice of Morse function.

This theory also works for functions which are not Morse but satisfy the following weaker condition: the critical point set consists of smooth submanifolds $\mathcal{R}$ such that the Hessian at every critical point $p \in \mathcal{R}$ is non-degenerate in the direction normal to the critical manifold through that point. Such functions are called *Morse–Bott functions.* The number of negative eigenvalues of the Hessian at $p \in \mathcal{R}$ only depends on $\mathcal{R}$ and not on the choice of p. It is called the *Morse index* of $\mathcal{R}$. Let $\mathcal{R}_i$ be the union of the critical submanifolds of Morse index i. There exists a spectral sequence with $E^1_{ij} = H_j(R_i, \mathbb{Z})$ converging to $H_*(M, \mathbb{Z})$, called the *Morse–Bott spectral sequence.* A detailed account of the above can be found in [20] and [108].

6.2.2 Floer Homology of Integral Homology Spheres

Let Σ be an integral homology 3–sphere and E a principal $SU(2)$–bundle over Σ. The bundle E is classified by its second Chern class $c_2(E) \in H^4(\Sigma, \mathbb{Z}) = 0$ which necessarily vanishes. Hence, E is a trivial bundle, and we fix a trivialization $E = \Sigma \times SU(2)$. Choose a Riemannian metric on Σ and let $\mathcal{B}^*(E)$ be the Banach manifold of gauge equivalence classes of irreducible connections in E, see Section 5.1, and $\mathbf{cs} : \mathcal{B}^*(E) \to \mathbb{R}/\mathbb{Z}$ the Chern–Simons function (5.1). The set of critical points of $\mathbf{cs}$ can be identified with the space $\mathcal{R}^*(\Sigma)$ of either irreducible flat connections on E or irreducible $SU(2)$–representations of $\pi_1\Sigma$, see Section 5.3.2. Note that all $SU(2)$–representations of $\pi_1\Sigma$ but the trivial one are irreducible, see Lemma 3.6.

A critical point α of $\mathbf{cs}$ is non-degenerate if and only if the group cohomology $H^1(\pi_1\Sigma, \operatorname{ad}\alpha)$ vanishes, see Lemma 5.7. For some homology spheres, all critical points of $\mathbf{cs}$ are non-degenerate, for others they are not. In the latter case, we perturb $\mathbf{cs}$ as described in Section 5.3.4 so that all critical points of the perturbed function are non-degenerate. For the sake of simplicity, we will *assume in the rest of this section that all critical points of* **cs** *are non-degenerate.* All the constructions remain valid after an admissible perturbation.

The Hessian of **cs** at a critical point α can be identified with either operator (5.8) or operator (5.9). In either case, the spectrum of Hessian is unbounded in both directions, and we cannot define the index of a critical point by simply counting negative (or positive) eigenvalues, as in the finite dimensional Morse theory. We can, however, compare the Hessians at two critical points by seeing how many eigenvalues change sign along a path connecting them. This is properly done by using the concept of spectral flow, see Section 5.3.2. This gives a relative index (in finite dimensional Morse theory, it corresponds to the difference between the Morse indices of the two critical points). To pick an overall normalization, we compare all critical points with the trivial connection θ. Define the index of a critical point α to be

$$\mu(\alpha) = \operatorname{ind} \mathcal{D}_A(\alpha, \theta) \bmod 8, \tag{6.10}$$

compare with (5.10) and Corollary 6.4. The index μ is called the *Floer index*; it is only well–defined modulo 8.

Remark 6.6. It should be mentioned that this is not the first instance that the Morse theory is extended to infinitely many dimensions. For instance, Milnor's book [216] contains an account of Morse theory for (infinite dimensional) loop spaces. Main difference is however that the spectra of Hessians in that situation are bounded from below, so that indices of critical points can be defined as usual, by counting negative eigenvalues.

As long as all critical points α of **cs** are non-degenerate, their number is finite. For every integer n between 0 and 7 we define $IC_n(\Sigma)$ to be a free abelian group generated by the α such that $\mu(\alpha) = n \bmod 8$. These groups will form a chain complex after we define a boundary operator.

Let E be the pull-back to the cylinder $\mathbb{R}\times\Sigma$ of the bundle $E = \Sigma\times SU(2)$, and view $\alpha, \beta \in \mathcal{R}^*(\Sigma)$ as irreducible flat connections. The moduli space of ASD connections in E with finite Yang–Mills action limiting to α and β at minus and plus infinity, respectively, will be denoted by $\mathcal{M}(\alpha,\beta)$. Note that the irreducibility of α (or β) implies that all ASD connections in $\mathcal{M}(\alpha,\beta)$ are irreducible so that $\mathcal{M}(\alpha,\beta) = \mathcal{M}^*(\alpha,\beta)$, see (6.5). Thus

$$\mathcal{M}(\alpha,\beta) = \left\{ A(t) \in \mathcal{A}(E) \;\middle|\; \frac{\partial A}{\partial t} + *F_{A(t)} = 0, \quad \|F_A\|^2_{L^2} < \infty, \right.$$
$$\left. \lim_{t\to-\infty} A(t) = \alpha, \; \lim_{t\to+\infty} A(t) = \beta \right\} \Big/ \sim,$$

where $\sim$ stands for the equivalence relation generated by gauge transformations constant in t. Even if $\mathcal{R}^*(\Sigma)$ is non–degenerate, the ASD equation may have to be perturbed to put $\mathcal{M}(\alpha,\beta)$ in general position, see Section 6.1.4.

Proposition 6.7. *For every α, $\beta \in \mathcal{R}^*(\Sigma)$, the moduli space $\mathcal{M}(\alpha,\beta)$ in general position is a smooth canonically oriented manifold of dimension*

$$\dim \mathcal{M}(\alpha,\beta) = \mu(\alpha) - \mu(\beta) \bmod 8.$$

Proof. That $\mathcal{M}(\alpha,\beta)$ is a smooth finite dimensional manifold follows from Section 6.1.4. Lemma 5.7 implies that $\dim\ker K_\theta = 3$ and $\dim\ker K_\alpha = 0$, therefore, $\operatorname{ind}\mathcal{D}_A(\theta,\alpha) + \operatorname{ind}\mathcal{D}_A(\alpha,\theta) = -3$, see Remark 6.16. This in turn implies that $\mu(\alpha)-\mu(\beta) = \operatorname{ind}\mathcal{D}_A(\alpha,\theta)+3+\operatorname{ind}\mathcal{D}_A(\theta,\beta) = \operatorname{ind}\mathcal{D}_A(\alpha,\beta) = \dim\mathcal{M}(\alpha,\beta) \bmod 8$. The manifold $\mathcal{M}(\alpha,\beta)$ is oriented by a choice of homology orientation, see Section 6.1.4. Since $H^1(\mathbb{R}\times\Sigma,\mathbb{R}) = H^2_+(\mathbb{R}\times\Sigma,\mathbb{R}) = H^1(\Sigma,\mathbb{R}) = 0$, this choice is canonical.

The real line acts on $\mathcal{M}(\alpha,\beta)$ by the rule $A(t)\mapsto A(t+T)$, $T\in\mathbb{R}$. Let $\hat{\mathcal{M}}(\alpha,\beta)$ be the oriented quotient of $\mathcal{M}(\alpha,\beta)$ by this translational symmetry then $\dim\hat{\mathcal{M}}(\alpha,\beta) = \mu(\alpha)-\mu(\beta)-1$. In particular, if $\mu(\beta)-\mu(\alpha) = 1$ then $\dim\hat{\mathcal{M}}(\alpha,\beta) = 0 \bmod 8$. According to [93], the zero dimensional part of $\hat{\mathcal{M}}(\alpha,\beta)$ is compact, hence consists of finitely many points with signs. Let

$$\partial(\alpha) = \sum_\beta \#\hat{\mathcal{M}}(\alpha,\beta)\cdot\beta,$$

where $\#\hat{\mathcal{M}}(\alpha,\beta)$ is the (signed) count of isolated points in $\hat{\mathcal{M}}(\alpha,\beta)$ if $\dim\hat{\mathcal{M}}(\alpha,\beta) = 0 \bmod 8$ and zero otherwise. If $\mu(\alpha) = n$ then $\dim\hat{\mathcal{M}}(\alpha,\beta) = 0$ if and only if $\mu(\beta) = n-1$ so that $\partial(\alpha)\in IC_{n-1}(\Sigma)$. The map ∂ extends by linearity to a homomorphism $\partial : IC_n(\Sigma)\to IC_{n-1}(\Sigma)$.

Proposition 6.8. *The map ∂ is a boundary operator, that is, $\partial^2 = 0$.*

Proof. Applying the operator ∂ twice to a flat connection α we obtain

$$\partial^2(\alpha) = \sum_\gamma\sum_\beta \#\hat{\mathcal{M}}(\alpha,\beta)\cdot\#\hat{\mathcal{M}}(\beta,\gamma)\cdot\gamma.$$

Therefore, to prove that $\partial^2(\alpha) = 0$, we need to show that

$$\sum_\beta \#\hat{\mathcal{M}}(\alpha,\beta)\cdot\#\hat{\mathcal{M}}(\beta,\gamma) = 0.$$

If $\mu(\alpha) = n$ then the latter sum counts pairs of gradient lines from α to β and from β to γ, where $\mu(\beta) = n-1$ and $\mu(\gamma) = n-2$, see Figure 6.1. According to Floer [93], for every such pair, there exist nearby gradient lines from α to γ, and they are parametrized by an open interval $(0,\varepsilon)$. This means that the 1-dimensional moduli space $\hat{\mathcal{M}}(\alpha,\gamma)$ has a boundary point corresponding to the factorization $\alpha\to\beta\to\gamma$. Moreover, Floer proved that all boundary points of $\hat{\mathcal{M}}(\alpha,\gamma)$ correspond to such factorizations. Thus every factorization $\alpha\to\beta\to\gamma$ arises as the end of an open interval in $\hat{\mathcal{M}}(\alpha,\gamma)$, and the other

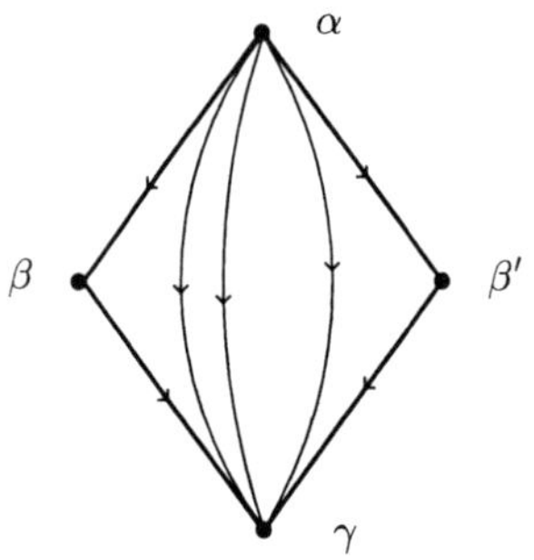

Fig. 6.1.

end of this interval corresponds to another factorization, $\alpha \to \beta' \to \gamma$. This proves that $\partial^2 = 0$ at least modulo 2. In fact, the orientations work out in such a way that this is also true over the integers.

An important point in the proof is that the factorization $\alpha \to \beta' \to \gamma$ cannot have β' equal to the trivial connection θ. If the trivial connection were to occur in the middle of the limiting ASD connection, the moduli space would have to be of dimension $\dim \mathcal{M}(\alpha,\gamma) = \dim \mathcal{M}(\alpha,\theta) + 3 + \dim \mathcal{M}(\theta,\gamma)$ which is at least 5. This is not the case for α and γ as above.

Remark 6.9. The part of the proof dealing with the description of the ends of the moduli space $\hat{\mathcal{M}}(\alpha,\gamma)$ is of much more general nature. It reflects the geometric principle which roughly states the following: if the metric on a 4–manifold $W = U \cup_\Sigma V$ is deformed to have a very long tube $[-T,T] \times \Sigma$ in the middle, then the ASD connections over W are obtained by gluing ASD connections over U and V with the same flat limit over Σ. This principle is one of the cornerstones of the Floer theory.

The groups $IC_n(\Sigma)$ together with the boundary operator $\partial : IC_n(\Sigma) \to IC_{n-1}(\Sigma)$ form a chain complex called the *Floer chain complex* of Σ. The index n in this complex is defined modulo 8 so the correct way to view $IC_*(\Sigma)$ is as a circular chain complex. The homology of the Floer chain complex,

$$I_n(\Sigma) = H_n(IC_*(\Sigma), \partial),$$

is called the *Floer homology* of Σ. We abuse notations here since it is not clear from the definition that the Floer homology only depends on the diffeomorphism type of Σ. *A priori* it may depend on the choice of Riemannian metric and perturbations. The proof of the fact that $I_*(\Sigma)$ is in fact well-defined is related to the functorial properties of Floer homology, which are discussed in the next section.

Remark 6.10. The boundary operator in the Floer chain complex is very difficult to compute. In most known computations of the Floer homology groups it vanishes due to a dimension count. For example, in the case of Seifert fibered homology spheres, generators in the Floer chain complex only occur in odd degrees, see Section 6.4.1, therefore, the boundary operator has to vanish. In fact, it is not easy to even find examples of homology spheres with non-vanishing boundary operator. Weiping Li [192] showed that $\Sigma = \Sigma(2,3,5) \# \Sigma(2,3,5)$ provides such an example. An irreducible example can be found in [276].

6.2.3 Functoriality with Respect to Cobordisms

Let Σ_1 and Σ_2 be integral homology spheres and suppose that $\mathcal{R}^*(\Sigma_1)$ and $\mathcal{R}^*(\Sigma_2)$ are non-degenerate: all the constructions below remain valid after an admissible perturbation. Given a smooth compact oriented cobordism W with boundary $\partial W = -\Sigma_1 \cup \Sigma_2$, make it into an open manifold by attaching ends $\mathbb{R}_- \times \Sigma_1$ and $\mathbb{R}_+ \times \Sigma_2$, and call the resulting manifold W again. Choose a Riemannian metric on W which is a product metric at the ends, and fix a homology orientation. Let $\mathcal{M}_W(\alpha,\beta)$ be the moduli space of finite Yang–Mills action ASD connections in an $SU(2)$–bundle E over W limiting to α and β at minus and plus infinity, respectively. The bundle E will be fixed for the rest of this section and will be suppressed in the notations.

Proposition 6.11. *If both α and β are irreducible and non-degenerate then $\mathcal{M}_W(\alpha,\beta)$ is a smooth oriented manifold (perhaps, after a generic perturbation) of dimension*

$$\dim \mathcal{M}_W(\alpha,\beta) = \mu(\alpha) - \mu(\beta) + 3\,(b_1 - b_2^+)(W) \bmod 8.$$

Proof. After the manifold $\mathcal{M}_W(\alpha,\beta)$ is put in general position, see Section 6.1.4, it is smooth. Since α and β are both irreducible and non-degenerate we have the following additivity formula, see Remark 6.16,

$$\dim \mathcal{M}_W(\theta,\theta) = \operatorname{ind} \mathcal{D}_A(\theta,\alpha) + \dim \mathcal{M}(\alpha,\beta) + \operatorname{ind} \mathcal{D}_A(\beta,\theta),$$

where $\dim \mathcal{M}_W(\theta,\theta) = -3(1 - b_1 + b_2^+)(W) \bmod 8$ and $\operatorname{ind} \mathcal{D}_A(\theta,\alpha) = -3 - \operatorname{ind} \mathcal{D}_A(\alpha,\theta) \bmod 8$, compare with Proposition 6.7.

In addition to all $\alpha \in \mathcal{R}^*(\Sigma_1)$ and $\beta \in \mathcal{R}^*(\Sigma_2)$ being non-degenerate, assume that the moduli spaces $\mathcal{M}_W(\alpha,\beta)$ are in general position so that each $\mathcal{M}_W(\alpha,\beta)$ is a smooth oriented manifold. Define

$$W_*(\alpha) = \sum_\beta \#\mathcal{M}_W(\alpha,\beta) \cdot \beta \tag{6.11}$$

where $\#\mathcal{M}(\alpha,\beta)$ is the number of isolated points in $\mathcal{M}_W(\alpha,\beta)$ counted with signs if $\dim \mathcal{M}_W(\alpha,\beta) = 0 \bmod 8$ (in which case the zero-dimensional part of $\mathcal{M}_W(\alpha,\beta)$ is compact), and $\#\mathcal{M}(\alpha,\beta) = 0$ otherwise. According to Proposition 6.11, dimension of $\mathcal{M}_W(\alpha,\beta)$ is equal to zero modulo 8 if and only if $\mu(\beta) = \mu(\alpha) + 3\,(b_1 - b_2^+)(W)$. Thus the map W_* extends by linearity to a homomorphism $W_* : IC_*(\Sigma_1) \to IC_*(\Sigma_2)$ of degree $3\,(b_1 - b_2^+)(W)$.

Proposition 6.12. *The homomorphism W_* is a chain homomorphism. The induced map $W_* : I_*(\Sigma_1) \to I_*(\Sigma_2)$ only depends on the cobordism W.*

Proof. The proof is based on roughly the same idea as that of Proposition 6.8. We need to prove that $W_* \circ \partial = \partial \circ W_*$ where the ∂ are respective boundary operators in the Floer chain complexes of Σ_1 and Σ_2. This amounts to showing that, for any $\alpha \in \mathcal{R}^*(\Sigma_1)$ and $\gamma \in \mathcal{R}^*(\Sigma_2)$ such that $\dim \mathcal{M}_W(\alpha, \gamma) = 1$, the following holds:

$$\sum_{\beta_1} \#\hat{\mathcal{M}}(\alpha, \beta_1) \cdot \#\mathcal{M}_W(\beta_1, \gamma) = \sum_{\beta_2} \#\mathcal{M}_W(\alpha, \beta_2) \cdot \#\hat{\mathcal{M}}(\beta_2, \gamma).$$

The two summations go over factorizations of ASD connections of the form $\alpha \to \beta_1 \to \gamma$ and $\alpha \to \beta_2 \to \gamma$ where $\beta_1 \in \mathcal{R}^*(\Sigma_1)$ and $\beta_2 \in \mathcal{R}^*(\Sigma_2)$. Floer shows in [93] that these factorizations correspond to the ends of the 1-dimensional moduli space $\mathcal{M}_W(\alpha, \gamma)$. Thus every factorization arises as the end of an open interval, whose other end corresponds to another factorization. This proves the result modulo 2, and in fact the orientations work out in such a way that this is also true over the integers.

Given two sets of metrics and perturbations on W, include them into a continuous family of parameters. A parametrized version of the above construction results in a chain homotopy proving that $W_* : I_*(\Sigma_1) \to I_*(\Sigma_2)$ only depends on W and not on the choices made in its definition.

Thus every pair (W, E), where W is a smooth compact oriented cobordism between oriented integral homology spheres Σ_1 and Σ_2 and E is an $SU(2)$–bundle over W, defines a homomorphism

$$W_* : I_*(\Sigma_1) \to I_*(\Sigma_2) \tag{6.12}$$

in the Floer homology of degree $3\,(b_1 - b_2^+)(W)$. It is functorial in the following sense.

Proposition 6.13. *For a composite cobordism $W = U \cup_\Sigma V$, we have $W_* = V_*\, U_*$. The map W_* is the identity for the product cobordism.*

Proof. Floer shows in [93] that, for 0–dimensional ASD moduli spaces on U, V, and W, the following holds:

$$\#\mathcal{M}_W(\alpha, \gamma) = \sum_\beta \mathcal{M}_U(\alpha, \beta) \cdot \#\mathcal{M}_V(\beta, \gamma).$$

This implies that the identity $W_* = V_*\, U_*$ already holds on the chain level. Given a cylinder $W = \mathbb{R} \times \Sigma$, choose a product metric and a translationally invariant perturbation, if necessary. If $\dim \mathcal{M}(\alpha, \beta) = 0$ then, due to the translational invariance, $\alpha = \beta$ and the only ASD connection on $\mathbb{R} \times \Sigma$ connecting α to β is the constant flat connection. After checking orientations, we conclude that $(\mathbb{R} \times \Sigma)_*$ is identity at the level of chain complexes, hence, $(\mathbb{R} \times \Sigma)_* = \mathrm{id} : I_*(\Sigma) \to I_*(\Sigma)$.

Theorem 6.14. *The Floer homology $I_*(\Sigma)$ is well-defined, that is, $I_*(\Sigma)$ only depends on Σ and not on the choice of metric and perturbations.*

Proof. The Floer homology groups we defined so far depend on a Riemannian metric g on Σ and a generic perturbation h so that in fact $I_*(\Sigma) = I_*(\Sigma, g, h)$. Given two choices of Riemannian metrics and perturbations, say (g_0, h_0) and (g_1, h_1), connect them by a path (g_t, h_t), $0 \le t \le 1$. This path defines a Riemannian metric and a perturbation over the cylinder $W = \mathbb{R} \times \Sigma$, constant away from $[0,1] \times \Sigma$, and hence a homomorphism $W_* : I_*(\Sigma, g_0, h_0) \to I_*(\Sigma, g_1, h_1)$ of degree zero. Reversing the $\mathbb{R}$–direction in W, we obtain a homomorphism $(-W)_* : I_*(\Sigma, g_1, h_1) \to I_*(\Sigma, g_0, h_0)$. The compositions $W \cup (-W)$ and $(-W) \cup W$ can be deformed to the products $\mathbb{R} \times \Sigma$ with product metrics and translationally invariant perturbations, (g_0, h_0) in the former case and (g_1, h_1) in the latter. These define the identity homomorphisms according to Proposition 6.13. Therefore, W_* and $(-W)_*$ are mutually inverse isomorphisms between $I_*(\Sigma, g_0, h_0)$ and $I_*(\Sigma, g_1, h_1)$.

One can generalize the above construction to cobordisms with any number of boundary components. For example, a cobordism W from a disjoint union $\Sigma_0 \cup \Sigma_1$ to Σ_2 induces a homomorphism from $I_*(\Sigma_0) \otimes I_*(\Sigma_1)$ to $I_*(\Sigma_2)$.

6.3 Spectral Flow Formulas

Calculation of the Floer homology groups $I_*(\Sigma)$ directly from the definition is a difficult problem. Best understood is the spectral flow part – we describe in this section several formulas for the spectral flow. One is a direct consequence of the Atiyah–Patodi–Singer theorem expressing the spectral flow as the index of an ASD operator on the cylinder $\mathbb{R} \times \Sigma$. The other two formulas express the spectral flow in terms of the pillowcase picture, and we recommend the paper [164] for an excellent exposition of these results.

It should be mentioned however that, even if one succeeds in describing the character variety $\mathcal{R}^*(\Sigma)$, perturbing it if necessary, and computing all of the Floer indices, one still needs to understand the boundary operator ∂ to complete the calculation of $I_*(\Sigma)$. Not much is known about ∂, compare with Remark 6.10. Unless ∂ vanishes by an obvious dimension count, it is practically untraceable. Seifert fibered homology spheres, for which ∂ vanishes, remain one of the very few classes of manifolds for which the calculation has been successfully completed, see Section 6.4.

6.3.1 The Atiyah–Patodi–Singer Formula

The spectral flow, or equivalently, the index of the ASD operator $\mathcal{D}_A$, see (6.10), can be computed with the help of the Atiyah–Patodi–Singer index theorem [12]. Let X be a smooth oriented 4–manifold with a single end $\mathbb{R}_+ \times$

M, and A a connection which limits to a flat connection α. Assume that A is in temporal gauge over the end, then $\mathcal{D}_A = \partial/\partial t - K_{A(t)} = \partial/\partial t + B_{A(t)}$, see (5.9) and (6.6). Application of the Atiyah–Patodi–Singer theorem [12] provides us with the formula

$$\operatorname{ind}\mathcal{D}_A(X) = -\int_X \hat{A}(X)\operatorname{ch}(V_+)\operatorname{ch}(\operatorname{ad}E) - \frac{1}{2}(h_\alpha - \eta_\alpha(0))(M) \qquad (6.13)$$

(our orientation convention is different from that of [12], hence we have a different sign for the contribution of the η–invariant, compare with [229]). Here, $h_\alpha = \dim\ker B_\alpha$, and $\eta_\alpha(0)$ is the spectral asymmetry of $B_\alpha = -K_\alpha$ defined as follows. The series

$$\eta_\alpha(s) = \sum_{\lambda\neq 0} \operatorname{sign}\lambda\,|\lambda|^{-s},$$

where the summation extends over all non-zero eigenvalues of B_α counted with multiplicities, converges absolutely for $\operatorname{Re}(s)$ large enough. It defines a function $\eta_\alpha(s)$ which extends meromorphically to the whole complex s–plane and has a finite value at $s = 0$. This value, $\eta_\alpha(0)$, is called the *spectral asymmetry* of B_α. The expression

$$\rho_\alpha(M) = \eta_\alpha(0) - \eta_\theta(0),$$

where θ is the trivial connection, is called the *ρ–invariant*, see Remark 6.17. Over a cylinder $\mathbb{R}\times M$, the formula (6.13) reduces to the following statement.

Theorem 6.15. *Let α and β be flat connections in an $SU(2)$–bundle E over a closed oriented 3–manifold M. Then*

$$\operatorname{ind}\mathcal{D}_A(\alpha,\beta) = 8\,(\mathbf{cs}(\beta) - \mathbf{cs}(\alpha)) - \frac{1}{2}(\rho_\alpha - \rho_\beta)(M) - \frac{1}{2}(h_\alpha + h_\beta) \bmod 8.$$

Proof. A simple calculation with the formula (6.13) shows that

$$\operatorname{ind}\mathcal{D}_A(\alpha,\beta) =$$
$$-\int_{\mathbb{R}\times M} \hat{A}(\mathbb{R}\times M)\operatorname{ch}(V_+)\operatorname{ch}(\operatorname{ad}E) - \frac{1}{2}(\rho_\alpha(M) - \rho_\beta(M)) - \frac{1}{2}(h_\alpha + h_\beta),$$

compare with Proposition 6.1. The integral term appearing in this formula can be evaluated as follows. First, the integrand can be split up,

$$-\hat{A}(\mathbb{R}\times M)\operatorname{ch}(V_+)\operatorname{ch}(\operatorname{ad}E) = -3\,\hat{A}(\mathbb{R}\times M)\operatorname{ch}(V_+) + 2\,c_2(\operatorname{ad}E),$$

with

$$c_2(\operatorname{ad}E) = \frac{1}{8\pi^2}\operatorname{tr}\left(F_{\operatorname{ad}A}\wedge F_{\operatorname{ad}A}\right),$$

where $F_{\operatorname{ad}A}$ stands for the curvature of the $SO(3)$-connection $\operatorname{ad}A$ in the adjoint bundle. Since α and β are $SU(2)$–connections,

$$\int_{\mathbb{R}\times M} c_2(\operatorname{ad} E) = 4\,(\mathbf{cs}(\beta) - \mathbf{cs}(\alpha)),$$

compare with Remark 5.6. The other term appearing in the integrand, $3\hat{A}(\mathbb{R}\times M)\operatorname{ch}(V_+)$, has zero integral modulo 8. This can be seen by considering the index $\operatorname{ind}\mathcal{D}_A(\theta,\theta)$. It equals sf (θ,θ), see Proposition 6.2, which in turn equals $-h_\theta$ mod 8. On the other hand, by the above formula, it equals

$$\operatorname{ind}\mathcal{D}_A(\theta,\theta) = 3\int_{\mathbb{R}\times M} \hat{A}(\mathbb{R}\times M)\operatorname{ch}(V_+) - h_\theta.$$

This completes the proof.

Remark 6.16. In general, the index of the operator $\mathcal{D}_A$ is not additive with respect to the gluing operation of 4–manifolds along common boundary. If $X = U\cup_M V$ then $\operatorname{ind}\mathcal{D}_A(X) = \operatorname{ind}\mathcal{D}_A(U) + h_\alpha + \operatorname{ind}\mathcal{D}_A(V)$, where α is the restriction of A onto M. The latter formula follows easily from (6.13).

Remark 6.17. Theorem 6.15 applied in the special case of $\beta = \theta$ implies that $-16\,\mathbf{cs}(\alpha) = \rho_\alpha(M)$ modulo $\mathbb{Z}$, so that the Chern–Simons function $\mathbf{cs}$ is completely determined by the ρ–invariant. Atiyah, Patodi and Singer [13] also show that, for an oriented smooth 4–manifold X with boundary M and a representation $\alpha:\pi_1(M)\to SU(2)$ which extends to a representation $A:\pi_1(X)\to SU(2)$, one has $\rho_\alpha(M) = 3\operatorname{sign}X - \operatorname{sign}_A X$. Here, $\operatorname{sign}_A X$ is the signature of the intersection form on the twisted cohomology $H^2(X,\operatorname{ad}A)$. In particular, if two representations $\alpha_M:\pi_1(M)\to SU(2)$ and $\alpha_N:\pi_1(N)\to SU(2)$ are flat cobordant via a flat cobordism (W,A), see Example 5.5, then $\rho_{\alpha_N}(N) - \rho_{\alpha_M}(M) = 3\operatorname{sign}W - \operatorname{sign}_A W$. David Auckly [17] gave a topological method to compute spectral flow based on this observation and the formula of Example 5.5.

Corollary 6.18. *Let Σ_1 and Σ_2 be h–cobordant integral homology spheres and suppose that $\mathcal{R}^*(\Sigma_1) = \mathcal{R}^*(\Sigma_2)$ are non-degenerate. Then $\lambda(\Sigma_1) = \lambda(\Sigma_2)$.*

Proof. Every irreducible representation $\alpha_1:\pi_1(\Sigma_1)\to SU(2)$ corresponds to a unique representation $\alpha_2:\pi_1(\Sigma_2)\to SU(2)$ under the canonical identification $\pi_1(\Sigma_1) = \pi_1(\Sigma_2)$. The Floer indices $\mu(\alpha_1)$ and $\mu(\alpha_2)$ can be computed using the formula of Theorem 6.15. Since Σ_1 and Σ_2 are flat homology cobordant, $\mathbf{cs}(\alpha_1) = \mathbf{cs}(\alpha_2)$, see Example 5.5, and $\rho_{\alpha_1}(\Sigma_1) = \rho_{\alpha_2}(\Sigma_2)$, see Remark 6.17. The terms h_{α_1} and h_{α_2} vanish because $\mathcal{R}^*(\Sigma_1)$ and $\mathcal{R}^*(\Sigma_2)$ are non-degenerate. The result now follows from the formula (5.11).

6.3.2 The Splitting Formula

Let $N = D^2\times S^1$ be the solid torus with boundary $\partial N = T^2$, and give T^2 the coordinates $m = S^1\times\{*\}$ and $\ell = \{*\}\times S^1$. Let $\hat{\mathcal{R}}(T^2)$ be $\mathcal{R}(T^2)$ minus the central representations; this is the pillowcase, see Figure 3.1, whose corners are removed. Let $\hat{\mathcal{R}}(N)$ be $\mathcal{R}(N)$ minus the central representations; this is an open arc mapping to the left edge of the pillowcase. Define the cohomology class

$$\mathfrak{s} \in H^1(\hat{\mathcal{R}}(T^2), \hat{\mathcal{R}}(N), \mathbb{Z}/8) \tag{6.14}$$

to be the Poincaré dual to the homology class z shown in Figure 6.2 : it equals two times the top edge plus two times the bottom edge plus four times the right edge, all oriented as indicated. Thus given a path γ in $\hat{\mathcal{R}}(T^2)$ with endpoints in $\hat{\mathcal{R}}(N)$, we have $\mathfrak{s}(\gamma) = \gamma \cdot z \bmod 8$.

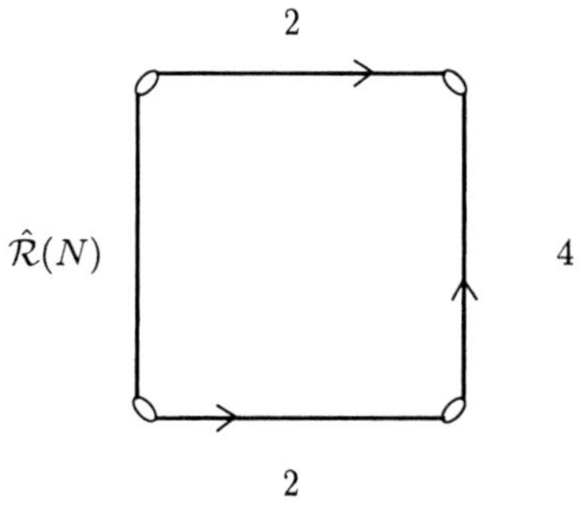

Fig. 6.2.

Another invariant we need to introduce is the Maslov index. Let $\mathcal{F}$ be the vertical line field on $\hat{\mathcal{R}}(T^2)$. Given an oriented immersed curve γ in $\hat{\mathcal{R}}(T^2)$ which is transversal to $\hat{\mathcal{R}}(N)$ at its endpoints, the *Maslov index* of γ is defined to be the winding number of the tangent vector $d\gamma/dt$ with respect to $\mathcal{F}$. It is denoted by $\mathfrak{m}(\gamma)$. Figure 6.3 shows how to compute $\mathfrak{m}(\gamma)$. It picks up $+1$ or -1 every time $d\gamma/dt$ is tangent to $\mathcal{F}$, depending on whether the tangency is on the right or on the left.

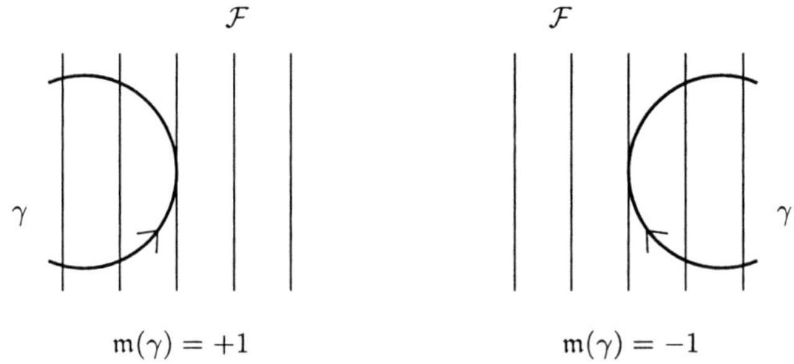

Fig. 6.3.

Theorem 6.19. *Let k be a knot in an integral homology sphere Σ, K its exterior, and $\alpha_0, \alpha_1 : \pi_1\Sigma \to SU(2)$ two irreducible representations. Suppose that there exists a path of representations $\alpha : [0,1] \to \mathcal{R}^*(K)$ connecting α_0 to α_1 with the following properties. First, $\mathcal{R}^*(K)$ is a smooth one–dimensional variety along α, that is, $H^1(\pi_1 K, \operatorname{ad}\alpha(t)) = \mathbb{R}$ for all t. Second, the image of α in the pillowcase lies in $\hat{\mathcal{R}}(T^2)$ and is transverse to $\mathcal{F}$ at the endpoints, that is, $H^1(\pi_1\Sigma, \operatorname{ad}\alpha_i) = 0$ for $i = 0, 1$. Denote by γ the image of α in the pillowcase. Then*

$$\operatorname{sf}(\alpha_0, \alpha_1) = \mathfrak{s}(\gamma) + \mathfrak{m}(\gamma).$$

This theorem is proved by Yoshida [308] with the help of his *splitting formula* [309] for the spectral flow. This formula recovers the spectral flow from decomposition data determined by a hypersurface which partitions the manifold. The splitting formula was later generalized by Nicolaescu [249] and Cappell, Lee and Miller [47] and [48]. It was transformed into an effective computational tool in [63].

Example 6.20. Let k be the singular fiber of degree 7 in the Brieskorn homology sphere $\Sigma(2,3,7)$. The character variety $\mathcal{R}(K)$ contains a single arc of irreducible representations limiting to two disjoint points on the abelian arc. Its image in the pillowcase is shown in Figure 6.4 (it is obtained from Figure 3.2 by change of coordinates). The two irreducible representations $\alpha_0, \alpha_1 : \pi_1\Sigma(2,3,7) \to SU(2)$ are shown on the left edge of the pillowcase. They are connected by a path γ in the pillowcase such that $\mathfrak{s}(\gamma) = 4 \bmod 8$. Since the vertical tangencies are missing, $\mathfrak{m}(\gamma) = 0$ and we conclude that $\mu(\alpha_0) - \mu(\alpha_1) = 4 \bmod 8$, compare with Example 6.23.

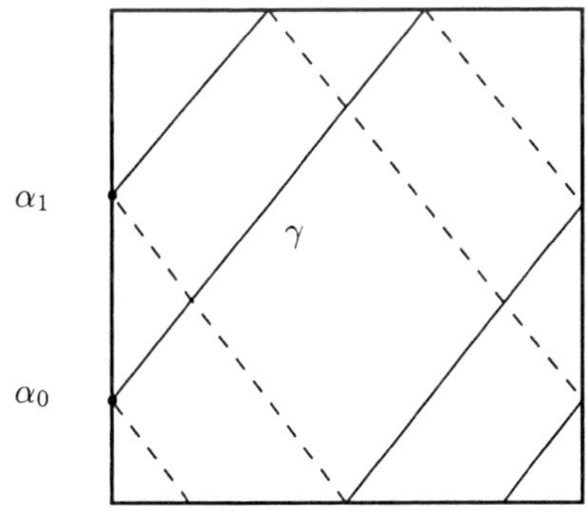

Fig. 6.4.

6.3.3 The Kirk–Klassen Formula

Let $\alpha(t)$ be a generic path of connections on a trivial $SU(2)$–bundle over an integral homology sphere Σ connecting two non-degenerate irreducible flat connections. The spectral flow of the family of operators $K_{\alpha(t)}$ counts the points along this path where $\ker K_{\alpha(t)} \neq 0$. Each of such points is counted with the weight equal the number of eigenvalues of $K_{\alpha(t)}$ crossing from negative to positive minus the number of those crossing from positive to negative, see Section 5.3.2. The points $\alpha(t)$ where $\ker K_{\alpha(t)} \neq 0$ can be described in purely topological terms due to the identification $\ker K_{\alpha(t)} = H^0(\pi_1\Sigma, \operatorname{ad}\alpha(t)) \oplus H^1(\pi_1\Sigma, \operatorname{ad}\alpha(t))$, see Lemma 5.7. The Kirk–Klassen formula determines in topological terms the weight of each of these points.

More precisely, suppose that the path $\alpha(t)$ is analytic. This can always be achieved, at least locally, by applying an appropriate gauge transformation. Then $K_{\alpha(t)}$ is an analytic path of self-adjoint operators, and its eigenvalues and eigenvectors vary analytically, see [167] and references therein. If $\lambda(t)$ is an eigenvalue of $K_{\alpha(t)}$ such that $\lambda(0) = 0$, one can tell if $\lambda(t)$ is crossing from positive to negative or from negative to positive by looking at the sign and the order of the first non-vanishing derivative of $\lambda(t)$ at $t = 0$.

Let $n = \dim\ker K_{\alpha(0)}$ and let $\lambda_i(t)$, $i = 1, \ldots, n$, are the paths of eigenvalues which are equal to zero at $t = 0$. Define a family of bilinear forms F_k which are diagonal forms with the kth derivatives of $\lambda_i(t)$ at $t = 0$ along the diagonal. Clearly, the contribution of $\alpha(0)$ to the spectral flow can be described in terms of the signatures of F_k. These signatures can be computed cohomologically as follows.

Let $\xi \in H^1(\Sigma, \operatorname{ad}\alpha(0))$ be the derivative of the path $\alpha(t)$ at $t = 0$. It is not difficult to show that the cup-product $[\xi, \xi] \in H^2(\Sigma, \operatorname{ad}\alpha(0))$ equals zero. Hence we have a derived complex whose chain groups are $H^*(\Sigma, \operatorname{ad}\alpha(0))$ and whose boundary operator

$$[\xi, -] : H^*(\Sigma, \operatorname{ad}\alpha(0)) \to H^*(\Sigma, \operatorname{ad}\alpha(0))$$

is given by the cup-product with ξ coupled with the Lie bracket on the coefficients. The following theorem is proved in [167].

Theorem 6.21. *Let $B_1 : H^1(\Sigma, \operatorname{ad}\alpha(0)) \times H^1(\Sigma, \operatorname{ad}\alpha(0)) \to \mathbb{R}$ be the symmetric bilinear form given by the formula $B_1(x, y) = -[\xi, x] \cdot y$ where $[\xi, -]$ is described above and*

$$\cdot : H^2(\Sigma, \operatorname{ad}\alpha(0)) \times H^1(\Sigma, \operatorname{ad}\alpha(0)) \to \mathbb{R}$$

is the cup product coupled with the bilinear form $(a, b) \mapsto -\operatorname{tr}(ab)$ on $\mathfrak{su}(2)$. Then the signature of B_1 equals the signature of F_1. Moreover, if the cohomology of the derived complex is zero then the signature of B_1 equals the contribution of $\alpha(0)$ to the spectral flow of the family $K_{\alpha(t)}$.

This takes care of the form F_1. Description of the forms F_k with $k \geq 2$ requires Massey products. The precise description (for all closed oriented 3–manifolds) together with some applications can be found in [167] and [169].

6.4 Floer Homology of Seifert Fibered and Graph Homology Spheres

Seifert fibered homology spheres are among very few known examples of homology spheres whose Floer homology can be computed directly from the definition. The main simplification for a Seifert fibered homology sphere Σ comes from the fact that all generators in $IC_*(\Sigma)$ have the same parity, hence the boundary operator vanishes and $I_*(\Sigma) = IC_*(\Sigma)$. The groups $I_*(\Sigma)$ are free abelian, and their ranks can be computed by using either an algorithm [83] or a closed form formula [272]. There has not been much progress in computing the Floer homology of graph homology spheres; some known results are outlined in the end of this section.

6.4.1 The Algorithm

Let $\Sigma(a_1,\dots,a_n)$ be a Seifert fibered homology sphere whose Seifert invariants b, $(a_1,b_1),\dots,(a_n,b_n)$ are chosen so that (1.5) holds and b is even, see Section 1.1.4. Moreover, if one of the a_i is even, assume it is a_1, and arrange Seifert invariants so that the b_i with $i \neq 1$ are even. Associated with every irreducible representation $\alpha : \pi_1\Sigma(a_1,\dots,a_n) \to SU(2)$ is an n–tuple $(\ell_1,\dots,\ell_n)$ of rotation numbers, $0 \le \ell_i \le a_i$, see Section 3.5.1 and Section 3.5.2. Denote

$$e = \sum_{i=1}^{n} \frac{a\ell_i}{a_i},$$

where $a = a_1 \cdots a_n$ is the product of the a_k. The following result can be found in Fintushel and Stern [83] and also in [238].

Proposition 6.22. *Let $\alpha : \pi_1\Sigma(a_1,\dots,a_n) \to SU(2)$ be an irreducible representation with rotation numbers $(\ell_1,\dots,\ell_n)$ such that $\pi\ell_i/a_i$ are different from 0 and π for $i = 1,\dots,m$ and are equal to either 0 or π for $i = m+1,\dots,n$. Then $\operatorname{ind}\mathcal{D}_A(\alpha,\theta)$ is always odd and, modulo 8, equals*

$$\frac{2e^2}{a} + 3 - m + \sum_{i=1}^{m} \frac{2}{a_i} \sum_{k=1}^{a_i-1} \cot\left(\frac{\pi a k}{a_i^2}\right) \cot\left(\frac{\pi k}{a_i}\right) \sin^2\left(\frac{\pi e k}{a_i}\right).$$

Proof. According to Theorem 6.15,

$$\operatorname{ind}\mathcal{D}_A(\alpha,\theta) = -8\,\mathbf{cs}(\alpha) - \frac{1}{2}\,\rho_\alpha(\Sigma(a_1,\dots,a_n)) - \frac{1}{2}\,h_\alpha - \frac{3}{2},$$

where $h_\alpha = 2m - 6$ is the dimension of the connected component of α in the representation variety, see Section 3.5.2, and $\mathbf{cs}(\alpha) = -e^2/4a$, see Example 5.2. The ρ–invariant $\rho_\alpha(\Sigma(a_1,\dots,a_n))$ can be computed as follows.

Let W be the mapping cylinder of the Seifert fibration $\Sigma(a_1,\dots,a_n) \to S^2$. It is an oriented 4–dimensional orbifold with boundary $\Sigma(a_1,\dots,a_n)$ and

n singular points whose neighborhoods are cones over the lens spaces $L(a_i, b_i)$. Let W_0 denote W with open cones around the singular points removed, then $\pi_1(W_0)$ is the quotient of $\pi_1\Sigma(a_1,\dots,a_n)$ by the center. In particular, the representation $\operatorname{ad}\alpha : \pi_1\Sigma(a_1,\dots,a_n) \to SO(3)$ extends to an $SO(3)$ representation of $\pi_1 W_0$. Therefore, W_0 is a flat cobordism and

$$\rho_\alpha(\Sigma(a_1,\dots,a_n)) - \sum_{i=1}^n \rho_{\alpha_i}(L(a_i,b_i)) = 3\operatorname{sign} W_0 - \operatorname{sign}_\alpha W_0,$$

see Remark 6.17. A calculation in [83] shows that $\operatorname{sign} W_0 = -1$ and $\operatorname{sign}_\alpha W_0 = 0$. The ρ–invariants of lens spaces can be computed with the help of Atiyah–Patodi–Singer G–signature theorem and the formula from [13] which expresses the ρ-invariants as the "Fourier transforms" of the G–signature defects. The resulting formula is

$$\rho_{\alpha_i}(L(a_i,b_i)) = -\frac{4}{a_i}\sum_{k=1}^{a_i-1} \cot\left(\frac{\pi a k}{a_i^2}\right)\cot\left(\frac{\pi k}{a_i}\right)\sin^2\left(\frac{\pi e k}{a_i}\right).$$

In particular, $\rho_{\alpha_i}(L(a_i,b_i))$ vanishes when $\ell_i = 0$ or a_i.

The fact that $\operatorname{ind}\mathcal{D}_A(\alpha,\theta)$ is always odd is proved in [87], Corollary 6.3. Another explanation of this fact, together with an alternative proof of the formula for $\operatorname{ind}\mathcal{D}_A(\alpha,\theta)$, can be found in [238]. It is based on the observation that the cylinder $\mathbb{R}\times\Sigma(a_1,\dots,a_n)$ is conformally equivalent to the complex surface $V_B(a_1,\dots,a_n)$ with the singular point removed, see (1.2). The ASD operator $\mathcal{D}_A$ on a complex surface is isomorphic to the complex operator $\bar\partial_A^*\oplus\bar\partial_A$, whose index is even, see [71]. A correction term which eventually makes the index odd is due to the discrepancy in the $L^p_{k,\delta}$–norms on the cylinder and those on the cone. Making this argument precise requires estimating smallest non-zero eigenvalues of the operators K_α and K_θ on $\Sigma(a_1,\dots,a_n)$.

The case of $n = 3$ is that of Brieskorn homology spheres $\Sigma(p,q,r)$. The representation space $\mathcal{R}^*(\Sigma(p,q,r))$ is non-degenerate, see [83], hence the representations $\alpha\in\mathcal{R}^*(\Sigma(p,q,r))$ generate the Floer chain complex $IC_*(\Sigma(p,q,r))$. Their Floer indices $\mu(\alpha) = \operatorname{ind}\mathcal{D}_A(\alpha,\theta)$ are given by Proposition 6.22. Since all of the $\mu(\alpha)$ are odd, the boundary operators in the Floer chain complex of $\Sigma(p,q,r)$ vanish and $I_*(\Sigma(p,q,r)) = IC_*(\Sigma(p,q,r))$. This gives an algorithm for computing the Floer homology of Brieskorn homology spheres.

Example 6.23. Poincaré homology sphere $\Sigma(2,3,5)$ has Seifert invariants $b = -2$, $b_1 = 1$, $b_2 = -2$ and $b_3 = -4$. There are only two triples (ℓ_1,ℓ_2,ℓ_3) which correspond to representations of $\pi_1\Sigma(2,3,5)$ into $SU(2)$. They are $(1,2,2)$ and $(1,2,4)$. The corresponding es are equal to 47 and 59 and the corresponding Floer indices are 5 and 1. Thus $I_*(\Sigma(2,3,5)) = (0,\mathbb{Z},0,0,0,\mathbb{Z},0,0)$, which is consistent with the fact that the Euler characteristic of $I_*(\Sigma(2,3,5))$

equals twice the Casson invariant, $\lambda(\Sigma(2,3,5)) = -1$. A similar calculation with $\Sigma(2,3,7)$ gives $I_*(\Sigma(2,3,7)) = (0,0,0,\mathbb{Z},0,0,0,\mathbb{Z})$, compare with Example 6.20.

Example 6.24. The Brieskorn homology sphere $\Sigma(4,3,5)$ has Seifert invariants $b = 0$, $b_1 = -1$, $b_2 = 2$, and $b_3 = -2$. The triples (ℓ_1, ℓ_2, ℓ_3) arising from representations are $(1,2,4)$, $(3,2,2)$, $(2,2,2)$, and $(2,2,4)$. The corresponding e s are 103, 109, 94, and 118, and the corresponding Floer indices are 1, 3, 5, and 7, respectively. Thus $I_*(\Sigma(4,3,5)) = (0,\mathbb{Z},0,\mathbb{Z},0,\mathbb{Z},0,\mathbb{Z})$.

Example 6.25. More calculations of the Floer homology of $\Sigma(p,q,r)$ can be found in [83], including several infinite series of examples. It should be noted that Fintushel and Stern [83] work with self-dual rather than anti-self-dual equations, and that their results match ours after one shifts the gradings from n to $-3-n$.

Let n now be an arbitrary integer greater than or equal to 3. An n–tuple $(\ell_1, \dots, \ell_n)$ of rotation numbers associated with an irreducible representation $\alpha : \pi_1 \Sigma(a_1, \dots, a_n) \to SU(2)$ describes a unique connected component $\mathcal{R}_\alpha$ in $\mathcal{R}^*(\Sigma(a_1, \dots, a_n))$ containing α. Let m be the number of the ℓ_i different from 0 and a_i then $\mathcal{R}_\alpha$ is a smooth closed orientable manifold of dimension $2m-6$ which admits a Morse function $\varphi_\alpha : \mathcal{R}_\alpha \to \mathbb{R}$ with critical points of even Morse indices, see Section 3.5.2. The manifolds $\mathcal{R}_\alpha$ are non-degenerate in the Morse–Bott sense. Therefore, the functions φ_α can be used to perturb the Chern–Simons function $\mathbf{cs}$ into $\mathbf{cs} + h$ so that the critical point set of $\mathbf{cs} + h$ is non-degenerate and consists entirely of the critical points of the functions $\varphi_\alpha : \mathcal{R}_\alpha \to \mathbb{R}$, see Section 5.3.5.

Proposition 6.26. *Let $a \in \mathcal{R}_\alpha$ be a critical point of $\varphi_\alpha : \mathcal{R}_\alpha \to \mathbb{R}$ of Morse index m_a, and let $\mu(a)$ be the Floer index of a. Then $\mu(a)$ is odd and, modulo 8, equals*

$$\begin{aligned}\mu(a) &= \operatorname{ind} \mathcal{D}_A(\alpha, \theta) + \dim \mathcal{R}_\alpha - m_a \\ &= \frac{2e^2}{a} - 3 + m + \sum_{i=1}^{m} \frac{2}{a_i} \sum_{k=1}^{a_i - 1} \cot\left(\frac{\pi a k}{a_i^2}\right) \cot\left(\frac{\pi k}{a_i}\right) \sin^2\left(\frac{\pi e k}{a_i}\right) - m_a .\end{aligned}$$

Proof. The Floer index $\mu(a)$ is defined as the index of the perturbed ASD operator, that is, $\mu(a) = \operatorname{ind}(\mathcal{D}_A + \pi)(a, \theta)$, where the perturbation π is a compact operator. According to [12], addition of a compact operator does not change the integral term in (6.13). In particular, the formula of Theorem 6.15 can be easily amended to give

$$\operatorname{ind}(\mathcal{D}_A + \pi)(a, \theta) = -8\, \mathbf{cs}(a) - \frac{1}{2}\, \rho'_\alpha(\Sigma(a_1, \dots, a_n)) - \frac{1}{2}\, h'_\alpha - \frac{3}{2},$$

where $\mathbf{cs}(a) = \mathbf{cs}(\alpha)$ (note that $\mathbf{cs}$ is constant at each of the connected components $\mathcal{R}_\alpha$) and $h'_\alpha = \dim \ker(B_\alpha - 4\pi^2 \operatorname{Hess}_\alpha h) = 0$. The ρ–invariant can be computed as follows:

$$\rho'_\alpha(\Sigma(a_1,\dots,a_n))$$
$$= \eta'_\alpha(0) - \eta_\theta(0) = \eta_\alpha(0) - \dim\mathcal{R}_\alpha + 2m_a - \eta_\theta(0)$$
$$= \rho_\alpha(\Sigma(a_1,\dots,a_n)) - \dim\mathcal{R}_\alpha + 2m_a,$$

where $\eta_\alpha(0)$ is the spectral asymmetry of B_α and $\eta'_\alpha(0)$ is the spectral asymmetry of $B_\alpha - 4\pi^2$ Hess$_\alpha$ h. Subtraction of $4\pi^2$ Hess$_\alpha$ h replaces $\dim\mathcal{R}_\alpha$ worth of zeroes in the spectrum of B_α with m_a positive and $\dim\mathcal{R}_\alpha - m_a$ negative eigenvalues of the Hessian of φ_α. From this point on we proceed as in the proof of Proposition 6.22.

Example 6.27. The representation variety $\mathcal{R}^*(\Sigma(2,3,5,7))$ was described in Example 3.31. Table 6.1 copied from [83] lists the isolated points corresponding to representations sending one of the x_i into ± 1, together with their Floer indices. Table 6.2 lists the 2–dimensional components $\mathcal{R}_\alpha$ of the representation space. Each of these components is a 2–dimensional sphere, with the standard Morse function $\varphi : S^2 \to \mathbb{R}$ having two critical points of indices 0 and 2. Corresponding to each of these critical points are two generators in the Floer homology. According to Proposition 6.26 their Floer indices are ind $\mathcal{D}_A(\alpha,\theta)$ and ind $\mathcal{D}_A(\alpha,\theta) + 2$.

$(\ell_1,\ell_2,\ell_3,\ell_4)$	Floer index	$(\ell_1,\ell_2,\ell_3,\ell_4)$	Floer index
(1,0,2,2)	7	(0,2,2,2)	1
(1,0,2,4)	5	(0,2,2,4)	7
(1,0,2,6)	1	(0,2,2,6)	3
(1,0,4,4)	7	(0,2,4,2)	3
(1,2,0,2)	7	(2,2,2,2)	5
(1,2,0,4)	5	(2,2,2,4)	3
(1,2,2,0)	7	(2,2,4,4)	5
(1,2,4,0)	1	(2,2,4,6)	1

Table 6.1.

We have total of 28 generators in $IC_*(\Sigma(2,3,5,7))$ and all of them have odd Floer indices as listed in the tables. Since the boundary operator necessarily vanishes we conclude that $I_*(\Sigma(2,3,5,7))$ is free abelian group of rank 7 in each odd dimension and is trivial in even dimensions.

More examples of explicit calculations of $I_*(\Sigma(a_1,\dots,a_n))$ for $n = 4$ and $n = 5$ can be found in [168].

6.4.2 The Closed Form Formula

A closed form formula for the Floer homology of Seifert fibered homology spheres can be found in [272] and [278]. Orient $\Sigma(a_1,\dots,a_n)$ as the link of sin-

$(\ell_1, \ell_2, \ell_3, \ell_4)$	Floer indices
(1,2,2,2)	3, 1
(1,2,2,4)	1, 7
(1,2,2,6)	5, 3
(1,2,4,2)	5, 3
(1,2,4,4)	3, 1
(1,2,4,6)	7, 5

Table 6.2.

gularity (1.2) and denote by λ the Casson invariant and by $\bar{\mu}$ the $\bar{\mu}$–invariant of Neumann and Siebenmann, see Section 7.2.3. Both $\lambda(\Sigma(a_1, \dots, a_n))$ and $\bar{\mu}(\Sigma(a_1, \dots, a_n))$ are easy to compute and in fact can be given by explicit formulas, see Section 3.5.2 and Section 7.2.3.

Theorem 6.28. *The Floer homology groups $I_k(\Sigma(a_1, \dots, a_n))$, $0 \le k \le 7$, are free abelian groups of ranks b_k such that $b_k = 0$ for even k and*

$$b_1 = b_5 = -\frac{1}{2}\,(\lambda(\Sigma(a_1, \dots, a_n)) + \bar{\mu}(\Sigma(a_1, \dots, a_n))),$$
$$b_3 = b_7 = -\frac{1}{2}\,(\lambda(\Sigma(a_1, \dots, a_n)) - \bar{\mu}(\Sigma(a_1, \dots, a_n))).$$

Proof. Let us assume for simplicity that $n = 3$ so that we deal with a Brieskorn homology sphere $\Sigma(p, q, r)$. A proof of the general case can be found in [278]. We saw in Section 6.4.1 that $I_k(\Sigma(p, q, r))$ are free abelian groups which vanish for even k. The four-periodicity theorem of Frøyshov, see Example 6.54, implies that $b_1 = b_5$ and $b_3 = b_7$. Since the Casson invariant is one half the Euler characteristic of the Floer homology we immediately obtain that $b_1 + b_3 = -\lambda(\Sigma(p, q, r))$. Thus in order to prove the theorem we need to show that $-b_1 + b_3 = \bar{\mu}(\Sigma(p, q, r))$.

As a link of complex surface singularity (1.1), manifold $\Sigma(p, q, r)$ admits an involution $\tau(x, y, z) = (\bar{x}, \bar{y}, \bar{z})$ which represents $\Sigma(p, q, r)$ as a double branched cover of S^3 with branch set a Montesinos knot k_τ, see Example 1.25. One checks by a direct calculation that the induced action of τ on the representation space $\mathcal{R}^*(\Sigma(p, q, r))$ is trivial, after which Theorem 5.18 implies that

$$\frac{1}{2} \sum_{\alpha \in \mathcal{R}^*(\Sigma(p,q,r))} (-1)^{\mu^\tau(\alpha)} = \frac{1}{8}\,\operatorname{sign} k_\tau. \tag{6.15}$$

The left hand side of (6.15) is the equivariant Casson invariant of $\Sigma(p, q, r)$ with respect to the involution τ, compare with Example 5.24. The right hand side of (6.15) can be identified with the $\bar{\mu}$–invariant by using methods of classical topology, see [272]. An equivariant version of the index calculation of Section 6.4.1 is used in [272] to show that $2\mu^\tau(\alpha) = \mu(\alpha) + 1 \bmod 4$. This identifies the left hand side of (6.15) as $-b_1 + b_3$, which completes the proof.

Example 6.29. Let $m \geq 0$ and consider the family $\Sigma(2,3,6m\pm 1)$ of Brieskorn homology spheres. Explicit calculations show that $\lambda(\Sigma(2,3,6m \pm 1)) = -m$ and $\bar{\mu}(\Sigma(2,3,6m \pm 1)) = \pm 1$ if m is odd and $\bar{\mu}(\Sigma(2,3,6m \pm 1)) = 0$ if m is even. Therefore, the ranks of the groups $I_*(\Sigma(2,3,6m \pm 1))$ are as follows:

$$\begin{aligned} (0,(m \mp 1)/2,0,(m \pm 1)/2,0,(m \mp 1)/2,0,(m \pm 1)/2), &\quad \text{if } m \text{ is odd},\\ (0,m/2,0,m/2,0,m/2,0,m/2), &\quad \text{if } m \text{ is even}. \end{aligned}$$

Remark 6.30. In general, let $\Sigma(a_1,\ldots,a_n)$ be a Seifert fibered homology sphere and let $a = a_1 \cdots a_{n-1}$ be the product of the first $(n-1)$ Seifert invariants a_i. Then $\bar{\mu}(\Sigma(a_1,\ldots,a_{n-1}, 2a \pm a_n)) = \pm\bar{\mu}(\Sigma(a_1,\ldots,a_n))$, see [241]. This results in a quasi-periodicity in the Floer homology of Seifert fibered homology spheres, which can also be deduced from the Floer exact triangle, see Example 6.41.

Corollary 6.31. *Given a Seifert fibered homology sphere $\Sigma(a_1,\ldots,a_n)$ let $q = a_1 \cdots a_j$ and $p = a_{j+1} \cdots a_n$ be products of the first j, respectively, last $n-j$, Seifert invariants, where j is any integer such that $2 \leq j \leq n-2$. Then*

$$I_*(\Sigma(a_1,\ldots,a_n)) = I_*(\Sigma(a_1,\ldots,a_j,p)) \oplus I_*(\Sigma(q,a_{j+1},\ldots,a_n)).$$

Proof. The homology sphere $\Sigma(a_1,\ldots,a_n)$ is the splice of $\Sigma(a_1,\ldots,a_j,p)$ and $\Sigma(q,a_{j+1},\ldots,a_n)$ along singular fibers of degrees p and q, see Example 1.10. Both λ– and $\bar{\mu}$–invariants are additive with respect to splicing, see Section 3.4.1 and Section 7.2.3, hence so are the Floer homology groups of $\Sigma(a_1,\ldots,a_n)$.

Example 6.32. The above corollary allows for a calculation of the Floer homology of $\Sigma(2,3,5,7)$ which is much easier than the one given in Example 6.27.

It should be mentioned that in general behavior of the Floer homology with respect to splicing is not known, see discussion in the next section.

6.4.3 Graph Homology Spheres

Graph homology spheres are integral homology spheres that can be obtained by splicing Seifert links, see Section 1.1.8. The Euler characteristic of Floer homology, which equals twice the Casson invariant, is additive with respect to splicing, see Section 3.4.1. Therefore, it is only natural to try to compute the Floer homology of a graph manifold from the Floer homology of its Seifert components. There has not been much progress in this direction, which is mostly due to the fact that the representation space of a manifold Σ obtained by splicing Σ_1 and Σ_2 may have odd-dimensional components even if the representation spaces of Σ_1 and Σ_2 are non-degenerate. This means that, after a generic perturbation, the Floer chain complex may have generators in both parities. This brings up the possibility of a non-trivial boundary operator, and computing such an operator is a difficult problem.

Example 6.33. Let Σ be the splice of two copies of the Brieskorn homology sphere $\Sigma(2,3,7)$ along the singular fibers of degree 7. The representation space $\mathcal{R}^*(\Sigma(2,3,7))$ consists of two non-degenerate representations, see Example 6.23. The representation variety $\mathcal{R}^*(\Sigma)$ consists of four non-degenerate points and four circles, which are non-degenerate in the Morse–Bott sense, see [274]. More examples of such behavior can be found in [170].

Example 6.34. Let Σ be the splice of $\Sigma(p,q,r)$ and $\Sigma(p',q',r')$ along the singular fibers k and k' of degrees r and r' such that $pqp'q' - rr' = \pm 1$. Then $\mathcal{R}^*(\Sigma)$ consists of finitely many non-degenerate representations and in fact can be identified with $\mathcal{R}^*(\Sigma(p,q,r)) \cup \mathcal{R}^*(\Sigma(p',q',r'))$. This is due to the fact that every irreducible representation of $\pi_1(\Sigma(p,q,r) \setminus N(k))$ can be uniquely extended to an irreducible representation of $\pi_1\Sigma$ so that its restriction to $\pi_1(\Sigma(p',q',r') \setminus N(k'))$ is reducible, and similarly for the other factor, see [271].

Once the representation space for a graph manifold is described, the spectral flow can be computed by using various splitting formulas, see Section 6.3.2. These formulas relate spectral flow on a graph manifold (or, more generally, on any manifold split along a hypersurface) to the spectral flows on the spliced components. They can be used to compute the Floer chain complex of the graph manifold and, if all generators happen to have the same parity, its Floer homology, see [170]. For some graph manifolds, their Floer homology groups can also be computed with the help of the Floer exact triangle, see Example 6.46.

6.5 Properties of the Floer Homology

The Floer homology has certain properties which are reminiscent, if only superficially, of those of the classical homology theories. Among these properties is the Floer exact triangle, which is an analogue of the long exact sequence in homology. Its relative term requires an extension of the Floer homology to homology handles which is outlined in Section 6.5.2. Other topics in this section include behavior of the Floer homology with the change of orientation, its functoriality with respect to cobordisms and diffeomorphisms, cup–product structures, mutation invariance, and the Floer homology of connected sums.

6.5.1 Orientation Reversal

Let Σ be an integral homology sphere and $-\Sigma$ the same integral homology sphere with reversed orientation. The representation varieties of Σ and $-\Sigma$ are identical, and the Floer indices are related by the formula $\mu_{-\Sigma}(\alpha) = -3 - \mu_\Sigma(\alpha)$. Therefore, one has a natural *pairing*

$$\langle\ ,\ \rangle : I_*(-\Sigma) \otimes I_{-3-*}(\Sigma) \to \mathbb{Z}. \tag{6.16}$$

This looks like Poincaré duality but that is deceptive: the 3 occurring in formula (6.16) is not the dimension of Σ but rather the dimension of $SU(2)$.

6.5.2 Floer Homology of Homology Handles

An oriented 3–dimensional manifold M is called a *homology handle*, or a *homology* $S^1 \times S^2$, if $H_*(M,\mathbb{Z}) = H_*(S^1 \times S^2, \mathbb{Z})$. For instance, the result of 0–surgery on a knot in an integral homology sphere is a homology handle. Principal $SO(3)$-bundles over a homology handle M are classified by their second Stiefel–Whitney class $w_2 \in H^2(M,\mathbb{Z}/2) = \mathbb{Z}/2$, hence there are only two isomorphism classes of such bundles.

Let E be an $SO(3)$-bundle over a homology handle M with $w_2(E) \neq 0$. Define the *gauge group* $\mathcal{G}$ as the space of sections of the bundle $E \times_{\mathrm{Ad}} SO(3)$ with the natural product structure, in other words, as the automorphism group of the bundle E. We get an exact sequence

$$\mathcal{G}_S \longrightarrow \mathcal{G} \stackrel{\eta}{\longrightarrow} H^1(M,\mathbb{Z}/2)$$

where η measures the obstruction to deforming a gauge transformation from $\mathcal{G}$ to the identity over the 1–skeleton of M, and $\mathcal{G}_S$ is defined as the kernel of η. The group $\mathcal{G}_S$ is called the *restricted gauge group*. From elementary obstruction theory one proves easily that η is onto and the group $\mathcal{G}_S$ consists of gauge transformations which lift to $SU(2)$–gauge transformations.

If $\mathcal{A}$ denotes the affine space of connections in E then there is an action of $H^1(M,\mathbb{Z}/2) = \mathbb{Z}/2$ on $\mathcal{A}/\mathcal{G}_S$ with quotient the orbit space of connections $\mathcal{A}/\mathcal{G}$. Let $\mathbf{cs} : \mathcal{A}/\mathcal{G}_s \to \mathbb{R}/\mathbb{Z}$ be the Chern–Simons function and assume for the sake of simplicity that its critical points, which are $\mathcal{G}_S$–orbits of flat connections in E, are non-degenerate. Given two flat connections, α and β, in the bundle E, we define $\mathrm{sf}\,(\alpha,\beta)$ to be the spectral flow of a family of operators (5.9) corresponding to a path of connections $A(t)$ in E connecting α to β. The spectral flow $\mathrm{sf}\,(\alpha,\beta)$ is well defined modulo 8 on the $\mathcal{G}_S$–orbits of α and β (but it is only well defined modulo 4 on the $\mathcal{G}$–orbits).

Since the trivial connection θ is not a connection in P, we cannot "compare" a flat connection α in P to the trivial connection to obtain an absolute grading the way we did it in Section 6.2.2. Instead there are only relative gradings. Let $IC_*(M)$ be the free chain complex generated by the $\mathcal{G}_S$–orbits of flat connections in E with the relative gradings.

The boundary operators $\partial : IC_n(M) \to IC_{n-1}(M)$ are defined exactly as in the case of integral homology spheres, see Section 6.2.2, and they have the property that $\partial^2 = 0$. The homology of the chain complex $(IC_*(M), \partial)$ is called the *Floer homology* of M and is denoted by $I_*(M)$. The groups $I_*(M)$ are relatively graded modulo 8.

Example 6.35. Let M be a homology handle and suppose that there is an embedded 2–sphere $S \subset M$ such that $w_2(E)$ evaluates non–trivially on S. Then there are no flat connections in E since the same is already true over S. Thus in this situation $I_*(M) = 0$. For example, $I_*(\Sigma\#(S^1 \times S^2)) = 0$ for any integral homology sphere Σ.

Remark 6.36. Let $M = \Sigma + 0 \cdot k$ be obtained by 0–surgery along a knot k in an integral homology sphere Σ. Then the Euler characteristic of $I_*(M)$ equals twice the Casson invariant $\pm\lambda'(k \subset \Sigma)$, see Section 3.1. This can be seen with the help of Theorem 5.9 and the Floer exact triangle, see Theorem 6.39. An independent proof can be found in [206].

Let γ be a knot in M representing a generator of $H_1(M, \mathbb{Z}) = \mathbb{Z}$, K its exterior and $m \subset \partial K$ its meridian. Then the $\mathcal{G}_S$–orbits of flat connections in E can be identified with the conjugacy classes of representations $\alpha : \pi_1 K \to SU(2)$ such that $\alpha(m) = -1$. The action of $H^1(M, \mathbb{Z}/2) = \mathbb{Z}/2$ on the $\mathcal{G}_S$–orbits translates into the involution $\alpha \mapsto \alpha \otimes \varepsilon$ on such representations, where $\varepsilon : \pi_1 K \to SU(2)$ is the central representation mapping m to -1. In the pillowcase picture, this involution corresponds to a 180° rotation about the vertical axis $\phi = \pi/2$, see Figure 3.1.

One can show that this involution is free on the $\mathcal{G}_S$–orbits of flat connections in E, and that it defines a free involution

$$g : I_*(M) \to I_{*+4}(M). \tag{6.17}$$

Its quotient is a $\mathbb{Z}/4$–graded Floer homology of M which can be defined directly from the $\mathcal{G}$–orbits of flat connections in E.

It should be also pointed out that after choosing an integral lift of $w_2(E)$ and a differential 2–form representing this lift, we can fix a $U(2)$–bundle $E^* \to M$ and uniquely lift flat connections from E to E^* once the central part is prescribed. The gauge transformations fixing this central part are the same as elements of $\mathcal{G}_S$ and the action of tensoring E^* with flat line bundle with monodromy in $\mathbb{Z}/2$ coincides with the action of $H^1(M, \mathbb{Z}/2)$.

To state the functoriality properties of this extended Floer homology, we need to make a slight adjustment in our definition of $I_*(\Sigma)$ for integral homology spheres Σ. Namely, we will replace $SU(2)$ with $SO(3)$ throughout the construction of $I_*(\Sigma)$ outlined in Section 6.2.2. Since Σ is an integral homology sphere, representations of its fundamental group in $SU(2)$ are in one-to-one correspondence with $SO(3)$–representations via the projection $\operatorname{ad} : SU(2) \to SO(3)$. All other appearances of $SU(2)$ or $SO(3)$ in the construction are through their Lie algebras $\mathfrak{su}(2) = \mathfrak{so}(3)$. Therefore, the resulting $SO(3)$ Floer homology is canonically isomorphic to $I_*(\Sigma)$.

The extended Floer homology groups are functorial in the following sense. Let W be a smooth oriented compact cobordism with $\partial W = -M_0 \cup M_1$ where each of M_0 and M_1 is either an integral homology sphere or a homology handle, and let E be an $SO(3)$–bundle over W which is non-trivial over each boundary component which is a homology handle. Given generators α and β in the Floer chain complexes of M_0 and M_1, consider the moduli spaces $\mathcal{M}_W(\alpha,\beta)$ of ASD connections in E interpolating between α and β and use them to define the homomorphism $W_* : IC_*(M_0) \to IC_*(M_1)$ by the formula

$$W_*(\alpha) = \sum_\beta \#\mathcal{M}_W(\alpha,\beta)\cdot\beta, \tag{6.18}$$

compare with (6.11). The homomorphism (6.18) is a chain homomorphism, therefore, it defines a homomorphism $W_* : I_*(M_0) \to I_*(M_1)$. This homomorphism can be used, in particular, to prove that the groups $I_*(M)$ are well–defined, compare with Theorem 6.14.

Remark 6.37. If both M_0 and M_1 are integral homology spheres then the degree of W_* equals $-2\,w_2(E)^2 - 3\,(b_2^+ - b_1)(W)$ mod 8, which can be seen as in Proposition 6.11 with the help of (6.4). If the bundle E lifts to an $SU(2)$–bundle, the homomorphism W_* coincides with the one defined in (6.11) by using $SU(2)$–connections.

Remark 6.38. There is a fine point in extending Proposition 6.13 to the $SO(3)$ Floer theory. Namely, given a cobordism U between M_0 and M_1 and a cobordism V between M_1 and M_2, we can glue them together to get a cobordism $W = U \cup_{M_1} V$. Suppose that M_1 is a homology handle. Since $H^1(M_1,\mathbb{Z}/2) = \mathbb{Z}/2$, there are two different ways to glue the bundles over U and V to form one over W. These may result in two different homomorphisms in the Floer homology.

6.5.3 The Floer Exact Triangle

Let $k \subset \Sigma$ be a knot in an integral homology sphere Σ. We introduce two other manifolds: the homology sphere $\Sigma' = \Sigma - k$ obtained from Σ by (-1)-surgery on k, and the homology handle $\overline{K} = \Sigma + 0\cdot k$ obtained by 0-surgery on k. Let Z be the trace of the (-1)–surgery, that is, the 4–manifold obtained by attaching a 2–handle to $[0,1]\times\Sigma$ along $k \subset \{1\}\times\Sigma$ with framing -1. Then Z is an oriented cobordism from Σ to Σ' whose intersection form is isomorphic to (-1). Likewise, we have a cobordism X with intersection form (0) from Σ' to $\overline{K}$, and a cobordism Y with intersection form (0) from $\overline{K}$ back to Σ. Together with the trivial $SO(3)$–bundle over Z and the non-trivial $SO(3)$–bundles over X and Y, these cobordisms induce homomorphisms in Floer homology which can be included in the following triangle

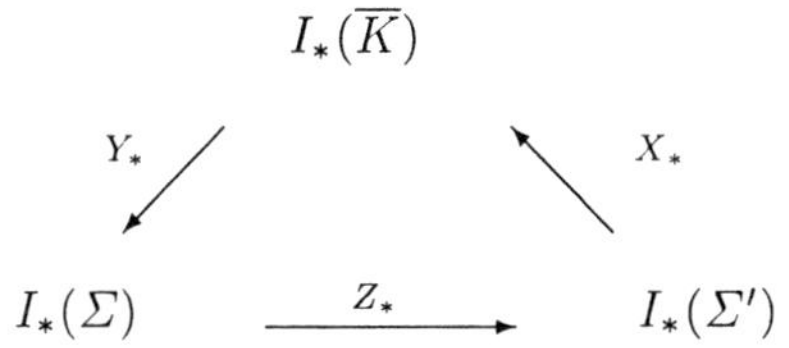

Note that the gradings on both $I_*(\Sigma)$ and $I_*(\Sigma')$ are well defined modulo 8 and that $\deg Z_* = 0$. We fix the relative grading of $I_*(\overline{K})$ in the triangle by the requirement that the homomorphism X_* preserve the grading.

The two different ways to glue the bundles over X and Y, see Remark 6.38, result in two different bundles over $Z \cup X \cup Y$ which restrict to the correct bundles over each of the cobordisms Z, X and Y. These bundles induce endomorphisms of $I_*(\Sigma)$ of degrees -1 and -5. It is common to use the bundle which induces the degree -1 endomorphism. With this choice, the triangle has total degree -1. A proof of the following result was outlined in [94] and its details were worked out in [38].

Theorem 6.39. *The above triangle is exact in that* $\operatorname{im} Z_* = \ker X_*$, $\operatorname{im} X_* = \ker Y_*$, *and* $\operatorname{im} Y_* = \ker Z_*$.

This is known as the *Floer exact triangle.* It induces a long exact sequence of Floer homology groups with the connecting homomorphism of degree -1.

Proof. We only need to prove exactness at one term, say $I_*(\Sigma')$, as exactness at the other two terms will follow by a similar argument. The easier part is to prove that $X_* Z_* = 0$, which is a counterpart of similar vanishing theorems for closed 4–manifolds, see Example 6.70. In the composite cobordism $Z \cup X$, there is an embedded 2–sphere S as illustrated in Figure 6.5. The sphere S is the union of the discs with common boundary the knot $k^* \subset \Sigma'$ which is the image in Σ' of the canonical longitude of k.

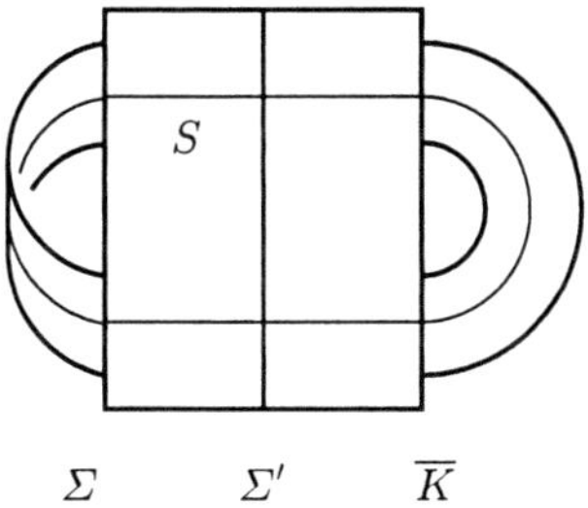

Fig. 6.5.

The homological self-intersection of S equals -1, therefore, we can write

$$Z \cup X = U \# P \quad \text{where} \quad P = -\mathbb{C}P^2$$

and U is another cobordism from Σ to $\overline{K}$. It is not important for this proof but one can use Kirby calculus to show that in fact U coincides with the cobordism Y with reversed orientation. The $SO(3)$–bundle over $Z \cup X$ restricts to an $SO(3)$–bundle over P with non-trivial w_2.

The composite $X_* \, Z_*$ is the same as the map defined by the cobordism $U \# P$ by using ASD connections for any choice of metric on $U \# P$. We consider a family of metrics g_λ, $\lambda \to 0$, in which the neck of the connected sum is made very small. For sufficiently small λ, all ASD connections over $U \# P$ are made up by gluing an ASD connection A over U to an ASD connection B, with non-zero w_2, over P. The ASD connection A is irreducible since it is already irreducible at the ends. The ASD connection B might be reducible, but only to S^1, since w_2 is non-zero. Gluing A and B together produces a family $A \# g^* B$ of ASD connections with $g \in SO(3)/S^1 = S^2$. Therefore, the moduli space of ASD connections over $U \# P$ is at least 2–dimensional. Since the 0–dimensional components are lacking, the map in Floer homology defined by $U \# P$ has to be zero.

We only briefly outline the more difficult part of Floer's proof, referring to [38] for a complete account. The first step is best understood in terms of the pillowcase image, see Section 3.3.1. Let K be the exterior of the knot $k \subset \Sigma$, and $\mathcal{R}(\overline{K})$ the space of $\mathcal{G}_S$–orbits of flat connections in the non–trivial $SO(3)$–bundle over $\overline{K} = \Sigma + 0 \cdot k$. Identify $\mathcal{R}(\overline{K})$ as in Section 6.5.2 with the space of conjugacy classes of representations $\pi_1(K) \to SU(2)$ sending the canonical longitude of k to -1. Then all the representation spaces $\mathcal{R}^*(\Sigma)$, $\mathcal{R}^*(\Sigma')$, and $\mathcal{R}(\overline{K})$ are contained in $\mathcal{R}(K)$. The inclusion $\partial K \to K$ induces a map from $\mathcal{R}(K)$ to $\mathcal{R}(\partial K)$, where $\mathcal{R}(\partial K)$ is the pillowcase shown in Figure 3.1. The image of $\mathcal{R}(K)$ in the pillowcase is generically a collection of curves, and the images of $\mathcal{R}^*(\Sigma)$, $\mathcal{R}^*(\Sigma')$, and $\mathcal{R}(\overline{K})$ are the intersections of these curves with the lines $\phi = 0$, $\theta = \phi$, and $\theta = \pi$, respectively. The Chern–Simons function $\mathbf{cs} : \mathcal{B}^*(\Sigma') \to \mathbb{R}/\mathbb{Z}$ can be perturbed into $\mathbf{cs} + h$ so that the set of its critical points is

$$\widetilde{\mathcal{R}}^*(\Sigma') = \mathcal{R}^*(\Sigma) \cup \mathcal{R}(\overline{K}).$$

In fact, $\mathbf{cs}$ can be deformed into $\mathbf{cs} + h$ through a continuous family $\mathbf{cs} + h_t$. In the pillowcase image, this deformation can be viewed as one deforming the diagonal $\theta = \phi$ of the pillowcase into the corner consisting of the arcs $\phi = 0$ and $\theta = \pi$, thus dragging $\mathcal{R}^*(\Sigma')$ into $\widetilde{\mathcal{R}}^*(\Sigma') = \mathcal{R}^*(\Sigma) \cup \mathcal{R}(\overline{K})$, see Figure 3.5.

The critical points and gradient flow lines of $\mathbf{cs} + h$ can be used to construct a chain complex $(\widetilde{IC}_*(\Sigma'), \widetilde{\partial})$ whose homology is isomorphic to $I_*(\Sigma')$. By construction, the chain group $\widetilde{IC}_*(\Sigma')$ is identified *as a group* with the direct sum

$$\widetilde{IC}_*(\Sigma') = IC_*(\Sigma) \oplus IC_*(\overline{K}). \tag{6.19}$$

If we choose any perturbations of the ASD equations over Z and over X compatible with the data on the ends, we obtain chain maps

$$IC_*(\Sigma) \xrightarrow{Z_*} \widetilde{IC}_*(\Sigma') \xrightarrow{X_*} IC_*(\overline{K})$$

which induce the maps Z_* and X_* in Floer homology. Now if it happens that (6.19) is an isomorphism of *chain complexes*, and that the chain maps Z_* and X_* can be taken to be the inclusion $IC_*(\Sigma) \to IC_*(\Sigma) \oplus IC_*(\overline{K})$ and the projection $IC_*(\Sigma) \oplus IC_*(\overline{K}) \to IC_*(\overline{K})$, respectively, then the exactness is immediate: the long exact sequence in homology will be induced by the short exact sequence of chain complexes. The above actually holds but only after one replaces Z_* with another map which is chain homotopic to Z_*. Some major work goes into the construction of such a chain homotopy, see [38].

Example 6.40. Let k be a trivial knot in an integral homology sphere Σ then the result of (-1)–surgery along k is again Σ, and the cobordism Z is diffeomorphic to the connected sum $([0,1] \times \Sigma)\#(-\mathbb{C}P^2)$. The result of the 0–surgery on Σ is $\Sigma\#(S^1 \times S^2)$, and we know from Example 6.35 that $I_*(\Sigma\#(S^1 \times S^2)) = 0$. The Floer exact triangle implies then that the map $Z_* : I_*(\Sigma) \to I_*(\Sigma)$ is an isomorphism. This is a special case of the more general *blow up formula* for Donaldson invariants, see Example 6.74.

Let Σ'' denote the homology sphere obtained from Σ by $(-1/2)$-surgery along k. Note that Σ'' can also be obtained by (-1)-surgery of Σ' along the image in Σ' of the canonical longitude of $k \subset \Sigma$. Then we have two Floer exact triangles

$$\begin{array}{ccccccccc}
 & & I_*(\overline{K}) & & & & I_*(\overline{K}') & & \\
 & \swarrow & & \nwarrow & & \swarrow & & \nwarrow & \\
I_*(\Sigma) & & \longrightarrow & & I_*(\Sigma') & & \longrightarrow & & I_*(\Sigma'')
\end{array}$$

An easy exercise in Kirby calculus shows that the manifolds $\overline{K}$ and $\overline{K}'$ are homeomorphic, and our grading conventions imply that $I_*(\overline{K}') = I_{*+2}(\overline{K})$.

Example 6.41. Let k_n be the singular fiber of degree a_n in a Seifert fibered homology sphere $\Sigma(a_1, \dots, a_n)$ then $(-1/m)$–surgery on k_n yields $\Sigma(a_1, \dots, a_n + m \cdot a)$ where $a = a_1 \cdots a_{n-1}$ is the product of the first $(n-1)$ Seifert invariants, compare with Example 1.5. Any two integral homology spheres $\Sigma(a_1, \dots, a_n)$ and $\Sigma(a_1, \dots, a_n + a)$ can be included into the Floer exact triangle whose relative term is the Floer homology of the homology handle $\overline{K}$ obtained by 0–surgery of $\Sigma(a_1, \dots, a_n)$ along k_n. The manifold $\overline{K}$ is Seifert fibered. The techniques of Section 6.4.1 can be generalized to show that the groups $I_q(\overline{K})$ are free abelian and that they vanish for even q, if the absolute grading on $I_*(\overline{K})$ is fixed as in the Floer exact triangle. The Floer exact triangle then splits into a family of short exact sequences implying that

$$I_*(\Sigma(a_1,\dots,a_n+a)) = I_*(\Sigma(a_1,\dots,a_n)) \oplus I_*(\overline{K}).$$

It should be mentioned that $I_*(\overline{K})$ depends on the a_n even though its Euler characteristic $\lambda'(k_n)$ does not, see (3.24). For example, if $\overline{K}$ is obtained by 0–surgery of the singular fiber of degree 13 in $\Sigma(2,7,13)$ then the groups $I_*(\overline{K})$ have ranks $(0,2,0,4,0,2,0,4)$. On the other hand, if $\overline{K}$ is obtained by 0–surgery on the singular fiber of degree 11 in $\Sigma(2,7,11)$ then the ranks of $I_*(\overline{K})$ are $(0,3,0,3,0,3,0,3)$. The above results generalize to all integers $m \geq 0$ as follows:

$$I_*(\Sigma(a_1,\dots,a_n+(m+2)\cdot a)) = \\ = I_*(\Sigma(a_1,\dots,a_n+m\cdot a)) \oplus I_*(\overline{K}) \oplus I_{*+2}(\overline{K}).$$

Since $I_{*+4}(\overline{K}) = I_*(\overline{K})$, this reflects the quasi-periodicity in the Floer homology of Seifert fibered homology spheres mentioned in Remark 6.30.

Example 6.42. Let k be the figure–eight knot in S^3 and, for any integer m, denote by Σ_m the integral homology sphere obtained by $(-1/m)$–surgery along k. A Kirby calculus argument shows that $\Sigma_1 = -\Sigma(2,3,7)$, compare with Example 1.4. This observation was used in [271] to compute the Floer homology groups of the Σ_m for all m. The argument is as follows. First we apply the Floer exact triangle to the manifold $\Sigma_1 = S^3 - k$. Since $I_*(S^3) = 0$ we conclude that $I_*(\overline{K}) = I_*(\Sigma_1) = I_*(-\Sigma(2,3,7)) = I_{-3-*}(\Sigma(2,3,7)) = (0,0,\mathbb{Z},0,0,0,\mathbb{Z},0)$, see Example 6.23. Application of the second Floer exact triangle to $\Sigma_2 = S^3 - (1/2)\cdot k$ gives that

$$I_*(\Sigma_2) = I_*(\Sigma_1) \oplus I_{*+2}(\Sigma_1) = (\mathbb{Z},0,\mathbb{Z},0,\mathbb{Z},0,\mathbb{Z},0).$$

Continuing by induction, we conclude that the Floer homology groups $I_*(\Sigma_m)$ for all positive m are free abelian groups of the following ranks:

$$((m-1)/2,0,(m+1)/2,0,(m-1)/2,0,(m+1)/2,0), \quad \text{if } m \text{ is odd,}$$
$$(m/2,0,m/2,0,m/2,0,m/2,0), \quad \text{if } m \text{ is even.}$$

The observation that $\Sigma_{-m} = -\Sigma_m$ takes care of the remaining m because $I_*(\Sigma_{-m}) = I_{-3-*}(\Sigma_m)$. The original approach to calculating $I_*(\Sigma_m)$ can be found in [309].

Example 6.43. The techniques used in Example 6.42 can be easily adapted as in [271] to calculate Floer homology of all integral homology spheres obtained by surgery on twist knots, compare with Example 1.4.

Example 6.44. Let k be a knot in S^3. The Floer homology groups of $S^3 - k$ and $S^3 + k$ are related by the following Floer exact triangles:

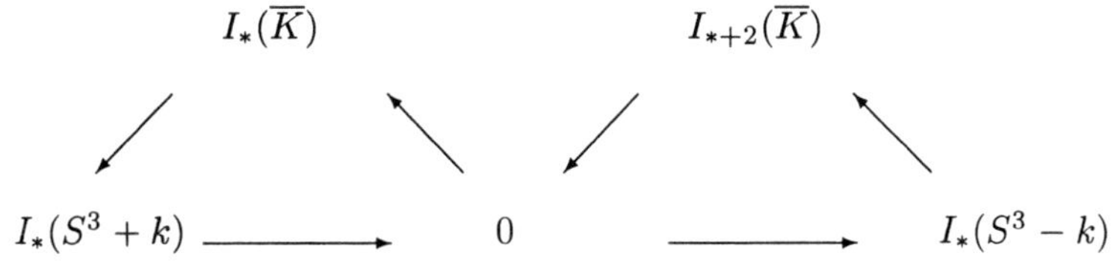

This implies that $I_n(S^3 - k) = I_{n+1}(S^3 + k)$ for any n.

Example 6.45. A knot $k \subset S^3$ is called *amphicheiral* if its mirror image k^* is isotopic to k. If k is amphicheiral then $S^3 - k = S^3 - k^* = -(S^3 + k)$, hence $I_n(S^3 - k)$ is dual to $I_{-3-n}(S^3 + k)$, see (6.16). Together with the result of Example 6.44 this implies that $I_n(S^3 - k)$ is dual to $I_{4-n}(S^3 - k)$ and $I_n(S^3 + k)$ is dual to $I_{-2-n}(S^3 + k)$, for any amphicheiral knot k.

Example 6.46. Let $k = T(p,q)\#T(p,q)^*$ be the connected sum of a (p,q)–torus knot $T(p,q)$ with its mirror image $T(p,q)^*$, and let $M_n(p,q)$ be the homology sphere obtained by $(-1/n)$–surgery of S^3 along k. For any n, this is a graph homology sphere. In particular, $M_1(p,q)$ is the splice of $\Sigma(p,q,pq-1)$ and $\Sigma(p,q,pq+1)$ along the singular fibers of degrees $pq-1$ and $pq+1$, respectively. The manifold $M_1(2,3)$ is shown in Figure 1.6. The Floer homology groups of $M_n(p,q)$ are as follows. For a positive n, the group $I_j(M_n(p,q))$ is trivial if j is even, and is isomorphic to a free abelian group of rank $n(p^2-1)(q^2-1)/24$ if j is odd. Since k is amphicheiral, $M_{-n}(p,q) = -M_n(p,q)$, which takes care of the negative n.

The groups $I_*(M_n(p,q))$ were computed in [290] for $p = 2$ and $q = 3$ directly from the definition of Floer homology. This calculation was later extended in [280] to cover $I_*(M_n(p,q))$ for all relatively prime p and q. A different approach used in [271] is as follows. The splicing decomposition for $M_1(p,q)$ mentioned above provides an identification

$$\mathcal{R}^*(M_1(p,q)) = \mathcal{R}^*(\Sigma(p,q,pq-1)) \cup \mathcal{R}^*(\Sigma(p,q,pq+1)), \tag{6.20}$$

see Example 6.34. In particular, $\mathcal{R}^*(M_1(p,q))$ is non-degenerate. The Casson invariant is additive with respect to splicing, see Section 3.4.1, therefore,

$$\lambda(M_1(p,q)) = \lambda(\Sigma(p,q,pq-1)) + \lambda(\Sigma(p,q,pq+1)). \tag{6.21}$$

Equations (6.20) and (6.21) together with Proposition 6.22 imply that the Floer indices of all the representations in $\mathcal{R}^*(M_1(p,q))$ are odd. The Floer homology groups $I_*(M_1(p,q))$ are 4–periodic, see Remark 6.53. Since k is amphicheiral, the groups $I_*(M_1(p,q))$ are in fact 2–periodic, see Example 6.45, which is enough to compute $I_*(M_1(p,q))$. Multiple application of the Floer exact triangle completes the calculation of $I_*(M_n(p,q))$ for all n.

Example 6.47. Let us consider a two-component link $\mathcal{L} = L_1 \cup L_2$ in S^3 such that $\mathrm{lk}(L_1, L_2) = \pm 1$ and the component L_1 is an unknot. Let $\Sigma_p(\mathcal{L})$ be the integral homology sphere obtained by surgery on the link $\mathcal{L}$ with L_1 framed by zero and L_2 framed by p. The manifolds $\Sigma_p(\mathcal{L})$ are referred to as *Mazur homology spheres* because they generalize the original Mazur's example [212] of a homology sphere bounding a smooth acyclic 4–manifold, see Section 7.1.1.

The Floer homology $I_*(\Sigma_p(\mathcal{L}))$ turns out to be independent of p. To see this, perform a (-1)–surgery on the meridian of L_2, see Figure 6.6 where the meridian is denoted by $\mathcal{O}$.

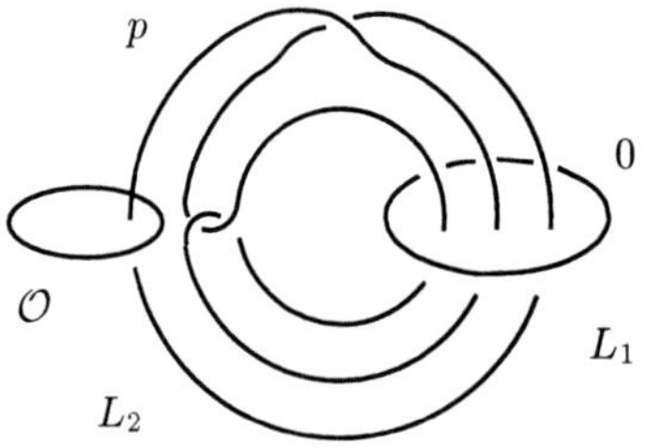

Fig. 6.6.

This surgery results in replacing framing p by framing $p+1$ while preserving the rest of the picture. On the other hand, 0-surgery along the same knot yields the manifold $\overline{K} = S^1 \times S^2$ with the trivial Floer homology, $I_*(\overline{K}) = 0$, see Example 6.35. The Floer exact triangle

$$\begin{array}{ccc} & I_*(\overline{K}) & \\ \swarrow & & \nwarrow \\ I_*(\Sigma_p) & \xrightarrow{\;Z_*\;} & I_*(\Sigma_{p+1}) \end{array}$$

now implies that the trace Z of the (-1)–surgery induces a degree zero isomorphism in Floer homology, which proves that $I_*(\Sigma_p(\mathcal{L}))$ is independent of p.

This observation allows to compute $I_*(\Sigma_p(\mathcal{L}))$ for links $\mathcal{L}$ having the property that $\Sigma_p(\mathcal{L})$ is Seifert fibered for at least one choice of p, see [271]. For example, the manifold shown in Figure 6.6 is homeomorphic to $\Sigma(2,5,7)$ when $p = 3$. Even when $I_*(\Sigma_p(\mathcal{L}))$ cannot be computed explicitly, one can often show that the groups $I_*(\Sigma_p(\mathcal{L}))$ are 2–periodic, see [271].

6.5.4 Special Boundary Maps

Let Σ be an integral homology sphere and assume for simplicity that the representation space $\mathcal{R}^*(\Sigma)$ is non-degenerate. For any $\alpha \in \mathcal{R}^*(\Sigma)$, consider the ASD moduli spaces $\mathcal{M}(\alpha,\theta)$ and $\mathcal{M}(\theta,\alpha)$ where θ stands for the trivial connection. Their dimensions can be computed as in Proposition 6.7 to be $\dim \mathcal{M}(\theta,\alpha) = -3 - \mu(\alpha) \bmod 8$ and $\dim \mathcal{M}(\alpha,\theta) = \mu(\alpha) \bmod 8$, where $\mu(\alpha) = \dim \mathcal{M}(\alpha,\theta) \bmod 8$ is the Floer index. Both $\mathcal{M}(\theta,\alpha)$ and $\mathcal{M}(\alpha,\theta)$ are oriented smooth manifolds with a free $\mathbb{R}$–action by translations. Let $\hat{\mathcal{M}}(\theta,\alpha) = \mathcal{M}(\theta,\alpha)/\mathbb{R}$ and $\hat{\mathcal{M}}(\alpha,\theta) = \mathcal{M}(\alpha,\theta)/\mathbb{R}$ and define maps

$$\delta : IC_1(\Sigma) \to \mathbb{Z} \quad \text{and} \quad \delta' : \mathbb{Z} \to IC_4(\Sigma) \tag{6.22}$$

as the homomorphisms such that

$$\delta(\alpha) = \#\,\hat{\mathcal{M}}(\alpha,\theta) \quad \text{and} \quad \delta'(1) = \sum_\beta \#\,\hat{\mathcal{M}}(\theta,\beta)\cdot\beta.$$

Here, $\#\hat{\mathcal{M}}(\alpha,\theta)$ and $\#\hat{\mathcal{M}}(\theta,\beta)$ stand for oriented counts of points in compact zero–dimensional spaces $\hat{\mathcal{M}}(\alpha,\theta)$ and $\hat{\mathcal{M}}(\theta,\beta)$ wit $\mu(\alpha) = 1$ and $\mu(\beta) = 4$. These maps satisfy the equations $\delta\partial = 0$ and $\partial\delta' = 0$ (where ∂ is the Floer boundary operator) for the same reason that $\partial^2 = 0$, see Proposition 6.8. Therefore, they define homomorphisms in the Floer homology, which are called *special boundary operators*,

$$\delta : I_1(\Sigma) \to \mathbb{Z} \quad \text{and} \quad \delta' : \mathbb{Z} \to I_4(\Sigma).$$

Example 6.48. The Poincaré homology sphere $\Sigma(2,3,5)$ is canonically oriented as the link of complex surface singularity (1.1). We saw in Example 6.23 that $I_1(\Sigma(2,3,5)) = I_5(\Sigma(2,3,5)) = \mathbb{Z}$ and $I_n(\Sigma(2,3,5)) = 0$ otherwise. The map δ' vanishes by a dimension count. On the contrary, $\delta : I_1(\Sigma(2,3,5)) \to \mathbb{Z}$ is an isomorphism. This result is due to Kronheimer (unpublished) and can also be found in [19] and [192]. With reversed orientation, $\delta' : \mathbb{Z} \to I_4(-\Sigma(2,3,5))$ is an isomorphism and δ is zero.

The maps δ and δ' play role in computing the Floer homology of connected sums of homology spheres, see Section 6.5.7. Frøyshov [104] used δ and δ' to define an invariant of rational homology cobordism now referred to as the *Frøyshov invariant.*

6.5.5 The u–map in Floer Homology

Let Σ be an integral homology sphere and assume for simplicity that its representation space $\mathcal{R}^*(\Sigma)$ is non-degenerate: all the constructions below are valid after an admissible perturbation. The u–map in Floer homology arises from an attempt to define a pairing $H^k(\mathcal{B}^*,\mathbb{Z}) \times I_n(\Sigma) \to I_{n-k}(\Sigma)$, which would be the intersection product if we dealt with a Morse function on a finite dimensional manifold. The cohomology $H^*(\mathcal{B}^*, \mathbb{Z})$ has a distinguished generator u in dimension 4 constructed as follows.

Let E be a trivial $SU(2)$–bundle over $\mathbb{R} \times \Sigma$ and denote by $\mathcal{G}$ the group of automorphisms of E in proper Sobolev completion. It acts as the gauge group on the space of irreducible connections $\mathcal{A}^*$ so that we can form the fiber product $\mathcal{A}^* \times_{\mathcal{G}} E$. The natural quotient map $\mathcal{A}^* \times_{\mathcal{G}} E \to \mathcal{B}^* \times (\mathbb{R} \times \Sigma)$ is a principal $SO(3)$–bundle which we call ξ. Then we have the first Pontryagin class $p_1(\xi)$, and slant product with this class defines a homomorphism $H_0(\mathbb{R} \times \Sigma, \mathbb{Z}) \to H^4(\mathcal{B}^*, \mathbb{Z})$ by sending $1 \in H_0(\mathbb{R} \times \Sigma, \mathbb{Z})$ to a class which we again call $p_1(\xi)$. Since $p_1(\xi) = w_2(\xi)^2 = 0 \mod 4$, this map is divisible by 4, and we define $u = (-1/4) \cdot p_1(\xi) \in H^4(\mathcal{B}^*, \mathbb{Z})$. Notice that this definition is an instance of a more general construction used to define Donaldson invariants, see Section 6.6.1.

Choose a lift of u which is a linear map from singular 4–chains in $\mathcal{B}^*$ to $\mathbb{Z}$ and call it u again. Special 4–chains occur in moduli spaces $\mathcal{M}(\alpha, \beta)$ with $\dim \mathcal{M}(\alpha, \beta) = \mu(\alpha) - \mu(\beta) = 4 \mod 8$. These chains define a map of abelian groups, $u : IC_n(\Sigma) \to IC_{n-4}(\Sigma)$, by the formula

$$u(\alpha) = \sum_{\mu(\beta)=\mu(\alpha)-4} \langle u, \mathcal{M}(\alpha, \beta) \rangle \cdot \beta, \tag{6.23}$$

where $\langle\ ,\ \rangle$ stands for the evaluation map. In order for (6.23) to be a chain map and as such define a map in Floer homology, the following diagram must commute:

$$\begin{array}{ccc} IC_n(\Sigma) & \xrightarrow{u} & IC_{n-4}(\Sigma) \\ \partial \downarrow & & \partial \downarrow \\ IC_{n-1}(\Sigma) & \xrightarrow{u} & IC_{n-5}(\Sigma) \end{array}$$

The commutativity of this diagram is equivalent to the equality

$$\sum_{\beta} \langle u, \mathcal{M}(\alpha, \beta) \rangle \cdot \#\hat{\mathcal{M}}(\beta, \gamma) = \sum_{\beta'} \#\hat{\mathcal{M}}(\alpha, \beta') \cdot \langle u, \mathcal{M}(\beta', \gamma) \rangle,$$

where $\mu(\beta) = \mu(\alpha) - 4$, $\mu(\beta') = \mu(\alpha) - 1$, and $\mu(\gamma) = \mu(\alpha) - 5$. Following the proof of Proposition 6.8, we need to check that the 4–dimensional part of the boundary of $\mathcal{M}(\alpha, \gamma)$ (this boundary may also contain low–dimensional strata) only arises from $\mathcal{M}(\alpha, \beta)$ in factorizations $\alpha \to \beta \to \gamma$ and from $\mathcal{M}(\beta', \gamma)$ in factorizations $\alpha \to \beta' \to \gamma$. This is true unless the trivial connection occurs in the middle, which may only happen if $\mu(\alpha) = \dim \mathcal{M}(\alpha, \theta) = 1 \mod 8$. Therefore, the map

$$u : I_n(\Sigma) \to I_{n-4}(\Sigma)$$

is well defined unless $n = 1$ or $n = 0$. In the latter two cases, only the maps $u : \ker \delta \to I_5(\Sigma)$ and $u : I_0(\Sigma) \to \operatorname{coker} \delta'$ are defined, where δ and δ' are special boundary maps defined in Section 6.5.4.

Theorem 6.49. *For any integral homology sphere Σ and for any $n \neq 0, 1$ modulo 8, the maps $u : I_n(\Sigma) \to I_{n-4}(\Sigma)$ are isomorphisms over any field of characteristic different from 2.*

Corollary 6.50. *Over any field of characteristic not equal to 2, Floer homology groups $I_*(\Sigma)$ are 4-periodic, $I_*(\Sigma) = I_{*+4}(\Sigma)$.*

Theorem 6.49 is proved in [104]. The reader should be warned that [104] uses different orientation conventions: the Floer indices of [104] are obtained from ours by switching from n to $-3-n$.

Example 6.51. An unpublished result of Kronheimer, see also [192], asserts that $u : I_5(\Sigma(2,3,5)) \to I_1(\Sigma(2,3,5))$ is multiplication by ± 2, hence it is an isomorphism over $\mathbb{Q}$ but not over the integers or $\mathbb{Z}/2$.

Example 6.52. The Floer homology need not be 4–periodic over a field of characteristic two. For example, Floer homology of $\Sigma(2,3,5) \# \Sigma(2,3,5)$ is not 4–periodic over $\mathbb{Z}/2$, see Example 6.58.

Remark 6.53. Precise relation between maps (6.17) and (6.23) is described in [104]. Unlike (6.23), the map (6.17) is well defined for all n and it is an isomorphism over the integers. This implies that the Floer homology of any integral homology sphere of the type $S^3 \pm k$ is 4–periodic over the integers. Indeed, $I_*(S^3 \pm k) = I_*(\overline{K})$ from the Floer exact triangle, and $I_*(\overline{K})$ is 4–periodic.

Example 6.54. The Floer homology groups of Seifert fibered homology spheres $\Sigma(a_1, \dots, a_n)$ are free abelian, see Section 6.4.1, therefore, 4–periodic over the integers.

In another application, Frøyshov [104] used the u–map to prove the simple type conjecture of Kronheimer and Mrowka concerning the structure of Donaldson invariants, see Example 6.73.

6.5.6 Integer Graded Floer Homology

Fintushel and Stern [84] extended the Floer homology theory to a theory with integer grading. Given an integral homology sphere Σ, assume for simplicity that all irreducible representations $\alpha : \pi_1\Sigma \to SU(2)$ are non-degenerate, and also assume that $\mathbf{cs}(\alpha) \neq 0 \bmod \mathbb{Z}$. These restrictions are satisfied for instance by all Brieskorn homology spheres, and in general can be removed.

Let $\alpha : \pi_1\Sigma \to SU(2)$ be an irreducible representation. It corresponds to a flat connection in the trivial bundle over Σ. Fix a trivialization of this bundle; this gives us a fixed trivial connection θ. Application of a gauge transformation $g : \Sigma \to SU(2)$ to α changes $\mathbf{cs}(\alpha)$ to $\mathbf{cs}(\alpha) + \deg g$. Therefore, there exists a unique flat connection, which we again call α, gauge equivalent to α and such that $\mathbf{cs}(\alpha) \in (-1, 0)$. Let

$$I(\alpha) = \dim \mathcal{M}(\alpha, \theta),$$

so that $I(\alpha) = \mu(\alpha) \bmod 8$ and it can be computed by using Theorem 6.15 in terms of $\mathbf{cs}(\alpha)$ and $\rho_\alpha(\Sigma)$.

For any $m \in \mathbb{Z}/8$ choose an integer s such that $s = m \bmod 8$ and define $F_s IC_m(\Sigma)$ to be abelian group generated by representations α with $I(\alpha) = s + 8j$, $j \geq 0$. Then

$$\ldots \subset F_s IC_m(\Sigma) \subset F_{s-8} IC_m(\Sigma) \subset \ldots \subset IC_m(\Sigma)$$

is a bounded filtration of the Floer chain group $IC_m(\Sigma)$. The fact that $\mathbf{cs}$ is non-decreasing along gradient trajectories, see Example 6.5, implies that Floer's boundary operator

$$\partial : F_s IC_m(\Sigma) \to F_{s-1} IC_{m-1}(\Sigma)$$

preserves the filtration. Thus Floer's $\mathbb{Z}/8$–graded chain complex $(IC_*(\Sigma), \partial)$ has a decreasing filtration $(F_s IC_*(\Sigma), \partial)$. For $m \in \mathbb{Z}/8$ and $s = m \bmod 8$, let $\mathcal{I}_{s,m}(\Sigma)$ be the homology of the complex

$$\ldots \to F_{s+1} IC_{m+1}(\Sigma) \xrightarrow{\ \partial\ } F_s IC_m(\Sigma) \xrightarrow{\ \partial\ } F_{s-1} IC_{m-1} \to \ldots$$

Then we have a bounded filtration of $I_m(\Sigma)$ defined by

$$F_s I_m(\Sigma) = \operatorname{im}\{\, \mathcal{I}_{s,m}(\Sigma) \to I_m(\Sigma)\,\}$$

with

$$\ldots \subset F_{s+8} I_m(\Sigma) \subset F_s I_m(\Sigma) \subset F_{s-8} I_m(\Sigma) \subset \ldots \subset I_m(\Sigma)$$

and the corresponding spectral sequence.

Theorem 6.55. *(Fintushel–Stern [84]) There is a convergent E^1–spectral sequence $(E^r_{s,m}(\Sigma), d^r)$ such that, for $m \in \mathbb{Z}/8$ and $s = m \bmod 8$, the groups $E^r_{s,m}$ are topological invariants and*

$$E^\infty_{s,m}(\Sigma) = F_s I_m(\Sigma)/F_{s+8} I_m(\Sigma).$$

The groups

$$\tilde{I}_s(\Sigma) = E^1_{s,m}(\Sigma), \quad s \in \mathbb{Z} \tag{6.24}$$

are called the *integer graded Floer homology*. These groups have the same functorial properties as the $\mathbb{Z}/8$–graded Floer homology. The special boundary maps and the u–map can be extended to the integer graded Floer homology, see [192].

Example 6.56. Fintushel and Stern show in [84] that the algorithm for computing $I_*(\Sigma(p,q,r))$, which we described in Section 6.4.1, can be adapted to compute the integer graded Floer homology of Brieskorn homology spheres. For example,

$$\begin{aligned}
P(2,3,5) &= t + t^5, \\
P(2,3,7) &= t^{-1} + t^3, \\
P(2,3,11) &= t + t^3 + t^5 + t^7 \\
P(3,4,13) &= t^{-5} + 2t^{-3} + 3t^{-1} + 2t + 2t^3,
\end{aligned}$$

where $P(p,q,r)(t)$ stands for the Poincaré polynomial of $\tilde{I}_s(\Sigma(p,q,r))$. In particular, the groups $\tilde{I}_s(\Sigma(p,q,r))$ are not 4–periodic as was the case for $I_*(\Sigma(p,q,r))$, see Example 6.54.

6.5.7 Floer Homology of Connected Sums

By the Mayer–Vietoris exact sequence, a connected sum $\Sigma_0 \# \Sigma_1$ of integral homology spheres Σ_0 and Σ_1 is again an integral homology sphere. The fundamental group of $\Sigma_0 \# \Sigma_1$ is isomorphic to the free product of the fundamental groups of the factors, $\pi_1(\Sigma_0 \# \Sigma_1) = \pi_1(\Sigma_0) * \pi_1(\Sigma_1)$. According to this splitting, there are four types of representations of $\pi_1(\Sigma_0 \# \Sigma_1)$ in $SU(2)$,

$$\theta * \theta, \quad \alpha_0 * \theta, \quad \theta * \alpha_1, \quad \alpha_0 * \alpha_1, \tag{6.25}$$

where $\alpha_k : \pi_1(\Sigma_k) \to SU(2)$ are non-trivial (hence irreducible) representations, $k = 0, 1$, and θ is one of the two trivial representations. The last three types of representations (6.25) are irreducible.

Suppose for simplicity that both $\mathcal{R}^*(\Sigma_0)$ and $\mathcal{R}^*(\Sigma_1)$ are non-degenerate: the general case will follow after applying a generic perturbation. Then the Mayer–Vietoris exact sequence in group cohomology implies that representations $\alpha_0 * \theta$ and $\theta * \alpha_1$ are non-degenerate in $\mathcal{R}^*(\Sigma_0 \# \Sigma_1)$. Moreover, for every pair $\alpha_0 \in \mathcal{R}^*(\Sigma_0)$, $\alpha_1 \in \mathcal{R}^*(\Sigma_1)$, there is a critical submanifold in $\mathcal{R}^*(\Sigma_0 \# \Sigma_1)$ which is homeomorphic to $SO(3)$ and is non-degenerate in the Morse–Bott sense, see Section 5.3.5. Therefore, even if $\mathcal{R}^*(\Sigma_0)$ and $\mathcal{R}^*(\Sigma_1)$ are non-degenerate, the space $\mathcal{R}^*(\Sigma_0 \# \Sigma_1)$ will contain higher-dimensional components and will need to be perturbed. This can be done with the help of the Morse–type perturbations, see Section 5.3.5. In particular, every $SO(3)$ component will contribute its Euler characteristic into the Casson invariant of $\Sigma_0 \# \Sigma_1$. Since the Euler characteristic of $SO(3)$ is zero, $\lambda(\Sigma_0 \# \Sigma_1) = \lambda(\Sigma_0) + \lambda(\Sigma_1)$, compare with Section 3.3.3.

Calculation of the Floer homology of $\Sigma_0 \# \Sigma_1$ is more involved and requires an analogue of the Morse–Bott spectral sequence, see Section 6.2.1. Due to some technical issues, such a spectral sequence only computes the integer graded Floer homology, see Section 6.5.6. In order to obtain the $\mathbb{Z}/8$–graded Floer homology of $\Sigma_0 \# \Sigma_1$, one should further use the spectral sequence of Theorem 6.55.

Theorem 6.57. *Let $\Sigma_0 \# \Sigma_1$ be a connected sum of integral homology spheres Σ_0 and Σ_1, and let $\tilde{I}_*$ denote the integer graded Floer homology, see (6.24). Then there is a spectral sequence $(E^r_{p,q}, d^r)$ which converges to $\tilde{I}_*(\Sigma_0 \# \Sigma_1)$.*

This theorem is proved in [192]. Its proof is based on some difficult analysis conducted in [194] which allows to represent ASD connections over $\mathbb{R} \times (\Sigma_0 \# \Sigma_1)$in terms of those over $\mathbb{R} \times \Sigma_0$ and $\mathbb{R} \times \Sigma_1$. The E^1–term of this spectral sequence and differentials d^r are explicitly described in [192] using the special boundary maps and the u–map in Floer homology. Differentials d^r with $r \geq 5$ all vanish, therefore, the spectral sequence collapses at $E^5_{p,q} = E^\infty_{p,q}$.

The above spectral sequence simplifies a great deal over the rational numbers. In particular, the terms $E^r_{p,q}$ vanish unless $r = 0$ and 3, so that the only non-trivial differentials are d^1 and d^4. This simplified spectral sequence can also be found in [108].

Example 6.58. Let $\Sigma = \Sigma(2,3,5) \# \Sigma(2,3,5)$ be the double of the Poincaré homology sphere. The $\mathbb{Z}/8$–graded Floer homology of Σ is computed in [192] to be as follows:

$$\begin{aligned} &I_0(\Sigma) = 0,\ I_1(\Sigma) = \mathbb{Z} \oplus \mathbb{Z} \oplus \mathbb{Z}/4,\ I_2(\Sigma) = 0,\ I_3(\Sigma) = \mathbb{Z}/2 \oplus \mathbb{Z}/2, \\ &I_4(\Sigma) = 0,\ I_5(\Sigma) = \mathbb{Z} \oplus \mathbb{Z}, \qquad\quad I_6(\Sigma) = 0,\ I_7(\Sigma) = \mathbb{Z}/2 \oplus \mathbb{Z}/2. \end{aligned}$$

In particular, $I_*(\Sigma)$ contains 2–torsion and $I_*(\Sigma)$ is 4–periodic over the rationals but not over the integers or over $\mathbb{Z}/2$, compare with Theorem 6.49 and Example 6.52.

Example 6.59. Let $\Sigma = \Sigma(2,3,5) \# - \Sigma(2,3,5)$ be a connected sum of two copies of the Poincaré homology sphere with two opposite orientations. Then $I_*(\Sigma) \otimes \mathbb{Q} = 0$, that is, $I_*(\Sigma)$ is all torsion, see [108] and [192].

Example 6.59 shows that in general $I_*(\Sigma_0 \# \Sigma_1)$ need not contain either $I_*(\Sigma_0)$ or $I_*(\Sigma_1)$ as a subgroup. However, in many instances this turns out to be the case, as demonstrated by Kawauchi [157].

Theorem 6.60. *Let Σ_1 be an integral homology sphere such that there exists a smooth embedding*

$$\Sigma_1 \subset n \cdot (-\mathbb{C}P^2) = (-\mathbb{C}P^2) \# \ldots \# (-\mathbb{C}P^2) \ (n \text{ times}).$$

Then $I_*(\Sigma_0) \subset I_*(\Sigma_0 \# \Sigma_1)$ for any integral homology sphere Σ_0.

Proof. We have a splitting $n \cdot (-\mathbb{C}P^2) = P \cup Q$ where P and Q are compact smooth manifolds such that $\partial P = \Sigma_1$, $\partial Q = -\Sigma_1$, and $b_2^+(P) = b_2^+(Q) = 0$. Let $U = ([-1,0] \times \Sigma_0) \natural P$ be the boundary connected sum along $D^3 \subset \{0\} \times \Sigma_0$ and $D^3 \subset \Sigma_1 = \partial P$, and similarly, $V = ([0,1] \times \Sigma_0) \natural Q$ the boundary connected sum along $D^3 \subset \{0\} \times \Sigma_0$ and $D^3 \subset (-\Sigma_1) = \partial Q$.

Then U is a smooth manifold with boundary $\partial U = -\Sigma_0 \cup (\Sigma_0 \# \Sigma_1)$. It induces a homomorphism $U_* : I_*(\Sigma_0) \to I_*(\Sigma_0 \# \Sigma_1)$ of degree zero. Similarly, V induces a homomorphism $V_* : I_*(\Sigma_0 \# \Sigma_1) \to I_*(\Sigma_0)$ of degree zero. According to Proposition 6.13, we have $V_* \circ U_* = (U \cup V)_*$ where $U \cup V = ([-1,1] \times \Sigma_0) \# n \cdot (-\mathbb{C}P^2)$ is a blow up of the cylinder $[-1,1] \times \Sigma_0$ in n points. As in Example 6.40 we conclude that $(U \cup V)_* = V_* \circ U_*$ is an isomorphism. Therefore, U_* is a monomorphism, which completes the proof.

Example 6.61. Theorem 6.60 and Example 6.59 together imply that the Poincaré homology sphere $\Sigma(2,3,5)$ cannot be embedded into $n \cdot (-\mathbb{C}P^2)$ for any n. There is of course a more direct way to see this: if $\Sigma(2,3,5)$ can be embedded into $n \cdot (-\mathbb{C}P^2)$ then there is a smooth manifold Q with boundary $-\Sigma(2,3,5)$ such that $b_2^+(Q) = 0$, compare with the proof of Theorem 6.60. Gluing V to the plumbed manifold W with negative definite intersection form E_8, see Example 1.16, gives us a closed smooth manifold with negative definite intersection form which is not diagonalizable over the integers. This contradicts Theorem 7.13.

Example 6.62. A smooth embedding $\Sigma(2,3,7) \subset (-\mathbb{C}P^2)\#(-\mathbb{C}P^2)$ can be constructed as follows. Consider a right–handed trefoil k in the boundary of D^4 and glue a 2–handle along k with framing (-1). The boundary of the resulting 4–manifold is $\Sigma(2,3,7)$. Attach another 2–handle as shown in Figure 6.7 and use Kirby calculus to show that the resulting 4–manifold is diffeomorphic to $(-\mathbb{C}P^2)\#(-\mathbb{C}P^2) \setminus D^4$. Filling in with a copy of D^4 provides the required embedding. Theorem 6.60 now implies that $I_*(\Sigma_0)$ is a subgroup of $I_*(\Sigma_0 \# \Sigma(2,3,7))$ for any integral homology sphere Σ_0.

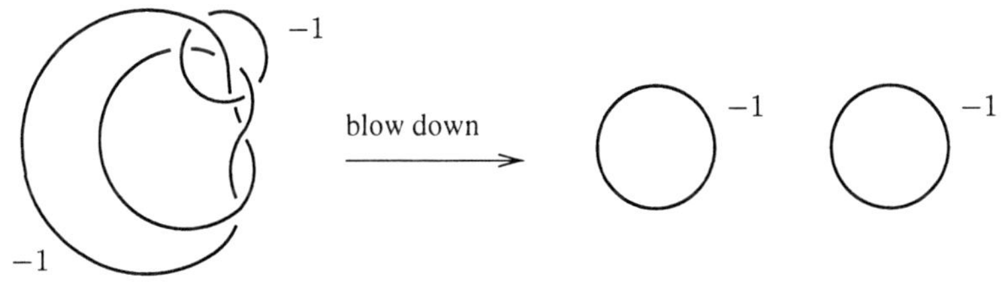

Fig. 6.7.

Example 6.63. Let $\Sigma \subset n \cdot (-\mathbb{C}P^2)$ then we have two embeddings $I_*(\Sigma) \to I_*(\Sigma \# \Sigma)$ corresponding to the two factors in $\Sigma \# \Sigma$. It is not difficult to show, see [271], that the images of these two embeddings have zero intersection, so that $I_*(\Sigma) \oplus I_*(\Sigma) \subset I_*(\Sigma \# \Sigma)$.

6.5.8 Functoriality with Respect to Diffeomorphisms

We saw in Section 6.2.3 and Section 6.5.2 that the Floer homology is functorial with respect to oriented cobordisms and $SO(3)$-bundles over them. The Floer homology is also functorial with respect to orientation preserving diffeomorphisms. Let M be either an integral homology sphere or a homology handle, and $f : M \to M$ an orientation preserving diffeomorphism. Let W_f be the mapping cylinder of f then W_f is an oriented cobordism between M and M such that the both inclusions $M \to W_f$ are homotopy equivalences. Any $SO(3)$–bundle over M extends uniquely over W_f, hence, we obtain a well defined homomorphism $(W_f)_* : I_*(M) \to I_*(M)$ of degree 0 which we denote by $f_* : I_*(M) \to I_*(M)$. This construction extends to orientation reversing diffeomorphisms f as long as the mapping cylinder of f is oriented.

The action of diffeomorphisms on the Floer homology is not very well understood. In fact, even finding examples of diffeomorphisms whose action on the Floer homology is non-trivial, is not easy. If a diffeomorphism $f : M \to M$ is such that $f_* \neq \mathrm{id}$, its mapping cylinder is a smooth 4–manifold which is homeomorphic but not diffeomorphic to the cylinder $[0,1] \times M$ relative to the boundary.

Example 6.64. For any diffeomorphism $f : \Sigma \to \Sigma$ of an integral homology sphere and any representation $\alpha : \pi_1(\Sigma) \to SU(2)$, we have $\mathbf{cs}(f^*\alpha) = \mathbf{cs}(\alpha)$. In particular, for any diffeomorphism of a Brieskorn homology sphere $\Sigma(p,q,r)$, the induced map on the Floer homology has to be the identity. This is due to the fact that, for any $\alpha, \beta \in \mathcal{R}^*(\Sigma(p,q,r))$, the equality $\mathbf{cs}(\alpha) = \mathbf{cs}(\beta)$ implies that $\alpha = \beta$, see Example 5.2.

Example 6.65. Let Σ be an integral homology sphere embedded into $n \cdot (-\mathbb{C}P^2)$, and consider the involution $\tau : \Sigma \# \Sigma \to \Sigma \# \Sigma$ permuting the two factors. According to Example 6.63, the Floer homology $I_*(\Sigma \# \Sigma)$ contains a subgroup $I_*(\Sigma) \oplus I_*(\Sigma)$ on which τ acts by permuting the factors. Therefore, the action of τ on $I_*(\Sigma \# \Sigma)$ is non-trivial.

Example 6.66. An irreducible integral homology sphere Σ with an involution acting non-trivially on $I_*(\Sigma)$ was constructed in [274]. The manifold Σ is a plumbed homology sphere obtained by (-1)–surgery on the right handed granny knot.

Example 6.67. Another irreducible example can be found in [276]. Let Σ be the integral homology sphere obtained by surgery on the link shown in Figure 7.3 with $p = 0$. The involution $\tau : \Sigma \to \Sigma$ is induced by the involution interchanging the link components (note that both components are unknots). The manifold Σ is a hyperbolic integral homology sphere. It has played an important role in many instances such as constructing Akbulut corks, see Example 7.40, or giving a counterexample to the Zeeman conjecture, see Section 7.3.3.

6.5.9 Mutation Invariance

Mutation is an operation on 3–manifolds described in Section 1.1.11. Given an integral homology 3–sphere Σ and an embedded Riemann surface F of genus 2, it produces another integral homology sphere denoted by Σ^τ.

Theorem 6.68. *(Ruberman [264]) For any choice of a genus two surface F in an integral homology sphere Σ, there exists a natural isomorphism $I_*(\Sigma) = I_*(\Sigma^\tau)$. The same is true for the integer graded Floer homology of Fintushel and Stern, see Section 6.5.6.*

The isomorphism $I_*(\Sigma) = I_*(\Sigma^\tau)$ of Theorem 6.68 is natural in the sense that it is induced by an (orbifold) cobordism between Σ and Σ^τ. It should be mentioned that Theorem 6.68 also provides an alternative proof of the mutation invariance of the Casson invariant, see Section 3.4.2.

6.6 Floer Homology in Donaldson Theory

In this section we briefly sketch the construction of Donaldson polynomial invariants of a smooth compact oriented 4–manifold X. If X is a closed manifold, its Donaldson invariants are polynomials with rational coefficients. If X has non-empty boundary, the Donaldson invariants of X are polynomials with coefficients in the Floer homology of ∂X, if at all defined. Gluing 4–manifolds along common boundary corresponds to pairing their Donaldson invariants via (6.16), at least heuristically. Making this statement precise requires either imposing certain technical conditions or using generalized Floer homology, see Section 6.7.

6.6.1 Donaldson Invariants of Closed 4–Manifolds

Our exposition in this section follows closely Morgan's lectures [227]. We assume throughout this section that X is a smooth closed oriented simply connected 4–manifold with a fixed orientation of $H^2_+(X,\mathbb{R})$ such that $b_2^+(X)$ is odd and not less than 3. For any non-negative integer d, the *Donaldson polynomial* $\gamma_d(X)$ is a symmetric multilinear function

$$\gamma_d(X) : \underbrace{H_2(X,\mathbb{Z}) \otimes \ldots \otimes H_2(X,\mathbb{Z})}_{d} \to \mathbb{Q},$$

or, equivalently, a homogeneous polynomial function

$$\gamma_d(X) : H_2(X,\mathbb{Z}) \to \mathbb{Q}$$

of degree d. The main property of the polynomials γ_d is that each of them is an orientation preserving diffeomorphism invariant of the pair $\{\, X$, orientation of $H^2_+(X,\mathbb{R}) \,\}$. The invariants γ_d are defined as follows.

Let E be a principal $SU(2)$–bundle over X and choose a Riemannian metric on X. Let $\mathcal{B}^*(E) = \mathcal{A}^*(E)/\mathcal{G}(E)$ be the Banach manifold of the gauge equivalence classes of irreducible connections in E, see Section 6.1.1. The group $\mathcal{G}(E)$ acts by automorphisms on E so that we can form the fiber product $\mathcal{A}^*(E) \times_{\mathcal{G}(E)} E$. The natural projection

$$\mathcal{A}^*(E) \times_{\mathcal{G}(E)} E \to \mathcal{B}^*(E) \times X$$

is a principal $SO(3)$–bundle which we denote by ξ. Then we have the first Pontryagin class $p_1(\xi)$, and slant product with this class defines a homomorphism

$$H_2(X,\mathbb{Z}) \to H^2(\mathcal{B}^*(E),\mathbb{Z}), \quad x \mapsto p_1(\xi)/x.$$

Restriction of this map to the moduli space $\mathcal{M}^*(E) \subset \mathcal{B}^*(E)$ of ASD connections, see Section 6.1.1, is a map $H_2(X,\mathbb{Z}) \to H^2(\mathcal{M}^*(E),\mathbb{Z})$. Since $p_1(\xi) = w_2(\xi)^2 = 0 \bmod 4$, we can define the *$\mu$–map*

$$\mu : H_2(X,\mathbb{Z}) \to H^2(\mathcal{B}^*(E),\mathbb{Z})$$

by the formula $\mu(x) = (-1/4) \cdot p_1(\xi)/x$. By restriction, it defines a map $\mu : H_2(X,\mathbb{Z}) \to H^2(\mathcal{M}^*(E),\mathbb{Z})$. There is a similar map

$$\mu : H_0(X,\mathbb{Z}) \to H^4(\mathcal{B}^*(E),\mathbb{Z})$$

given by sending $1 \in H_0(X,\mathbb{Z})$ to $(-1/4)\cdot p_1(\xi)$. This map is also integral. The significance of the μ–map is that its image generates the rational cohomology $H^*(\mathcal{B}^*(E),\mathbb{Q})$. More precisely,

$$H^*(\mathcal{B}^*(E),\mathbb{Q}) = S^*(\mu(H_2(X,\mathbb{Q}))) \otimes S^*(\mu(H_0(X,\mathbb{Q}))),$$

where $S^*(V)$ denotes the symmetric algebra generated by V. The usual grading of this algebra is induced by defining the vector space $\mu(H_i(X,\mathbb{Q}))$ to be homogeneous of degree $4 - i$.

Let d be a non-negative integer such that $d = (-3/2) \cdot (1 + b_2^+(X)) \bmod 4$ then there exists a principal $SU(2)$–bundle E over X such that the formal dimension of the moduli space $\mathcal{M}(E)$ equals $2d$, compare with (6.3). For a generic metric on X, the moduli space $\mathcal{M}(E) = \mathcal{M}^*(E)$ is a smooth oriented manifold of dimension $2d$. For any $x_1, \dots, x_d \in H_2(X,\mathbb{Z})$ we define

$$\gamma_d(X)(x_1,\dots,x_d) = (\mu(x_1) \cup \ldots \cup \mu(x_d))\,[\mathcal{M}(E)].$$

Because $\mathcal{M}(E)$ is non-compact in general, the above formula needs to be correctly interpreted before it can be regarded as well defined. A recipe for evaluation producing integral valued invariants was given in [69] subject to the constraint $2d > 3(1 + b_2^+(X))$. The construction has later been extended so as to remove this restriction, see [227]. Due to the nature of this extension, the resulting invariant takes rational values.

If d is a non-negative integer such that $d = 2 - (3/2) \cdot (1 + b_2^+(X)) \bmod 4$, we choose a bundle E such that $\dim \mathcal{M}(E) = 2d + 4$ and define

$$\gamma_d(X)(x_1, \ldots, x_d) = \frac{1}{2}\left(\mu(x_1) \cup \ldots \cup \mu(x_d) \cup \mu(1)\right)[\mathcal{M}(E)]$$

where $1 \in H_0(X, \mathbb{Z})$. If d is not equal to $(-3/2) \cdot (1 + b_2^+(X)) \bmod 2$ we define $\gamma_d(X) = 0$. All the polynomials γ_d can be combined into one analytic function or formal power series $\gamma : H_2(X, \mathbb{R}) \to \mathbb{R}$ by defining

$$\gamma(h) = \sum_d \frac{\gamma_d(X)(h)}{d!}. \tag{6.26}$$

This function is well-defined if $b_2^+(X) \geq 1$ and is an invariant of orientation preserving diffeomorphism if $b_2^+(X) \geq 3$. In particular, it is independent of the choice of Riemannian metric.

The Donaldson polynomials γ_d can be extended to symmetric multilinear functions

$$\gamma_{d,p}(X) : \underbrace{H_2(X, \mathbb{Z}) \otimes \ldots \otimes H_2(X, \mathbb{Z})}_{d} \otimes \underbrace{H_0(X, \mathbb{Z}) \otimes \ldots \otimes H_0(X, \mathbb{Z})}_{p} \to \mathbb{Q}$$

by defining

$$\gamma_{d,p}(X)(x_1, \ldots, x_d, 1, \ldots, 1) = \left(\mu(x_1) \cup \ldots \cup \mu(x_d) \cup \mu(1)^p\right)[\mathcal{M}(E)],$$

where $\dim \mathcal{M}(E) = 2d + 4p$.

Remark 6.69. The group $SU(2)$ can be replaced with $SO(3)$ throughout the above construction. The resulting invariants are called *$SO(3)$-Donaldson polynomials*, see [71]. It should also be mentioned that, with some extra precautions, this theory can be extended to non-simply connected manifolds.

Donaldson polynomials are one of the very few tools available to distinguish between homeomorphic but non-diffeomorphic 4–manifolds. They have been extensively studied over the years and a wealth of information about them has been amassed, see [178], [227], [288] etc. Listed below are just a few results which will be used later on.

Example 6.70. Let X_1 and X_2 be closed oriented simply connected 4–manifolds with $b_2^+(X_1) > 0$ and $b_2^+(X_2) > 0$. If X is orientation preserving diffeomorphic to a connected sum $X_1 \# X_2$ then all Donaldson polynomials of X vanish, $\gamma_d(X) = 0$ for $d \geq 0$. This is known as the *connected sum theorem*, see [71]. It can be viewed as a special case of the gluing formula for relative Donaldson polynomials, see Theorem 6.85.

Example 6.71. Let X be a simply connected algebraic surface. Then for all sufficiently large integers $d > 0$ such that $d = 3(1 + b_2^+)(X) \bmod 4$ the Donaldson polynomial $\gamma_d(X)$ is non-trivial. This result is usually referred to as the *non-vanishing theorem* for algebraic surfaces, see [71]. Together with the connected sum theorem, it implies that simply connected algebraic surfaces cannot split into connected sum of factors with positive b_2^+.

Example 6.72. Let X be the *Kummer K3 surface,*

$$X = \{\, [z_0 : z_1 : z_2 : z_3] \in \mathbb{C}P^3 \mid z_0^4 + z_1^4 + z_2^4 + z_3^4 = 0 \,\}.$$

This is a smooth closed simply connected 4–manifold. Its intersection form Q is isomorphic over the integers to $2{\cdot}E_8 \oplus 3{\cdot}H$ where E_8 is the negative definite E_8–form associated with the tree in Figure 1.14, and H is the hyperbolic two-by-two form,

$$H = \begin{pmatrix} 0 & 1 \\ 1 & 0 \end{pmatrix}. \tag{6.27}$$

The Donaldson polynomials for X are as follows, see [178],

$$\gamma_{2k} = \frac{(2k)!}{2^k\, k!}\, Q^k \quad \text{or} \quad \gamma = \exp\left(\frac{Q}{2}\right).$$

Example 6.73. A manifold X is said to have *simple type* if, for any choice of $x_1, \ldots, x_d \in H_2(X, \mathbb{Z})$ and for any $p \geq 0$,

$$\gamma_{d,p+2}(X)(x_1, \ldots, x_d, \underbrace{1, \ldots, 1}_{p+2}) = 4 \cdot \gamma_{d,p}(X)(x_1, \ldots, x_d, \underbrace{1, \ldots, 1}_{p}),$$

see Kronheimer and Mrowka [178]. All manifolds X with $b_2^+(X) \geq 3$ have simple type, see [104], [234], and [305]. The proof in [104] is based on the study of the u–map in Floer homology.

The simple type condition has the following significance, see [178]. If X is a simply connected 4–manifold of simple type, then there exist finitely many cohomology classes $K_1, \ldots, K_m \in H^2(X, \mathbb{Z})$ and non-zero rational numbers $a_1, \ldots, a_m$ such that the series (6.26) is given by the formula

$$\gamma(X) = \exp\left(\frac{Q}{2}\right) \cdot \sum_{s=1}^{m} a_s \exp(K_s)$$

as an analytic function on $H_2(X, \mathbb{R})$. Here, Q is the intersection form of X. The classes K_s are called *basic classes* and each of them is an integral lift of $w_2(X)$. For example, the only basic class of the Kummer $K3$ surface is zero, see Example 6.72.

Example 6.74. If X has simple type then the $SU(2)$ Donaldson series of $X\#(-\mathbb{C}P^2)$ is given by $\gamma(X\#(-\mathbb{C}P^2)) = \gamma(X) \cdot \exp(e^2/2) \cdot \cosh(e)$ where e

is dual to the generator of $H_2(\mathbb{C}P^2, \mathbb{Z})$ represented by $\mathbb{C}P^1 \subset \mathbb{C}P^2$, see [78]. A similar formula for $SO(3)$ Donaldson polynomials is obtained by replacing $\cosh(e)$ with $\sinh(e)$, see [78]. These formulas are known as the *blow up formulas* for Donaldson polynomials.

Example 6.75. Donaldson invariants provide many examples of smooth manifolds which are homeomorphic but not diffeomorphic to each other. For example, let $X_1 = K3 \# (-\mathbb{C}P^2)$ be a blow up of the Kummer surface and $X_2 = 3\,\mathbb{C}P^2 \# 20\,(-\mathbb{C}P^2)$. Then the intersection forms of X_1 and X_2 are isomorphic over the integers hence X_1 and X_2 are homeomorphic by Freedman's theorem [98]. However they are not diffeomorphic because X_1 is an algebraic surface and $\gamma(X_1) \neq 0$ by the non-vanishing theorem, see Example 6.71, and $\gamma(X_2) = 0$ by the connected sum theorem, see Example 6.70.

6.6.2 Relative Donaldson Polynomials

Let X be a compact oriented smooth simply connected 4–manifold whose boundary is an integral homology sphere Σ. Under suitable circumstances, one can associate with X relative Donaldson polynomials, which are homogeneous polynomial functions on $H_2(X, \mathbb{Z})$ with values in the Floer homology of Σ. Moreover, for any closed smooth oriented 4–manifold $W = X \cup_\Sigma Y$ split along Σ into smooth simply connected manifolds X and Y such that $\partial X = \Sigma$ and $\partial Y = -\Sigma$, the Donaldson polynomials of W can be computed from the relative Donaldson polynomials of X and Y via pairing (6.16). Atiyah [9] and Witten [306] realized that this pattern constitutes a $3+1$ topological quantum field theory, and formalized it in a functorial way. Despite this, it remains a challenging problem to implement the above scheme rigorously. The book [70] is the standard reference on the topic, see also [20], [39], [82] and [104].

Let E be a principal $SU(2)$–bundle over X, and use the same symbol X to denote the open manifold $X \cup (\mathbb{R}_+ \times \Sigma)$. Suppose for the sake of simplicity that all flat connections on Σ are non-degenerate; the general case will follow by applying an admissible perturbation. Let $\alpha \in \mathcal{R}^*(\Sigma)$ and denote by $\mathcal{M}_X(\alpha)$ the moduli space of ASD connections in E with finite Yang–Mills action limiting to α at the end $\mathbb{R}_+ \times \Sigma$. Notice that $\mathcal{M}_X(\alpha)$ consists of irreducible connections because their limit at infinity, α, is irreducible.

Proposition 6.76. *For any $\alpha \in \mathcal{R}^*(\Sigma)$, the moduli space $\mathcal{M}_X(\alpha)$ is generically a collection of smooth manifolds of dimensions $-\mu(\alpha) - 3(1 + b_2^+(X))$ mod 8 where $\mu(\alpha)$ is the Floer index of α.*

Proof. The additivity of the index implies that $\dim \mathcal{M}_X(\theta) = \dim \mathcal{M}_X(\alpha) + \mu(\alpha)$ hence $\dim \mathcal{M}_X(\alpha) = -\mu(\alpha) + \dim \mathcal{M}_X(\theta) = -\mu(\alpha) - 3(1 + b_2^+(X))$ modulo 8, compare with (6.3).

Let d be a non-negative integer and choose d homology classes $x_1, \ldots, x_d$ in $H_2(X, \mathbb{Z})$. Let $n = -2d - 3(1 + b_2^+)(X) \bmod 8$ so that $\dim \mathcal{M}_X(\alpha) = 2d \bmod 8$. The formula

$$\gamma_d(X)(x_1, \ldots, x_d) = \sum_{\alpha \in \mathcal{R}^*(\Sigma)} (\mu(x_1) \cup \ldots \cup \mu(x_d)) \, [\mathcal{M}_X(\alpha)] \cdot \alpha, \tag{6.28}$$

where the $2d$–dimensional cohomology class $\mu(x_1) \cup \ldots \cup \mu(x_d)$ is evaluated on the components of $\mathcal{M}_X(\alpha)$ of dimension $2d$, defines a chain in the Floer chain group $IC_n(\Sigma)$. It is known that great care has to be taken to make sense of the formula (6.28), see [39] and [229].

Proposition 6.77. *The chain $\gamma_0(X)$ is a cycle, that is, $\partial \gamma_0(X) = 0$. If $b_2^+(X) \geq 1$ then, for any $d \geq 0$ and for any choice of $x_1, \ldots, x_d$, the chain $\gamma_d(X)(x_1, \ldots, x_d)$ is a cycle.*

Proof. Let us first consider the case of $d = 0$. The formula (6.28) reduces to

$$\gamma_0(X) = \sum_{\alpha} \#\mathcal{M}_X(\alpha) \cdot \alpha,$$

where $\#\mathcal{M}_X(\alpha)$ is a signed count of points in the 0–dimensional part of $\mathcal{M}_X(\alpha)$. Proving that $\partial \gamma_0(X) = 0$ amounts to showing that

$$\sum_{\alpha} \#\mathcal{M}_X(\alpha) \cdot \#\hat{\mathcal{M}}(\alpha, \beta) = 0 \tag{6.29}$$

for any $\beta \in \mathcal{R}^*(\Sigma)$ of Floer index $\mu(\beta) = n - 1$. Consider the moduli space $\mathcal{M}_X(\beta)$ whose dimension is $1 \bmod 8$. The boundary points of the 1–dimensional components of $\mathcal{M}_X(\beta)$ correspond to factorizations of ASD connections on X limiting to β through flat connections α of Floer index $\mu(\alpha) = n$. Since the number of such boundary points is even, this shows that at least modulo 2 the coefficient (6.29) vanishes, compare with the proof of Proposition 6.8. In fact, the orientations work in such a way that this is also true over the integers.

An important point in the above argument is that the trivial connection θ does not appear as a boundary point of any of the 1–dimensional components of $\mathcal{M}_X(\beta)$. This follows from the dimension count: were θ to occur, the dimension of the moduli space $\mathcal{M}_X(\beta)$ would be at least $\dim \mathcal{M}_X(\theta) + 3 + \dim \mathcal{M}(\theta, \beta) \geq 3$.

The trivial connection may however occur as a limiting point for higher dimensional components, and this is why the condition $b_2^+ \geq 1$ is needed if we are to deal with $d \geq 1$.

Suppose that $d \geq 1$. For every $i = 1, \ldots, d$, consider a codimension two submanifold $V_i \subset \mathcal{B}^*(E)$ which is Poincaré dual to $\mu(x_i) \in H^2(\mathcal{B}^*(E), \mathbb{Z})$. That such a manifold exists away from reducible connections is guaranteed by the condition $b_2^+(X) \geq 1$, see [69] or [71]. Choose the surfaces V_i so that they intersect each other and $\mathcal{M}_X(\alpha)$ transversely. Then

$$(\mu(x_1) \cup \ldots \cup \mu(x_d))\,[\mathcal{M}_X(\alpha)] = \#\,(V_1 \cap \ldots \cap V_d \cap \mathcal{M}_X(\alpha)),$$

where # stands for a signed count of points. Since the surfaces V_i avoid the trivial connection θ, essentially the same argument as we used in the case $d = 0$ shows that $\partial\gamma_d(X)(x_1, \ldots, x_d) = 0$.

Let d be a non-negative integer and suppose that $b_2^+(X) \geq 1$ if $d \geq 1$. Then the formula (6.28) defines a symmetric bilinear function

$$\gamma_d(X) : \underbrace{H_2(X, \mathbb{Z}) \oplus \ldots \oplus H_2(X, \mathbb{Z})}_{d} \to I_n(\Sigma)$$

where $n = -2d - 3(1 + b_2^+)(X) \bmod 8$, which is called the *relative Donaldson polynomial* of degree d. It is an invariant of orientation preserving diffeomorphism of (X, Σ) (and a homology orientation).

The above definition of Donaldson polynomials can be extended to allow cup-products with $p \geq 0$ copies of the class $\mu(1)$:

$$\gamma_{d,p}(X)(x_1, \ldots, x_d, \underbrace{1, \ldots, 1}_{p}) = \sum_{\alpha \in \mathcal{R}^*(\Sigma)} (\mu(x_1) \cup \ldots \cup \mu(x_d) \cup \mu(1)^p)\,[\mathcal{M}_X(\alpha)] \cdot \alpha, \quad (6.30)$$

where $\mu(\alpha) = -2d - 4p - 3(1 + b_2^+) \bmod 8$. The formula (6.30) defines an element in the Floer chain complex. This element in general need not be a cycle, see Example 6.80 below, and hence need not define an element in the Floer homology. It will, however, define an element in the equivariant Floer homology $I_*^{\mathcal{G}}(\Sigma)$, see Section 6.7.1.

Remark 6.78. The group $SU(2)$ in the above constructions can be replaced with $SO(3)$ resulting in relative $SO(3)$ Donaldson polynomials which we call again $\gamma_d(X)$. The $SO(3)$ Donaldson polynomial $\gamma_d(X)$, when it is defined, takes values in $I_n(\Sigma)$ where $n = -2w_2(E)^2 - 3(1 + b_2^+)(X) - 2d \bmod 8$. The definition of Donaldson polynomials can also be extended to include non-simply connected manifolds.

Example 6.79. Let W be an oriented smooth cobordism between integral homology 3–spheres Σ_1 and Σ_2 so that $\partial W = -\Sigma_1 \cup \Sigma_2$. The degree zero Donaldson polynomial is just a class $\gamma_0(W) \in I_n(-\Sigma_1 \cup \Sigma_2)$ where $n = -3(1 - b_1 + b_2^+(W)) \bmod 8$. The formula $I_*(-\Sigma_1 \cup \Sigma_2) = I_*(-\Sigma_1) \otimes I_*(\Sigma_2)$ together with the pairing (6.16) allows to identify $\gamma_0(W)$ with the homomorphism $W_* : I_*(\Sigma_1) \to I_*(\Sigma_2)$ of degree $3(b_1 - b_2^+)(W)$, see (6.12).

Example 6.80. Let $X = \mathbb{R} \times \Sigma$ be an infinite cylinder over an integral homology sphere Σ. The relative Donaldson polynomial $\gamma_{0,1}(X)(1) \in IC_1(-\Sigma \cup \Sigma)$ can be identified with the u–map $u : IC_*(\Sigma) \to IC_{*-4}(\Sigma)$. We saw in Section 6.5.5 that this is not necessarily a chain map, which reflects the fact that $\gamma_{0,1}(X)(1)$ in general is not a cycle.

Example 6.81. According to Example 6.23 we have $I_n(\Sigma(2,3,5)) = \mathbb{Z}$ if $n = 1, 5$, and $I_n(\Sigma(2,3,5)) = 0$ otherwise. Let X be the smooth simply connected 4–manifold obtained by plumbing on the negative definite E_8–graph shown in Figure 1.14, then $\partial X = \Sigma(2,3,5)$. Let $e \in H^2(X,\mathbb{Z})$ be a cohomology class with square $e \cdot e = -2$ and let $E = L \oplus \mathbb{R}$ be the reducible $SO(3)$-bundle over X such that the Euler class of L is e. Corresponding to the pair (X, E) is a degree zero $SO(3)$ Donaldson polynomial $\gamma_0(X) \in I_n(\Sigma(2,3,5))$ where $n = -2\, w_2^2(E) - 3\,(1 + b_2^+)(X) = 1 \bmod 8$. Fintushel and Stern show in [91] that $\gamma_0(X) = \pm\alpha$ where α is a generator in $I_1(\Sigma(2,3,5))$.

Example 6.82. Let M be the Milnor fiber of singularity (1.2) so that M is a smooth simply-connected 4–manifold with boundary a Seifert fibered homology sphere $\Sigma(a_1,\dots,a_n)$. Assume that $\Sigma(a_1,\dots,a_n)$ is different from $\Sigma(2,3,5)$ then $b_2^+(M) > 0$. Ebeling and Okonek [74] show that $\gamma_d(M) = 0$ if d is odd, and

$$\gamma_d(M) = a_d\, Q^{d/2} \quad \text{for some} \quad a_d \in I_*(\Sigma(a_1,\dots,a_n)) \otimes \mathbb{Q}$$

if d is even. Here, Q is the intersection form of M viewed as a homogeneous quadratic function on $H_2(M,\mathbb{Z})$. In the special case of $\Sigma(2,3,6m \pm 1)$ with $6m \pm 1 > 5$, the coefficients a_d are non-trivial for all even d. Similar results hold for Milnor fibers of all isolated complete intersection singularities whose link is an integral homology sphere $\Sigma \neq \Sigma(2,3,5)$ such that $I_*(\Sigma)$ has no torsion, see [74].

Example 6.83. An example of contractible 4–manifold W with non-trivial Donaldson invariant $\gamma_0(W) \in I_5(\partial W)$ can be found in [276].

Relative Donaldson polynomials have the following functoriality property. Suppose that $\partial X = \Sigma$ and W is a cobordism with $\partial W = -\Sigma \cup \Sigma'$. Then, for any choice of $x_1,\dots,x_d \in H_2(X,\mathbb{Z})$,

$$\gamma_d(X \cup_\Sigma W)(x_1,\dots,x_d) = W_*\, \gamma_d(X)(x_1,\dots,x_d),$$

where $W_* : I_*(\Sigma) \to I_*(\Sigma')$ is the homomorphism induced by the cobordism W, see (6.12).

Example 6.84. If $\gamma_d(X)(x_1,\dots,x_d) \neq 0$ then $\gamma_d(X\#(-\mathbb{C}P^2))(x_1,\dots,x_d) \neq 0$ because

$$X\#(-\mathbb{C}P^2) = X \cup_\Sigma (([0,1] \times \Sigma)\#(-\mathbb{C}P^2)),$$

and $([0,1] \times \Sigma)\#(-\mathbb{C}P^2)$ induces an isomorphism in Floer homology, see Example 6.40.

6.6.3 The Gluing Formula

Let X be a smooth oriented simply connected closed 4–manifold which is decomposed into two pieces, $X = U \cup V$, where U and V are compact simply connected smooth manifolds with $\partial U = \Sigma$ and $\partial V = -\Sigma$. Suppose that Σ is an integral homology sphere then $H_2(X, \mathbb{Z}) = H_2(U, \mathbb{Z}) \oplus H_2(V, \mathbb{Z})$. The following is known as the *gluing formula* for Donaldson polynomials. We refer the reader to [70] for all the details.

Theorem 6.85. *Let $\gamma_d(X)$ be a degree d Donaldson polynomial of X corresponding to a bundle E over X, and let $u_1, \dots, u_r \in H_2(U, \mathbb{Z})$ and $v_1, \dots, v_{d-r} \in H_2(V, \mathbb{Z})$. If*

(a) $b_2^+(U) > 0$, or $w_2(E_U) \neq 0$ and $r = 0$, and
(b) $b_2^+(V) > 0$, or $w_2(E_V) \neq 0$ and $r = d$,

then

$$\gamma_d(X)(u_1, \dots, u_r, v_1, \dots, v_{d-r}) = \langle \gamma_r(U)(u_1, \dots, u_r), \gamma_{d-r}(V)(v_1, \dots, v_{d-r}) \rangle$$

where $\langle\ ,\ \rangle$ stands for the pairing (6.16) of $I_*(\Sigma)$ with $I_{-3-*}(-\Sigma)$.

Proof. This formula can be viewed as a version of the Mayer–Vietoris principle described in [232]. The general set up is as follows. Assume that a Riemannian metric g on X is a product metric in a neighborhood tube $[-1, 1] \times \Sigma$. We then study the effect of changing g in a family $\{ g_t \mid t \geq 1 \}$ where g_t agrees with g off $[-1, 1] \times \Sigma$ but has stretched the tube to $[-t, t] \times \Sigma$. According to [299], any sequence of connections A_t which are ASD with respect to the metric g_t, converges away from finitely many points to an ASD connection A_U on U and an ASD connection A_V on V as $t \to +\infty$. Both connections A_U and A_V limit to a flat connection on Σ. Conversely, any two ASD connections A_U and A_V limiting to the same flat connection over Σ can be glued together as in [70] to produce an ASD connection over X.

An essential part of the proof is to show that the trivial connection θ over Σ fails to show up as a limiting connection. This can be achieved by using conditions (a) and (b) of the theorem. The same conditions guarantee that both $\gamma_r(U)$ and $\gamma_{d-r}(V)$ are cycles in the respective Floer chain groups.

Example 6.86. Let $X = X_1 \# X_2$ be connected sum of two smooth simply connected manifolds with $b_2^+(X_1) > 0$ and $b_2^+(X_2) > 0$. The gluing formula can then be applied to $X = U \cup V$ where U and V are punctured X_1 and X_2 with $\partial U = \partial V = S^3$. Since $I_*(S^3) = 0$, this proves that $\gamma(X) = 0$, compare with Example 6.70.

Example 6.87. Let X be the Kummer surface, see Example 6.72. Brieskorn homology sphere $\Sigma(2, 3, 7)$ embeds into X splitting it into two submanifolds, $X = U \cup V$, with intersection forms $E_8 \oplus 2H$ and $E_8 \oplus H$, respectively. In fact, U is the Milnor fiber of singularity at zero of $x^2 + y^3 + z^7 = 0$, and V

is a plumbed manifold. Let $u_1, \ldots, u_6 \in H_2(U, \mathbb{Z})$ be homology classes such that the pair $u_1, u_2 \in E_8$ satisfies $u_1^2 = u_2^2 = -2$ and $u_1 \cdot u_2 = 1$, the pair $u_3, u_4 \in H$ satisfies $u_3^2 = u_4^2 = 0$ and $u_3 \cdot u_4 = 1$, and the pair $u_5, u_6 \in H$ satisfies $u_5^2 = u_6^2 = 0$ and $u_5 \cdot u_6 = 1$. Similarly, choose $v_1, \ldots, v_4 \in H_2(V, \mathbb{Z})$ so that v_1, v_2 form a pair in E_8 and v_3, v_4 form a pair in H as above. Let α be a generator of $I_3(\Sigma(2,3,7)) = I_2(-\Sigma(2,3,7)) = \mathbb{Z}$, see Example 6.23. The relative $SU(2)$ Donaldson polynomials are of the form $\gamma_6(U)(u_1, \ldots, u_6) = m \cdot \alpha$ and $\gamma_4(V)(v_1, \ldots, v_4) = n \cdot \alpha$ for some integers m and n. According to the gluing formula, $\gamma_{10}(X)(u_1, \ldots, u_6, v_1, \ldots, v_4) = m \cdot n$. A direct calculation with the degree 10 Donaldson polynomial of X, ses Example 6.72, shows that $m \cdot n = 1 \bmod 2$. In particular, both m and n are odd.

Example 6.88. Every smooth 4–manifold X having a homotopy type of the $K3$ surface is called a *homotopy K3 surface.* Fintushel and Stern show in [82] that any homotopy $K3$ surface which admits an embedding of $\Sigma(2,3,7)$ has a non-trivial $SU(2)$ Donaldson polynomial of degree 10, very much like the Kummer surface itself, see Example 6.87. Their proof goes as follows. If $\Sigma(2,3,7)$ embeds into a homotopy $K3$ surface X, it splits X into smooth oriented submanifolds $X = U \cup V$ with intersection forms $E_8 \oplus 2H$ and $E_8 \oplus H$, respectively. Let α be a generator of $I_3(\Sigma(2,3,7)) = \mathbb{Z}$. A direct calculation shows that, for a choice of homology classes as in Example 6.87, $\gamma_6(U)(u_1, \ldots, u_6) = m \cdot \alpha$ and $\gamma_4(V)(v_1, \ldots, v_4) = n \cdot \alpha$ for some odd integers m and n. By the gluing formula, $\gamma_{10}(X)(u_1, \ldots, u_6, v_1, \ldots, v_4) = m \cdot n = 1 \bmod 2$, hence, $\gamma_{10}(X)$ is non-trivial. This result has been an essential ingredient in the original construction of Akbulut corks [1], compare with Example 7.40.

Example 6.89. Another generalization of Example 6.87 can be found in [74]. View a Seifert fibered homology sphere $\Sigma(a_1, \ldots, a_n)$ as the link of singularity (1.2) and let U be the corresponding Milnor fiber. There exists a natural compactification of U as a complete intersection in a weighted projective space, see [66]. Let X be the minimal resolution of the (quotient) singularities of this compactification. The curve at infinity of X consists of smooth rational curves whose intersection graph is tree, and a suitable linear combination of these curves is a canonical class K_X of X. Let V be a tubular neighborhood of the curve at infinity, then V is a plumbed 4–manifold with $b_2^+(V) = 1$. Thus we have a splitting $X = U \cup V$ along $\Sigma(a_1, \ldots, a_n)$. Ebeling and Okonek [74] show that $\gamma_d(U) = a_d\, Q^{d/2}$ where Q is the intersection form of U, see Example 6.82. They also show that, under certain technical restrictions, $\gamma_d(V)$ are polynomials in the intersection form of V and the canonical class K_X. The gluing formula then implies that the Donaldson polynomial of X can be expressed in terms of the intersection forms of U and V and the canonical class K_X. This is in agreement with Kronheimer and Mrowka's structure theorem for Donaldson polynomials, see Example 6.73.

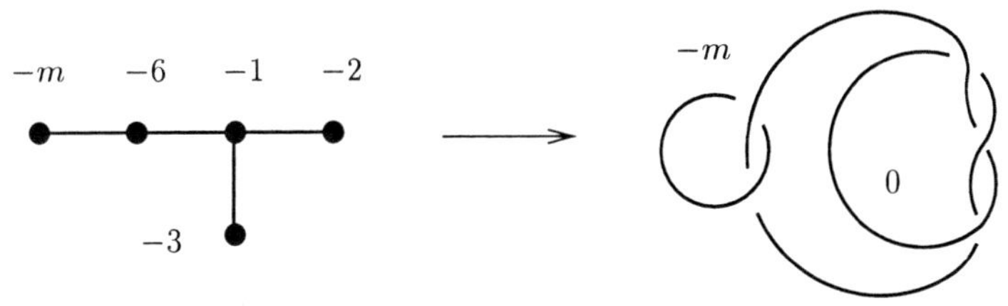

Fig. 6.8.

Example 6.90. In the special case of $\Sigma(2,3,6m-1)$, the resolution X from the above example is an elliptic surface. The manifold V is obtained by three blow downs in the plumbing graph shown in Figure 6.8. The manifold V is otherwise known as the *Gompf nucleus*, see [124]. Its relative Donaldson polynomials are computed in [200].

Example 6.91. Fintushel and Stern [91] used the above gluing formula and the calculations outlined in Example 6.81 to compute degree zero Donaldson polynomial of the Kummer $K3$ surface. More applications of the gluing formula can be found in [90].

6.7 Extending Floer Homology

The Floer homology theory, which was originally defined in [93] for integral homology 3–spheres, has been extended in many directions. There exists a Floer type theory for rather general 3–manifolds which (conjecturally) is uniquely determined by a set of axioms reminiscent of the Eilenberg–MacLane axioms for homology of CW complexes. More generally, the orbifold Floer homology of Collin and Steer is defined for pairs (Σ, k) where k is a knot in an integral homology 3–sphere Σ. Several theories have been specifically crafted to accomodate gluing of Donaldson polynomials. Among these are the equivariant Floer homology of Austin and Braam and the Fukaya–Floer homology. The (instanton) Floer homology has also been related to the symplectic Floer homology defined in [95]. Another Floer type theory arising from Seiberg–Witten gauge theory is conjecturally isomorphic to the Ozsváth–Szabó Floer homology.

6.7.1 Equivariant Floer Homology

Theorem 6.85 suggests that, in order for a general gluing formula for Donaldson polynomials to work, the space of gluing parameters should account for all flat connections on an integral homology sphere, including the trivial one. This leads to the concept of *equivariant Floer homology*, see [21].

Let Σ be an integral homology sphere and assume for simplicity that $\mathcal{R}^*(\Sigma)$ is non-degenerate: this condition can be removed by using admissible perturbations. Consider the representation space $R(\Sigma) = \mathrm{Hom}(\pi_1\Sigma, SU(2))$ acted upon by the group $SO(3)$. Roughly speaking, the equivariant Floer homology $I_*^{\mathcal{G}}(\Sigma)$ is the $SO(3)$–equivariant cohomology of $R(\Sigma)$ with real coefficients. More precisely, let us consider the equivariant cohomology groups

$$H^*_{SO(3)}(R(\Sigma),\mathbb{R}) = H^*(ESO(3)\times_{SO(3)} R(\Sigma),\mathbb{R})$$

where $ESO(3)\to BSO(3)$ is the universal $SO(3)$–bundle. Given a representation α let R_α be the connected component of $R(\Sigma)$ containing α. If α is irreducible then $R_\alpha = SO(3)$ and

$$H^*_{SO(3)}(R_\alpha,\mathbb{R}) = H^*(R_\alpha/SO(3),\mathbb{R}) = H^*(\mathrm{pt},\mathbb{R}) = \mathbb{R}.$$

If α is the trivial representation θ then $R_\theta = \mathrm{pt}$ and

$$H^*_{SO(3)}(R_\theta,\mathbb{R}) = H^*(BSO(3),\mathbb{R}) = \mathbb{R}[u],$$

where u is the universal Pontryagin class of degree 4. These cohomology groups together form the E_1–term of a Morse–Bott spectral sequence which, in favorable circumstances, converges to the equivariant Floer homology $I_*^{\mathcal{G}}(\Sigma)$.

The E_1–term has only one non–zero row and one non–zero column and is described by the formula $E_1^{*,*} = E_1^{*,0}\oplus E_1^{0,*}$, where

$$E_1^{*,0} = \bigoplus_\alpha H^*_{SO(3)}(R_\alpha,\mathbb{R}) = \bigoplus_\alpha \mathbb{R}$$

graded by the Floer indices $\mu(\alpha)$, and $E_1^{*,0} = \mathbb{R}[u]$. The differential d_1 of this spectral sequence coincides on the irreducible representations with the boundary operator in the Floer chain complex. The remaining differential d_1 and higher differentials measure the interaction between the trivial representation and irreducible representations, and are described in full detail in [21] in terms of the u–map in Floer homology, see Section 6.5.5.

Example 6.92. The equivariant Floer homology groups for S^3 are especially simple since there is only one flat connection, the trivial one. Thus $I_*^{\mathcal{G}}(S^3) = \mathbb{R}[u]$.

Example 6.93. It is shown in [21] that $I_*^{\mathcal{G}}(\Sigma(2,3,5)) = \mathbb{R}[u]$ as an $\mathbb{R}[u]$–module with the generator $1-u^2$.

Austin and Braam [21] generalize the above construction and define equivariant Floer homology for all 3–manifolds M such that either $H_1(M,\mathbb{Q})=0$ or $H_1(M,\mathbb{Z})$ is torsion free. These include, in particular, rational homology spheres and homology handles. Explicit calculations of $I_*^{\mathcal{G}}(M)$ for spherical space forms M are contained in [19]. Gluing formula for Donaldson polynomials employing the equivariant Floer homology can be found in [21].

6.7.2 Fukaya–Floer Homology

Let X be a smooth closed 4–manifold split into $X = U \cup_M V$ by a 3–dimensional submanifold M. In Section 6.6.3 we discussed how to recover Donaldson polynomials of X from those of U and V when M is an integral homology sphere. To understand more general splittings one needs to overcome two significant further problems. One is the effect of reducible connections, which is the main concern of the equivariant Floer theory. The other is the decomposition of homology classes in X by M: when M has non-trivial first homology, one may encounter the situation that a second homology class in X cannot be realized as combination of classes in U and V, but rather arises from relative classes in $H_2(U, M, \mathbb{Z})$ and $H_2(V, M, \mathbb{Z})$. This last problem is addressed by the *Fukaya–Floer homology*, for which we refer the reader to [39] and [107].

6.7.3 Floer's Category and Functor

In this section we describe a functorial approach to the Floer homology outlined by Floer in [94] and later developed in [38]. Let $E \to M$ be a principal $SO(3)$–bundle over an oriented 3–manifold M. We call E *admissible* if either M is an integral homology sphere or else E is free of reducible connections. For example, the unique non-trivial $SO(3)$–bundle over a homology handle discussed in Section 6.5.2 is admissible.

Denote by $\mathcal{C}^*$ the category whose objects are admissible $SO(3)$–bundles $E \to M$. A morphism between $E_1 \to M_1$ and $E_2 \to M_2$ consists of the following data:

- An $SO(3)$–bundle $Q \to W$ over an oriented 4–manifold W such that $\partial W = -M_1 \cup M_2$ and the restriction of Q onto M_1 and M_2 coincides with E_1 and E_2, respectively,
- A lift of $w_2(Q) \in H^2(W, \mathbb{Z}/2)$ to $H^2(W, \mathbb{Z})$, and
- A choice of orientations of moduli spaces of ASD connections on Q.

Floer [94] assigned to every object $E \to M$ in category $\mathcal{C}^*$ an abelian group $I_*(M, E)$ which is relatively graded modulo 8. Moreover, he assigned to every morphism $Q \to W$ between $E_1 \to M_1$ and $E_2 \to M_2$ a morphism $(W, Q)_* : I_*(M_1, E_1) \to I_*(M_2, E_2)$ of graded abelian groups (the bundles will normally be suppressed in our notations). This defines a functor I_* from the category $\mathcal{C}^*$ to the category of abelian groups with relative $\mathbb{Z}/8$–grading and their homomorphisms. This functor is called the *Floer functor.* On the full subcategory of $\mathcal{C}^*$ consisting of integral homology spheres and homology handles, this functor agrees with the functor I_* defined in Section 6.5.2.

The Floer exact triangle, see Theorem 6.39, extends to a Floer exact triangle in the category $\mathcal{C}^*$, see [38] and [94]. In addition to this, the following excision property holds.

First note that there exists only one non–trivial $SO(3)$–bundle over a 2-torus. Let $E_1 \to M_1$ and $E_2 \to M_2$ be objects in $\mathcal{C}^*$, and suppose that each contains an embedded 2–torus $T_k \subset M_k$ such that the restriction of E_k onto T_k is non–trivial, $k = 1, 2$. Note that T_k necessarily represents a non–trivial class in $H_2(M_k, \mathbb{Z})$.

Let $M = M_1 \cup M_2$ be the 3–manifold obtained by cutting M_1 and M_2 open along the tori T_1 and T_2 and gluing together the resulting boundary components in pairs, in a way determined by the orientations of M_1 and M_2. There is a standard cobordism W from the disjoint union of M_1 and M_2 to M. This gluing operation extends to the bundles given a choice of attaching maps. This operation results is an admissible $SO(3)$–bundle over M and an $SO(3)$–bundle over W.

Theorem 6.94. *For any choice of attaching maps, the cobordism W induces an isomorphism $I_*(M) = (I_*(M_1) \otimes I_*(M_2))/u$ over the rational numbers.*

Here, $u : I_*(M_1) \otimes I_*(M_2) \to I_*(M_1) \otimes I_*(M_2)$ is an involution induced by the u–map, see Section 6.5.5. Theorem 6.94 is proved in [38]. A stronger version of this result which holds over the integers is proved in [195]. It is usually referred to as the *Künneth formula* or the *excision property* for Floer homology.

Example 6.95. Let $E \to T^2$ be a non-trivial $SO(3)$–bundle over T^2 and $F \to T^3$ its pull-back to T^3 via the projection $T^3 = T^2 \times S^1 \to T^2$. An easy calculation in [38] shows that the Floer homology of $F \to T^3$ has two free generators in dimensions differing by 4. According to Theorem 6.94, the Floer homology $I_*(M)$ is independent of the choice of the attaching map, hence, all torus bundles over S^1 will have the same Floer homology as the bundle $F \to T^3$.

Example 6.96. Let $k \subset S^3$ be a fibered knot of genus one then the homology handle $\overline{K}$ obtained by 0–surgery along k is a torus bundle over S^1. Therefore, $I_*(\overline{K}) = \mathbb{Z} \oplus \mathbb{Z}$ in dimensions differing by 4, see Example 6.95. The Floer exact triangle then implies that $I_*(S^3 \pm k) = I_*(\overline{K}) = \mathbb{Z} \oplus \mathbb{Z}$ in dimensions differing by 4. This is consistent with the calculation of Example 6.23 if k is a trefoil and with that of Example 6.42 if k is the figure–eight knot.

Example 6.97. Let E be a non-trivial $SO(3)$–bundle over a closed oriented Riemann surface F_g of genus g, and let $\pi : F_g \times S^1 \to F_g$ be the projection. The Floer homology groups $I_*(F_g \times S^1, \pi^* E)$ are explicitly computed in [235].

Floer conjectured in [94], see also [38], that the functor I_* on the category $\mathcal{C}^*$ can be completely determined by a set of axioms similar to those of Eilenberg and MacLane for homology of CW complexes. More precisely, the following conjecture has been proposed. Let J_* be a functor on $\mathcal{C}^*$ such that

- The Floer surgery triangle is exact for J_*,

- The Künneth formula is true for J_*, and
- $I_*(S^3) = J_*(S^3)$ and $I_*(T^3, \pi^* E) = J_*(T^3, \pi^* E)$, where $\pi^* E$ is the pull–back of the unique non–trivial $SO(3)$–bundle E over T^2 via the projection $T^3 = T^2 \times S^1 \to T^2$.

Then $J_* = I_*$. Discussion on this conjecture and possible approaches to its proof can be found in [38] and [94].

6.7.4 The Atiyah–Floer Conjecture

Let M be a closed symplectic manifold of dimension $2n$ and L_1, L_2 two Lagrangian submanifolds. Floer [95] associated with the triple (M, L_1, L_2) a homology theory called *symplectic Floer homology.* Since $\dim L_1 = \dim L_2 = n$, the intersection $L_1 \cap L_2$ generically consists of finitely many points. These points generate the symplectic Floer chain complex. The boundary operator counts pseudoholomorphic discs with boundary in $L_1 \cup L_2$ connecting intersection points.

Let Σ be an integral homology sphere with Heegaard splitting $\Sigma = M_1 \cup M_2$ along a Riemann surface F of genus g. Away from the singularities, $\mathcal{R}(F)$ is a symplectic manifold of dimension $6g - 6$ and $\mathcal{R}(M_1), \mathcal{R}(M_2) \subset \mathcal{R}(F)$ are Lagrangian submanifolds. The intersection $\mathcal{R}(M_1) \cap \mathcal{R}(M_2)$ is naturally identified with $\mathcal{R}(\Sigma)$, so that the symplectic Floer chain complex has essentially the same generators as $IC_*(\Sigma)$.

The following is known as the *Atiyah–Floer conjecture*: symplectic Floer homology can be adapted to handle singularities in the representations spaces arising from a Heegaard splitting, and so defined symplectic Floer homology equals the instanton Floer homology.

A proof of this conjecture has been outlined in preprints [188] and [189], see also [310]. Dostoglou and Salamon [72] proved an easier version of the Atiyah–Floer conjecture, with the integral homology sphere replaced by the mapping torus of an automorphism of a Riemann surface F.

6.7.5 Floer Homology of Knots

Let k be a knot in an integral homology 3–sphere Σ and $\Delta_k(t)$ its Alexander polynomial. For every $0 \le \varphi \le \pi$ such that $\Delta(e^{2i\varphi}) \ne 0$, Collin and Steer [59] defined four abelian groups $I_n^{\varphi}(\Sigma, k)$, $n = 0, 1, 2, 3$, called *Floer homology of the knot* $k \subset \Sigma$. These groups refine the Herald–Lin invariant $h_{\varphi}(\Sigma, k)$, see Theorem 5.17, in that the Euler characteristic of $I_*^{\varphi}(\Sigma, k)$ equals $h_{\varphi}(\Sigma, k)$.

The groups $I_*^{\varphi}(\Sigma, k)$ are defined as the homology groups of a chain complex $IC_*^{\varphi}(\Sigma, k)$ generated by irreducible representations $\alpha : \pi_1 K \to SU(2)$ such that $\operatorname{tr} \alpha(m) = 2 \cos \varphi$. Here, K is the exterior of the knot k and $m \subset \partial K$ is its meridian. A generic admissible perturbation as described in Section 5.4 may be necessary to assure non–degeneracy. The Floer index and the boundary operator are defined as in the (regular) Floer theory after one overcomes analytical difficulties related to the fact that the manifold K has boundary.

These difficulties are resolved in [59] by making K into a (compact) orbifold (Σ, k) singular along the knot k. Suppose first that $\varphi = \pi m/n$ for some rational number m/n. Then all analysis can be done equivariantly on a smooth compact 3–manifold M obtained as an n–fold branched cover of (Σ, k) with branch set the knot k. In particular, the Floer index of a generator in $IC_*^{\varphi}(\Sigma, k)$ can be computed as the equivariant index of an appropriate ASD operator on M. The definition is then extended to the remaining values of φ by continuity. This orbifold approach is the reason why the Floer homology $I_*^{\varphi}(\Sigma, k)$ is also referred to as the *orbifold Floer homology.*

The groups $I_*^{\varphi}(\Sigma, k)$ have the following properties, see [59]. Let $\mathcal{Z} = \{\varphi \in [0,\pi] \mid \Delta_k(e^{2i\varphi}) = 0\}$, then the Floer homology $I_*^{\varphi}(\Sigma, k)$ is constant on each of the connected components of $[0,\pi]\setminus\mathcal{Z}$. The groups $I_*^{0}(\Sigma, k)$ are isomorphic to (twice) the regular Floer homology $I_*(\Sigma)$, with grading reduced modulo 4.

The Floer homology $I_*^{\varphi}(\Sigma, k)$ is functorial in the following sense. A cobordism between (Σ_1, k_1) and (Σ_2, k_2) is a pair (W, F) where W is a cobordism between Σ_1 and Σ_2 and $F \subset W$ is an oriented embedded surface such that $F \cap \Sigma_1 = k_1$ and $F \cap \Sigma_2 = k_2$. For every φ away from the roots of the Alexander polynomials of k_1 and k_2, the pair (W, F) induces a homomorphism $(W, F)_* : I_*^{\varphi}(\Sigma_1, k_1) \to I_*^{\varphi}(\Sigma_2, k_2)$ of degree $3(b_1 - b_2^+)(W) + e(F)$, where $e(F)$ is the Euler characteristic of F. This homomorphism satisfies the usual functorial properties.

Example 6.98. Let k be a knot in S^3 such that $\Delta_k(t) = 1$ then all $0 \le \varphi \le \pi$ belong to the connected component of 0 in $[0,\pi]\setminus\mathcal{Z}$ and hence $I_*^{\varphi}(S^3, k) = 0$. More generally, the same holds for knots whose Alexander polynomials have no roots on the unit complex circle; for example, the figure eight knot is in this category so its Floer homology groups vanish for all values of φ.

Example 6.99. Torus knots $k_{p,q}$ may have non-trivial Floer homology. In fact, one can show as in [59] that the generators in $IC_*^{\varphi}(S^3, k_{p,q})$ only occur in one parity and hence $I_*^{\varphi}(S^3, k_{p,q}) = IC_*^{\varphi}(S^3, k_{p,q})$. If $\varphi = \pi m/n$ where n is relatively prime with p and q then the cyclic branched cover of S^3 with branch set $k_{p,q}$ is the Brieskorn homology sphere $\Sigma(p, q, n)$, see Example 1.22. One can use this observation to explicitly describe the generators in $I_*^{\varphi}(S^3, k_{p,q})$, see [58].

Remark 6.100. The theory of orbifold Floer homology is believed to be closely related to the work of Kronheimer and Mrowka on connections singular along embedded surfaces in 4–manifolds, see [176]. The orbifold Floer homology could play here a role similar to that played by the original Floer homology in the Donaldson theory.

Remark 6.101. Weiping Li [191] and [193] constructed the so called *symplectic Floer homology* for knots in S^3. The Euler characteristic of this Floer homology equals up to an overall constant the (equivariant) knot signature.

He also conjectured that this symplectic theory matches the instanton Floer homology of knots defined above, possibly in the spirit of the Atiyah–Floer conjecture, see Section 6.7.4.

Remark 6.102. Let M be a closed oriented 3–manifold with a $\mathbb{Z}/n$–action $\tau : M \to M$ making it into a cyclic branched cover of an integral homology sphere Σ with branch set a knot k. The Euler characteristic of the Floer homology of (Σ, k) equals the Herald–Lin invariant of (Σ, k), which in turn equals (up to an overall constant) the equivariant Casson invariant of (M, τ), see Section 5.5.3. One may wonder if there exists "equivariant Floer homology" of (M, τ) isomorphic to the orbifold Floer homology of (Σ, k).

6.7.6 Seiberg–Witten Floer Homology

The Floer homology groups were introduced in Section 6.2 as the homology groups associated with the Chern–Simons function $\mathbf{cs}$. There appears to be no serious obstacles to adapting this construction to the Chern–Simons–Dirac function $\mathbf{cs}_D$, see Section 5.3.6, and defining the *Seiberg–Witten Floer homology*. There is not yet a complete account of such a construction in the literature, see however [175] and [205].

Another Floer homology theory was defined by Ozsváth and Szabó in a series of preprints starting with [252]. Conjecturally, it is equivalent to the Seiberg–Witten Floer homology but its definition is combinatorial, starting with a Heegaard splitting of a closed oriented three–manifold.

7 The Homology Cobordism Group

7.1 Homology Cobordisms

Let Σ_0 and Σ_1 be integral homology 3–spheres. They are said to be *homology cobordant*, or *H–cobordant*, if there exists a smooth compact oriented 4–manifold W with boundary $\partial W = -\Sigma_0 \cup \Sigma_1$ such that the inclusions $\Sigma_i \to W$ induce isomorphisms $H_*(\Sigma_i, \mathbb{Z}) \to H_*(W, \mathbb{Z})$, $i = 0, 1$. Equivalently, Σ_0 and Σ_1 are homology cobordant provided $(-\Sigma_0) \# \Sigma_1$ bounds a smooth acyclic 4-manifold, that is a smooth compact oriented 4–manifold W with $H_*(W, \mathbb{Z}) = H_*(D^4, \mathbb{Z})$. Note that we do not require that $\pi_1(W) = 0$, only that $H_1(W, \mathbb{Z}) = 0$. A homology sphere Σ is said to be *homology cobordant to zero* if it is homology cobordant to S^3 or, equivalently, bounds a smooth acyclic 4–manifold. Note that the requirement that the homology cobordism be *smooth* is essential – it is known [98] that any integral homology 3-sphere bounds a *topological* compact oriented 4–manifold which is acyclic.

Example 7.1. Any homology sphere Σ is homology cobordant to itself via the product cobordism $W = [0, 1] \times \Sigma$. There may exist other manifolds W providing homology cobordism of Σ to itself. Sometimes, such a W can be chosen to be simply–connected, but this is not always the case. For example, the Poincaré homology sphere $\Sigma(2, 3, 5)$ is not homology cobordant to itself via any simply–connected homology cobordism, see [293].

Example 7.2. Fintushel and Stern [83] found many examples of Seifert fibered homology spheres $\Sigma(a_1, \dots, a_n)$ which are not homology cobordant via simply connected homology cobordism to any other integral homology sphere.

7.1.1 Mazur Homology Spheres

In this section we describe a rather large class of integral homology spheres homology cobordant to zero. These manifolds are sometimes referred to as *Mazur homology spheres*, after Mazur [212].

Example 7.3. The following is the original construction [212] of a non-simply connected integral homology sphere which is homology cobordant to zero. Let us consider the manifold $S^1 \times D^3$ obtained by attaching a 1–handle $D^1 \times D^3$

to D^4. The boundary of this manifold is $S^1 \times S^2$, which can be viewed as the manifold glued from two copies of $S^1 \times D^2$ along the identity map of the boundaries. Let us pick a knot k in a copy of $S^1 \times D^2$ as shown in Figure 7.1.

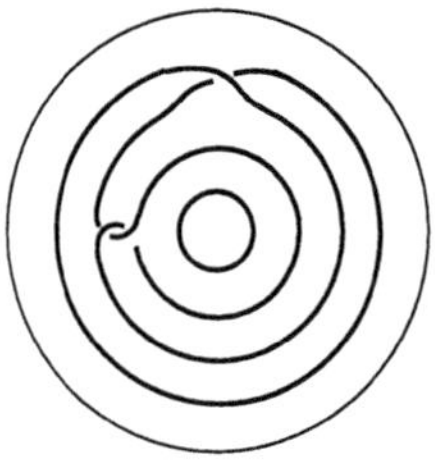

Fig. 7.1.

Use the knot k to attach a 2–handle to $S^1 \times D^3$ with framing 3. We obtain a smooth 4–manifold W. One can easily see from the Mayer–Vietoris exact sequence that $H_*(W, \mathbb{Z}) = H_*(D^4, \mathbb{Z})$. In a situation like this we say that the 1– and 2–handles *cancel each other* in the homology of W. Moreover, one can show that $\pi_1(W) = 0$ so the manifold W is in fact contractible (by the Whitehead Theorem).

The link diagram shown in Figure 7.2 describes the manifold W, and the one in Figure 7.3 describes its boundary, an integral homology sphere Σ (in both cases $p = 3$). One can use Kirby calculus to show that in fact Σ is orientation preserving homeomorphic to $\Sigma(2,5,7)$, see [4]. In particular, $\pi_1(\Sigma)$ is not trivial.

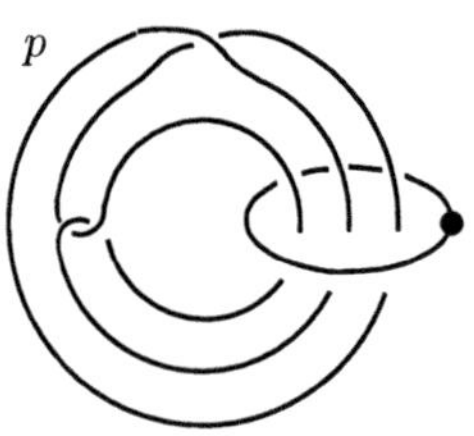

Fig. 7.2.

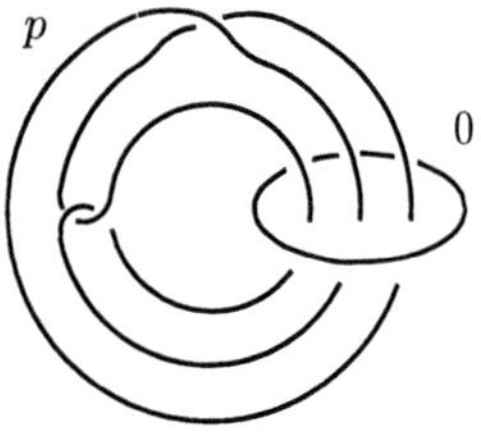

Fig. 7.3.

Example 7.4. One can modify this construction to get many more examples of integral homology spheres homology cobordant to zero. The framing 3 can be replaced by any integer p. Among the homology spheres obtained by surgery on the link in Figure 7.3 for various p are $\Sigma(3,4,5)$ and $\Sigma(2,3,13)$, see [4], and also the splice of $\Sigma(2,3,5)$ with $\Sigma(2,3,7)$ along the fibers of degrees 5 and 7. The knot we started with, see Figure 7.1, can also be replaced by any knot in $S^1 \times D^2$ which is homologous to $S^1 \times \{0\}$. If it is homotopic to $S^1 \times \{0\}$, we end up with S^3.

Example 7.5. The construction from Example 7.4 can be generalized to produce homology cobordisms with more than one pair of canceling 1– and 2–handles. More precisely, given a connected oriented smooth 4–manifold V with connected boundary $\partial V = M$ such that $H_1(V,\mathbb{Z}) = \mathbb{Z}^s$ and $H_2(V,\mathbb{Z}) = H_3(V,\mathbb{Z}) = 0$, one can easily see that $H_1(M,\mathbb{Z}) = \mathbb{Z}^s$. Represent an integral basis of $H_1(M,\mathbb{Z})$ by s disjoint embedded circles, and glue s copies of a 2-handle $D^2 \times D^2$ to V along these circles. The resulting 4–manifold is clearly an integral homology cobordism. Applied to $V = S^1 \times D^3$, this construction results in the homology cobordisms described in Example 7.4.

Example 7.6. Several infinite series of Brieskorn homology spheres homology cobordant to zero can be found in [52] and [289]. All these examples are obtained via the construction of Example 7.5, with one or two pairs of canceling 1– and 2–handles. There are many other examples scattered through the literature. However, we do not know of any examples of Seifert fibered homology spheres with four or more singular fibers which are homology cobordant to zero.

7.1.2 Knot Cobordisms and Homology Cobordisms

Another source of examples of homology cobordisms is surgery along knot cobordisms (also called knot concordances). We outline this topic rather briefly here and refer the reader for a comprehensive survey to [127] and [130].

A *knot cobordism*, or *concordance*, between oriented knots k_0 and k_1 in S^3 is an oriented annulus $A = [0,1] \times S^1$ smoothly embedded into $[0,1] \times S^3$ so that $k_i = A \cap (\{\,i\,\} \times S^3)$, $i = 0,1$, with matching orientations. Here, the embedding $A \subset [0,1] \times S^3$ is required to be *proper*, that is, $A \cap \partial([0,1] \times S^3) = \partial A$. Cobordism classes of oriented knots form an abelian group $\mathcal{C}^{3,1}$ with respect to connected sum of knots, the zero element being the class of the trivial knot, and the inverse being given by the reversal of orientation. The group $\mathcal{C}^{3,1}$ is called the (classical) *knot concordance group.*

Given a knot cobordism $A \subset [0,1] \times S^3$ between knots k_0 and k_1, cut a tubular neighborhood $A \times D^2 = [0,1] \times S^1 \times D^2$ out of $[0,1] \times S^3$. The resulting manifold has boundary $K_0 \cup ([0,1] \times S^1 \times \partial D^2) \cup K_1$ where K_0 and K_1 are the knot exteriors of k_0 and k_1, respectively. Now pick an integer q and glue the tubular neighborhood back in via the orientation reversing diffeomorphism $\mathrm{id} \times h : [0,1] \times S^1 \times \partial D^2 \to [0,1] \times S^1 \times \partial D^2$ of the boundaries, where h is the diffeomorphism corresponding to the $(1/q)$–surgery on a knot. The following is an immediate consequence of the Mayer–Vietoris exact sequence.

Proposition 7.7. *The smooth 4–manifold resulting from the above construction is an integral homology cobordism between integral homology spheres obtained by $(1/q)$–surgery along the knots k_0 and k_1, respectively.*

A knot $k \subset S^3$ representing zero element in the group $\mathcal{C}^{3,1}$ is called *slice.* In other words, a knot $k \subset S^3$ is slice if there is a smoothly embedded 2–disc $D^2 \subset D^4$ with $\partial D^2 = k$. Every knot in S^3 bounds some disc in D^4 (for example, the cone over k), thus the requirement that the disc be smoothly embedded is an essential part of the definition.

Corollary 7.8. *For any integer q, the integral homology sphere obtained by $(1/q)$–surgery on a slice knot in S^3 is homology cobordant to zero.*

Example 7.9. Let ℓ be an oriented knot in S^3 and $k = \ell \# \ell^*$ a connected sum of the knot ℓ with its mirror image ℓ^*. Then k is slice. This follows because k can be situated in $\mathbb{R}^3$ so as to be symmetric with respect to a plane, and a slice disc can be obtained as the trace of 180° rotation of ℓ in $\mathbb{R}^4$ with respect to the plane. Shown in Figure 7.4 is the *square knot* $k = \ell \# \ell^*$ where ℓ is a trefoil. The manifold Σ resulting from the (-1)–surgery on the square knot is homology cobordant to zero. One can use Kirby calculus to show that in fact Σ equals the splice of $\Sigma(2,3,5)$ and $\Sigma(2,3,7)$ along the singular fibers of degrees 5 and 7, see Example 1.9.

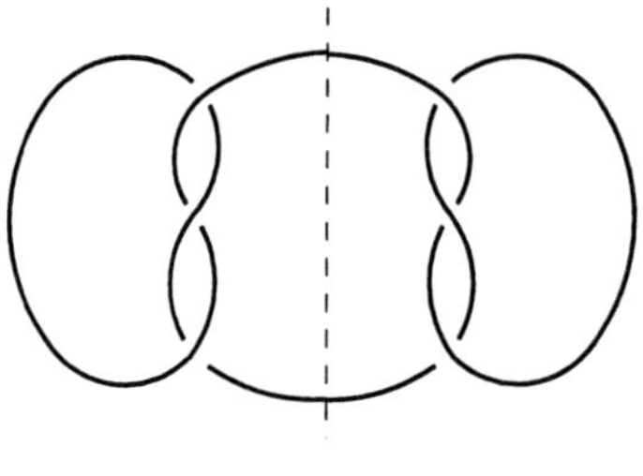

Fig. 7.4.

The above constructions of homology cobordisms can be generalized in several directions, see [127]. First, one can replace the product $[0,1] \times S^3$ by *any* homology cobordism W between integral homology spheres, and think of the knot cobordism as an annulus properly embedded into W. The result of a $(1/q)$-surgery along such a knot cobordism results again in an integral homology cobordism. Second, knots and their cobordisms can be replaced with links and their cobordisms. A *link cobordism* is a family of disjoint annuli in $[0,1] \times S^3$ (or a homology cobordism) restricting to the two given links at the ends. Finally, instead of doing surgery on a knot cobordism, one can glue together exteriors of two knot cobordisms to get a homology cobordism, if the gluing map is chosen properly.

7.1.3 Ribbon Concordances and Ribbon Knots

Ribbon concordance is a special case of the knot cobordism, see [129]. Given a cobordism $A \subset [0,1] \times S^3$ between knots $k_i \subset \{i\} \times S^3$, $i = 0, 1$, the embedding of A can be adjusted by small isotopy so that the restriction to A of the projection map $[0,1] \times S^3 \to [0,1]$ is a Morse function. This function will typically have critical points of index 0, 1 and 2. If there are no critical points of index 2 (local maxima), we say that A is a *ribbon concordance* from k_0 to k_1.

In practical terms, this means the following, see [296]. Let $L = L_0 \cup L_1$ be an oriented link of two components in $\mathbb{R}^3$ and $b : I \times I \to \mathbb{R}^3$, where $I = [0,1]$, an embedding such that $b(I \times I) \cap L = b(I \times \partial I)$, $b(I \times 0) \subset L_0$, $b(I \times 1) \subset L_1$, and the orientations inherited from L on $b(I \times \partial I)$ induce the same orientation on $b(I \times I)$. In this case, the link

$$(L \setminus b(I \times \partial I)) \cup b(\partial I \times I),$$

with inherited orientation, is denoted $L_0 +_b L_1$. A 2–disc immersed into S^3 is called a *ribbon* if each of its singularities consists of two sheets intersecting in an arc which is interior to one of the sheets, see Figure 7.5.

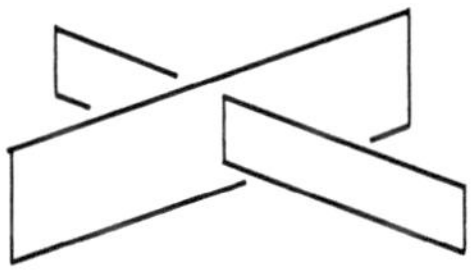

Fig. 7.5.

A knot $k \subset \mathbb{R}^3$ is *ribbon concordant* to a knot k' if there exist ribbons D_i^2, $i = 1, \ldots, n$, in the knot k exterior such that $\partial D_i^2 \cap \partial D_j^2 = \emptyset$ for $i \neq j$ and

$$k' = (\ldots((k +_{b_1} \partial D_1^2) +_{b_2} \partial D_2^2) +_{b_3} \cdots +_{b_n} \partial D_n^2).$$

This definition easily generalizes to links of more that one component, see [296].

Example 7.10. The *stevedore's knot* shown in Figure 7.6 is ribbon concordant to a trivial knot by means of the ribbon shown in Figure 7.7.

A knot $k \subset S^3$ is called *ribbon* if it is ribbon concordant to a trivial knot. This means that k is the boundary of an immersed disc in $\mathbb{R}^3$ whose only self-intersections are of the form shown in Figure 7.5. The stevedore's knot shown in Figure 7.6 is a ribbon knot. It is immediate from the definition that all ribbon knots $k = \partial D^2$ are slice knots. The slice disc is obtained by pushing the self-intersections of the ribbon slightly into the D^4.

An old problem of Fox, which is still unresolved, asks whether every slice knot is ribbon. For example, all slice knots of Example 7.9 can be shown to be ribbon.

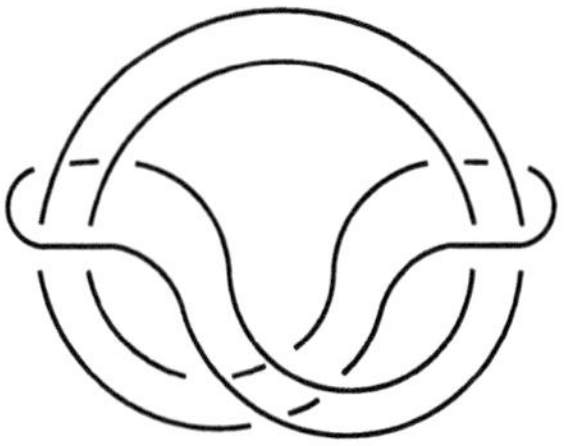

Fig. 7.6.

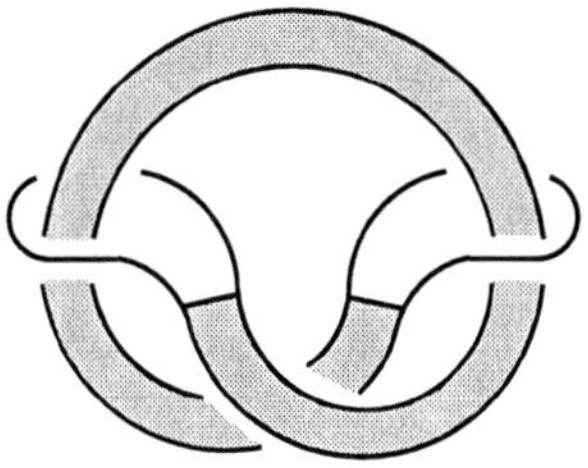

Fig. 7.7.

7.1.4 Branched Coverings

Let k_0 and k_1 be knots in S^3 concordant to each other via an embedded annulus $A \subset [0,1] \times S^3$. Given a positive integer n, let W_n be the n–fold cyclic branched cover of $[0,1] \times S^3$ with branch set A. Then W_n is an oriented cobordism between 3–manifolds Σ_0 and Σ_1, which are n–fold cyclic branched covers of S^3 with branch sets k_0 and k_1, respectively. If n is a power of a prime then $H_*(M_n, \mathbb{Q}) = H_*([0,1] \times S^3, \mathbb{Q})$, see for example [130], hence W_n is a rational homology cobordism between the rational homology spheres Σ_0 and Σ_1. In particular, if k_1 is slice then Σ_1 is rational homology cobordant to zero. In certain circumstances this construction results in an integral homology cobordism.

Example 7.11. The link shown in Figure 7.3 is strongly invertible. Therefore, the integral homology sphere Σ obtained by surgery on this link (for any choice of p) is a double branched covering of S^3. The algorithm of Section 1.1.12 can be used to show that the branch set of this covering is a ribbon knot k. The surgery on the link in Figure 7.2 results in a contractible 4–manifold W with boundary Σ. One can easily see that W is a double branched covering of D^4 with branch set a ribbon spanning k.

7.2 The Structure of $\Theta^3_{\mathbb{Z}}$

In Section 2.3.6, we defined the homology cobordism group $\Theta^3_{\mathbb{Z}}$ of integral homology 3–spheres and the Rokhlin homomorphism

$$\mu : \Theta^3_{\mathbb{Z}} \to \mathbb{Z}/2. \tag{7.1}$$

These objects have been studied since the 1950's but, for a few decades, very little was known about either of them. It was not even known if $\Theta^3_{\mathbb{Z}}$ is finite or infinite, and a reasonable conjecture in the 1970's was that μ is an isomorphism. Some major progress was made in the 1980's with the introduction of gauge-theoretical methods. These methods allowed to prove that $\Theta^3_{\mathbb{Z}}$ is infinite and, in fact, infinitely generated. It is still not known whether $\Theta^3_{\mathbb{Z}}$ has any torsion, or whether modulo torsion it is free abelian.

In particular, it is not known whether $\Theta^3_{\mathbb{Z}}$ has elements of second order with non-trivial Rokhlin invariant. Existence of such an element would solve in positive the triangulation conjecture, see Section 7.3.1. It is known however that the non-triviality of $\mu(\Sigma)$ implies that Σ has infinite order in $\Theta^3_{\mathbb{Z}}$ for certain classes of integral homology spheres Σ, including all Seifert fibered homology spheres.

7.2.1 A Classical Approach to $\Theta^3_{\mathbb{Z}}$

Among results about $\Theta^3_{\mathbb{Z}}$ obtained by "classical" methods, that is, without the gauge theory, are the following theorems. Livingston proved in [201] that every integral homology 3–sphere Σ is homology cobordant to an irreducible integral homology sphere. This result was improved by Myers [236] who showed that every Σ is in fact homology cobordant to a hyperbolic integral homology sphere. Both theorems also hold for more general 3-manifolds.

The results of Livingston and Myers were strengthened by Ruberman, who showed in [267] that any closed orientable 3–manifold has an invertible homology cobordism to a hyperbolic 3–manifold. A cobordism W from M to N is called *invertible* if there is another cobordism W', so that $W \cup W' = [0,1] \times M$.

It does not seem likely that every integral homology sphere is homology cobordant to a Seifert fibered homology sphere, although we do not know of any such examples.

Remark 7.12. It is not true that every integral homology sphere M has an invertible homology cobordism to a Seifert fibered homology sphere N. One way to see this is by using Gromov's norm, compare with [267]; another is as follows. If this were true we would have homomorphisms $I_*M \to I_*N \to I_*M$ induced by W and W' whose composition is an isomorphism, see Proposition 6.13. This would mean that I_*M embeds into I_*N hence $I_nM = 0$ for even n. On the other hand, there exist many homology spheres M which have non–zero Floer homology groups in both parities, see Example 7.19.

7.2.2 Infinite Order Elements in $\Theta^3_{\mathbb{Z}}$

The fact that $\Theta^3_{\mathbb{Z}}$ is infinite is an immediate consequence of the following *Donaldson theorem* [67], which was historically the first major application of gauge theory to the 4–dimensional topology.

Theorem 7.13. *If the intersection form Q_X of a closed smooth oriented 4-manifold X is definite then Q_X is diagonalizable over the integers, and hence isomorphic to either $p \cdot (+1)$ or $p \cdot (-1)$.*

This result is in a sharp contrast with Freedman's theorem [98] which asserts that every unimodular integral form can be realized as the intersection form of a closed *topological* 4–manifold.

Proof. We sketch a proof in the case when X is simply-connected. Many aspects of this proof were simplified in [79]. A proof of the general case can be found in [68]. A proof using the Seiberg–Witten gauge theory is given in [228].

Without loss of generality we may suppose that the intersection form of X is negative definite. Let E be an $SU(2)$–bundle over X with $c_2(E) = 1$ and consider the moduli space $\mathcal{M}$ of ASD connections on E. The formal dimension of $\mathcal{M}$ is $8c_2(E) - 3(1 - b_1 + b_2^+)(X) = 5$. Let m denote half the number of solutions to $Q_X(a,a) = -1$; note that $m \le \operatorname{rank} Q_X$. According to [67], for almost all metrics on X, there exist points $p_1, \dots, p_m \in \mathcal{M}$ such that $\mathcal{M} \setminus \{p_1, \dots, p_m\}$ is a smooth orientable 5-dimensional manifold. Each point $p_i \in \mathcal{M}$ has a neighborhood homeomorphic to a cone over $\mathbb{C}P^2$ with either orientation. Moreover, $\mathcal{M} \setminus \{p_1, \dots, p_m\}$ is not empty, and there exists a collar $(0,\lambda] \times X$ inside $\mathcal{M}$ such that attaching $\{0\} \times X$ to $\mathcal{M}$ makes it into a compact manifold $\bar{\mathcal{M}}$.

Thus if we remove open neighborhoods of the p_i from $\bar{\mathcal{M}}$, we will obtain an oriented cobordism between X and m copies of $\mathbb{C}P^2$. Since $\operatorname{sign} Q_X$ is an oriented cobordism invariant, we conclude that $\operatorname{rank} Q_X = -\operatorname{sign} Q_X \le m$. But $\operatorname{rank} Q_X \ge m$ therefore $\operatorname{rank} Q_X = m$ and the form Q_X is diagonalizable.

Example 7.14. The Poincaré homology sphere has infinite order in $\Theta^3_{\mathbb{Z}}$. To prove this, recall that $\Sigma(2,3,5)$ bounds a plumbed manifold M with the intersection form E_8, see Example 1.16. For any $n \ge 1$, the n-multiple of Σ bounds the manifold $n \cdot M$ obtained as the boundary connected sum of n copies of M. The intersection form of $n \cdot M$ is isomorphic to $n\,E_8$. Now suppose that there is an integer $n \ge 1$ such that $n\,\Sigma$ is homology cobordant to zero via a homology cobordism W, and form the manifold $n\,M \cup_{n\,\Sigma} (-W)$. This is a smooth closed oriented manifold with the intersection form $n\,E_8$, which is definite and non-diagonalizable over the integers. This contradiction with the Donaldson theorem proves the result.

Remark 7.15. Curiously enough, if $\Sigma(2,3,5)$ is the boundary of a smooth compact oriented 4–manifold X with negative definite intersection form then the intersection form of X must be isomorphic over the integers to $E_8 \oplus n \cdot (-1)$, $n \ge 0$, see [105]. On the other hand, any negative definite 4–manifold with boundary $\Sigma(2,3,7)$ has diagonalizable intersection form, see [247]. In particular, Theorem 7.13 cannot be applied to show that $\Sigma(2,3,7)$ has infinite order in $\Theta^3_{\mathbb{Z}}$, compare with Example 7.27. These and other results in [105] and [247] rely on the study of the so called *Frøyshov invariant* originating in the Seiberg–Witten gauge theory.

Example 7.16. For any integers $p \ne 0$ and q, the integral homology sphere $p\cdot\Sigma(2,3,5) \# q\cdot\Sigma(2,3,7)$ has infinite order in the homology cobordism group. This follows from Donaldson's theorem after observing that $\Sigma(2,3,7)$ bounds a smooth manifold with intersection form (-1) (as (-1)–surgery on the right-handed trefoil) and another smooth manifold with intersection form $(+1)$ (as $(+1)$–surgery on the figure-eight knot).

Example 7.17. Whenever an integral homology sphere Σ bounds a smooth 4–manifold whose intersection form is definite and non–diagonalizable over the integers, Σ has infinite order in the homology cobordism group $\Theta^3_{\mathbb{Z}}$ (and also in the rational homology cobordism group $\Theta^3_{\mathbb{Q}}$). A method for constructing such homology spheres was proposed by Cochran and Gompf [55]. Given two knots, k_1 and k_2, we say that $k_1 \geq k_2$ if k_1 is *concordant* to k_2 inside of a 4–manifold with negative definite intersection form. This generalizes the relation of k_2 being obtained from k_1 by change of a negative crossing, see Figure 7.8 (for example, the left-handed trefoil admits a planar projection with all negative crossings).

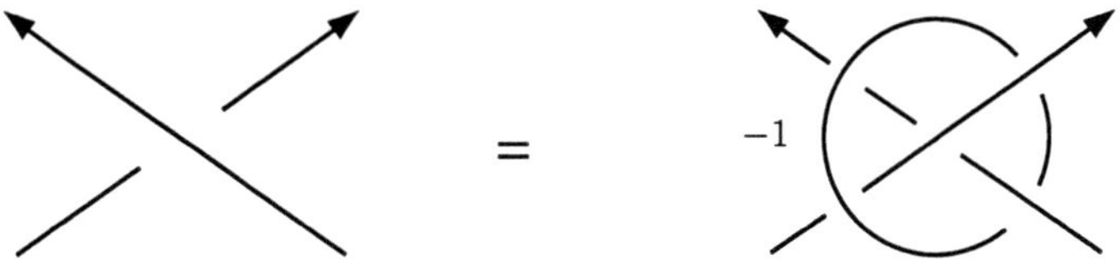

Fig. 7.8.

Perform a $(-1/q)$–surgery, $q > 0$, along the concordance relating k_1 and k_2 to obtain a negative definite cobordism between the integral homology spheres $S^3 - (1/q) \cdot k_1$ and $S^3 - (1/q) \cdot k_2$. Therefore, if the former bounds a smooth manifold with negative definite intersection form which is not diagonalizable over the integers, so does the latter. One can easily see that if k is a knot in S^3 admitting a planar projection with all negative crossings then $k \geq$ the left-handed trefoil. Since the result of (-1)–surgery on the left-handed trefoil is the Poincaré homology sphere, any (-1)–surgery on k (and more generally, any $(-1/q)$–surgery with q positive) has infinite order in $\Theta^3_{\mathbb{Z}}$.

Remark 7.18. The Donaldson Theorem 7.13 holds as well for manifolds X with boundary an integral homology sphere Σ as long as $I_*(\Sigma)$ is trivial. Integral homology spheres Σ having $I_*(\Sigma)$ are called *Floer spheres.* No examples of Floer spheres other than S^3 are currently known.

Example 7.19. Let k be a knot in S^3 with all negative crossings, and $D(k)$ its untwisted Whitehead double, see Section 3.4.3. Cochran and Gompf [55] show that the homology sphere $\Sigma = S^3 - D(k)$ bounds a smooth manifold with negative definite intersection form which is not diagonalizable over the integers. By Donaldson's theorem, $I_*(\Sigma)$ must be non–trivial. On the other hand, $\lambda(\Sigma) = 0$ because the Alexander polynomial of $D(k)$ is one. Therefore, the Floer homology groups $I_*(\Sigma)$ are non–trivial in each parity.

Fintushel and Stern in [87] initiated the study of homology cobordisms of Seifert fibered homology spheres via equivariant gauge theory on orbifolds.

Theorem 7.20. *Let $\Sigma(a_1,\dots,a_n)$ be a Seifert fibered homology sphere oriented as the link of singularity (1.2), and define*

$$R(a_1,\dots,a_n) = \frac{2}{a} - 3 + n + \sum_{i=1}^{n} \frac{2}{a_i} \sum_{k=1}^{a_i-1} \cot\left(\frac{\pi a k}{a_i^2}\right) \cot\left(\frac{\pi k}{a_i}\right) \sin^2\left(\frac{\pi k}{a_i}\right),$$

where $a = a_1 \cdots a_n$. Then $R(a_1,\dots,a_n)$ is an odd integer. If $R(a_1,\dots,a_n) > 0$ then $\Sigma(a_1,\dots,a_n)$ does not bound an oriented smooth 4-manifold V whose intersection form is positive definite and whose first homology $H_1(V,\mathbb{Z})$ contains no 2–torsion.

Proof. Let W be the mapping cylinder of Seifert fibration $\Sigma(a_1,\dots,a_n) \to S^2$. It is an oriented 4–dimensional V–manifold with boundary $\Sigma(a_1,\dots,a_n)$ and n singular points whose neighborhoods are cones over the lens spaces $L(a_i,b_i)$, where (a_i,b_i) are Seifert invariants of $\Sigma(a_1,\dots,a_n)$. The usual concepts of topology and differential geometry extend to V–manifolds via equivariant setting, see for instance [269]. In particular, the rational intersection form of W is given by $w^2 = -1/a_1 \cdots a_n$ where w is the canonical generator in $H_2(W,\mathbb{Z}) = H_2(S^2,\mathbb{Z}) = \mathbb{Z}$.

Suppose that $\Sigma(a_1,\dots,a_n)$ bounds a smooth manifold V with positive definite intersection form and assume for simplicity that V is simply connected. Consider the closed V–manifold $X = W \cup (-V)$, which has negative definite intersection form, and form the $SO(3)$ V–bundle $E_w = L_w \oplus \mathbb{R}$ over X where L_w is the linear V–bundle with Euler class dual to w. The moduli space $\mathcal{M}$ of ASD–connections in E_w is a compact manifold (perhaps after a compact perturbation) except at a unique singular point at the single reducible ASD–connection. The singular point has a neighborhood homeomorphic to a cone on a complex projective space $\mathbb{C}P^m$. An index calculation shows that, if $R(a_1,\dots,a_n)$ is positive, it equals the dimension of $\mathcal{M}$. This means that $\mathbb{C}P^m$, which is the link of the unique singularity of $\mathcal{M}$, is null-cobordant. If m is even, this gives an immediate contradiction; if m is odd, a similar contradiction is obtained after some extra work.

Corollary 7.21. *If $R(a_1,\dots,a_n) > 0$ then $\Sigma(a_1,\dots,a_n)$ has infinite order in the homology cobordism group.*

Proof. Suppose that there exists $m \geq 1$ such that $m \cdot \Sigma(a_1,\dots,a_n)$ is homology cobordant to zero via a homology cobordism W. As a link of singularity, $\Sigma(a_1,\dots,a_n)$ bounds a negative-definite smooth manifold X which is a resolution of singularity (1.2), see Section 1.1.10. Cap off all but one copy of $\Sigma(a_1,\dots,a_n)$ in the boundary of W with $m-1$ copies of the manifold $(-X)$ to get a smooth manifold Z with boundary $\Sigma(a_1,\dots,a_n)$. The intersection form of Z is positive definite and $H_1(Z,\mathbb{Z}) = 0$, which contradicts Theorem 7.20.

Example 7.22. An easy calculation shows that $R(2,3,6m-1) = 1$ for any $m \geq 1$ so all $\Sigma(2,3,6m-1)$ have infinite order in the homology cobordism group. Brieskorn homology spheres $\Sigma(2,3,12m-1)$ have infinite order although the Rokhlin invariant vanishes for all of them. At the same time, $R(\Sigma(2,3,6m+1)) = -1$ for all $m \geq 1$ so that Theorem 7.20 gives no information about homology cobordisms of $\Sigma(2,3,6m+1)$.

Remark 7.23. The invariant $R(a_1,\dots,a_n)$ does not carry the Rokhlin invariant: for instance, $R(2,3,7) = -1$ so that Theorem 7.20 does not prohibit $\Sigma(2,3,7)$ from bounding an acyclic manifold. On the other hand, $\mu(\Sigma(2,3,7)) = 1 \bmod 2$ and hence $\Sigma(2,3,7)$ is not homology cobordant to zero.

The *Fintushel–Stern invariant* $R(a_1,\dots,a_n)$ was interpreted by Neumann and Zagier [244] as follows. Recall that $\Sigma(a_1,\dots,a_n)$ is the boundary of a negative definite manifold obtained by plumbing according to the graph in Figure 1.7 with weights $b \leq -1$ and $b_{ij} \leq -2$ determined by the equations (1.10) and (1.11). Neumann and Zagier prove that $R(a_1,\dots,a_n) = -2b-3$. In particular, $R(a_1,\dots,a_n)$ is always an odd integer greater than or equal to -1.

Corollary 7.24. *If $\Sigma(a_1,\dots,a_n)$ has finite order in the homology cobordism group then $b=-1$, where b is the integer (1.11).*

More elements of infinite order in $\Theta^3_{\mathbb{Z}}$, including many Seifert fibered homology spheres left out by the Fintushel–Stern invariant, can be detected by using the so called 10/8–*theorem* of Furuta [115], see also [2].

Theorem 7.25. *Let X be a smooth closed oriented spin 4–manifold and let Q be its intersection form (2.1). If Q is not definite then*

$$\operatorname{rank} Q > 5/4 \cdot |\operatorname{sign} Q|.$$

Note that if the Q is definite then $\operatorname{rank} Q = \operatorname{sign} Q = 0$ by Donaldson's theorem. Theorem 7.25 can be reformulated as follows. The intersection form Q is isomorphic over the integers to a form $a \cdot E_8 \oplus b \cdot H$ where E_8 is the unimodular form associated with the plumbing graph in Figure 1.14, and H is the hyperbolic two-by-two form (6.27). Theorem 7.25 asserts then that $|a| < |b|$. Conjecturally, a stronger conclusion holds that $\operatorname{rank} Q \geq 11/8 \cdot |\operatorname{sign} Q|$. This last statement is known as the 11/8–*conjecture* (hence the name 10/8–theorem).

Corollary 7.26. *Suppose that an integral homology sphere Σ is the boundary of a smooth oriented spin 4–manifold X whose intersection form Q is such that*

$$\operatorname{rank} Q \leq 5/4 \cdot |\operatorname{sign} Q|.$$

Then Σ has infinite order in the homology cobordism group.

Example 7.27. Brieskorn homology sphere $\Sigma(2,3,7)$ bounds a plumbed manifold with even intersection form E_{10} isomorphic over the integers to $E_8 \oplus H$. It does not satisfy the 10/8–inequality of Theorem 7.25, therefore, $\Sigma(2,3,7)$ has infinite order in the homology cobordism group. In fact, for any relatively prime integers $p, q \geq 2$ and an odd $k \geq 1$, one can construct a smooth simply connected 4–manifold with boundary $\Sigma(p,q,pqk+1)$ and with even intersection form for which the 10/8–inequality fails, see [270]. This implies that all Brieskorn homology spheres $\Sigma(p,q,pqk+1)$ with odd k have infinite order in $\Theta^3_{\mathbb{Z}}$. This result need not be true for even k as it is known that, for example, $\Sigma(2,3,13)$ is homology cobordant to zero, see Example 7.4.

More integral homology spheres having infinite order in $\Theta^3_{\mathbb{Z}}$ can be detected with the help of the $\bar{\mu}$–invariant, see Section 7.2.3.

7.2.3 The $\bar{\mu}$–invariant

Let Γ be a plumbing graph and $P(\Gamma)$ the 4–manifold obtained by plumbing disc bundles according to Γ, see Section 1.1.9. Suppose that the boundary Σ of $P(\Gamma)$ is an integral homology sphere, that is, $\det A(\Gamma) = \pm 1$ where $A(\Gamma)$ is the matrix (1.8). Then there exists a unique homology class $w \in H_2(P(\Gamma), \mathbb{Z})$ satisfying the following two conditions. First, w is *characteristic*, that is (dot represents intersection number)

$$w \cdot x = x \cdot x \bmod 2 \quad \text{for all } x \in H_2(P(\Gamma), \mathbb{Z}), \tag{7.2}$$

and second, all the coordinates of $w = (w_1, \ldots, w_s)$ are either 0 or 1 in the natural basis of $H_2(P(\Gamma), \mathbb{Z})$ represented by zero-sections of the plumbed bundles. We call w the *Wu class* for $P(\Gamma)$, and the set of vertices v_i with $w_i = 1$ the *Wu set* of Γ, compare with Section 2.4.1. It is proved in [241] that the integer

$$\bar{\mu}(\Sigma) = \frac{1}{8}\,(\operatorname{sign} P(\Gamma) - w.w) \tag{7.3}$$

only depends on $\Sigma(\Gamma)$ and not on Γ. It is called the *$\bar{\mu}$–invariant* of Σ. The equation (7.2) implies that no two adjacent vertices of Γ belong to the Wu set. Therefore, the class w can be chosen to be spherical, which implies that $\bar{\mu}(\Sigma) = \mu(\Sigma) \bmod 2$ where $\mu(\Sigma)$ is the Rokhlin invariant, see (2.3).

An equivalent definition of the $\bar{\mu}$–invariant is due to Siebenmann [286]. According to Montesinos, on any plumbed homology sphere Σ, there exists a preferred class I_Σ of orientation preserving involutions such that, for $\sigma \in I_\Sigma$, the orbit space Σ/σ is S^3 and the fixed point set is a knot k_σ in S^3. The knot k_σ is called a *knot of plumbing type*, see Example 1.26. If σ' is also in I_Σ, the knot $k_{\sigma'}$ is related to k_σ by a sequence of mutations. The knot signature is a mutation invariant, thus the integers $\operatorname{sign} k_\sigma$, $\sigma \in I_\Sigma$, all coincide. Their common value is an invariant of Σ which coincides with $8 \cdot \bar{\mu}(\Sigma)$. The latter is not difficult to prove, see [272].

Keeping this identification in mind, one can also view $\bar{\mu}(\Sigma)$ as the equivariant Casson invariant of Σ, see Example 5.25.

Remark 7.28. Either definition of the $\bar{\mu}$–invariant easily extends to all plumbed $\mathbb{Z}/2$–homology spheres Σ, defining $\bar{\mu}(\Sigma)$ which need not be an integer but which still reduces modulo 2 to the Rokhlin invariant of Σ.

The $\bar{\mu}$–invariant of graph homology spheres is easy to compute. It is additive with respect to splicing of graph links, hence, one only needs to calculate it for Seifert fibered homology spheres. The formulas (2.8) and (2.9) for the Rokhlin invariant of Seifert manifolds which are $\mathbb{Z}/2$–homology spheres are actually formulas for the $\bar{\mu}$–invariant, if one does not reduce modulo 2. A surgery formula for the $\bar{\mu}$–invariant can be found in [279].

W. Neumann conjectured in [241] that the $\bar{\mu}$–invariant is an invariant of homology cobordism. The following result from [277] weighs for a positive solution to this conjecture.

Theorem 7.29. *Let an integral homology sphere Σ be a link of algebraic singularity with the complex orientation. If Σ has a finite order in the homology cobordism group then $\bar{\mu}(\Sigma) \geq 0$.*

Proof. Since Σ is an algebraic link, it bounds a plumbed manifold P with negative intersection form, see Section 1.1.10. If Σ bounds a smooth acyclic manifold W, then there is a smooth closed manifold X with negative definite intersection form obtained by gluing the plumbed manifold P to $-W$. According to Donaldson's Theorem 7.13, the intersection form of X is diagonalizable over the integers. In the diagonalizing basis, all coordinates of the characteristic vector w are odd, therefore, $-w \cdot w \geq \operatorname{rank} H_2(P, \mathbb{Z})$ and $\bar{\mu}(\Sigma) \geq 0$. If an n–multiple of Σ bounds a smooth acyclic manifold, the argument above should be repeated with n copies of the manifold P.

All Seifert fibered homology spheres are links of singularities, therefore, Theorem 7.29 applies to them. We will show in the next section that, in fact, a stronger result holds, namely, if a Seifert fibered homology sphere has finite order in the homology cobordism group then its $\bar{\mu}$–invariant vanishes.

7.2.4 The Fukumoto-Furuta w–invariants of Integral Homology Spheres

Let X be a V-manifold with finitely many singular points, each singular point having a neighborhood isomorphic to the cone over a lens space or other spherical space form. The usual concepts of topology and differential geometry extend to V–manifolds after defining everything in an equivariant setting, see [269]. We are interested in a spin structure on X and the index of the associated Dirac operator $\mathcal{D}(X)$. Fukumoto and Furuta proved in [112] the following result about this index, extending Theorem 7.25 from smooth manifolds to V–manifolds.

Theorem 7.30. *Let X be an oriented closed spin V–manifold with $b_1(X) = 0$, and let c be a spin structure on X. Then* $\operatorname{ind}\mathcal{D}(X)$ *is even, and either* $\operatorname{ind}\mathcal{D}(X) = 0$ *or* $b_-(X) < \operatorname{ind}\mathcal{D}(X) < b_+(X)$.

Fukumoto and Furuta [112] used this theorem to define a family of w–invariants. Let Σ be an integral homology sphere, and let X be a compact oriented spin V–manifold with only isolated singularities such that $b_1(X) = 0$ and $\partial X = \Sigma$. Such a V–manifold always exists, moreover, it can be chosen to be smooth and simply connected. Let c be a spin structure on X. For every triple (Σ, X, c), the *w–invariant* is defined as

$$w(\Sigma, X, c) = \operatorname{ind}\mathcal{D}(X \cup Y) + \operatorname{sign}(Y)/8,$$

where Y is any smooth simply connected spin 4–manifold with $\partial Y = -\Sigma$. Here, $\mathcal{D}(X \cup Y)$ is the Dirac operator associated with the spin structure on $X \cup Y$ induced by c and the spin structure on Y. Of course, if X is smooth we simply get $w(\Sigma, X, c) = -\operatorname{sign}(X)/8$.

The integer $w(\Sigma, X, c)$ does not depend on the choice of Y. The reduction of $w(\Sigma, X, c)$ modulo 2 equals the Rokhlin invariant of Σ for any choice of X and c. Moreover, $w(\Sigma, X, c)$ is additive with respect to connected sums of the triples (Σ, X, c). Theorem 7.30 implies the following important property of the w–invariants.

Theorem 7.31. *Let Σ be an integral homology sphere which has finite order in the homology cobordism group $\Theta^3_{\mathbb{Z}}$. Then, for any choice of X and c, either $w(\Sigma, X, c) = 0$ or*

$$-b_-(X) < w(\Sigma, X, c) < b_+(X).$$

For example, if Σ happens to bound a spin V–manifold X such that $b_+(X) \leq 1$, $b_-(X) \leq 1$ and $w(\Sigma, X, c) \neq 0$, then Σ cannot be of finite order in the homology cobordism group. Several explicit constructions of spin V–manifolds with small $b_\pm$ are given in [113] together with various ramifications of Theorem 7.31.

The w–invariants turned out to be very useful in the study of the homology cobordisms of Seifert fibered and plumbed homology spheres, see [113] and [273]. Given a plumbed manifold P with boundary an integral homology sphere Σ, one can view P as a cobordism between a disjoint union of lens spaces and Σ. Capping off the boundary lens spaces with cones, one gets the so called plumbed V–manifold with boundary Σ. The mapping cone W of any Seifert fibration $\Sigma(a_1, \ldots, a_n) \to S^2$ is an example of a plumbed V–manifold.

Theorem 7.32. *Let (X, c) be a spin plumbed V–manifold with boundary an integral homology sphere Σ. Then $w(\Sigma, X, c) = -\bar{\mu}(\Sigma)$ where $\bar{\mu}$ is the invariant of Neumann and Siebenmann.*

This was proved in several special cases in [113] and in general in [273]. The following is an application of the w–invariants to the study of homology cobordisms of Seifert fibered homology spheres. An extension to certain classes of plumbed homology spheres can be found in [111] and [273].

Theorem 7.33. *Any Seifert fibered homology sphere $\Sigma(a_1, \dots, a_n)$ with non-vanishing $\bar{\mu}$–invariant has infinite order in the homology cobordism group.*

Proof. Suppose that one of the invariants a_i of $\Sigma(a_1, \dots, a_n)$ is even. Then the mapping cylinder W of $\Sigma(a_1, \dots, a_n) \to S^2$ is a spin V–manifold with $b_+ = 0$ and $b_- = 1$. According to Theorem 7.31, $\Sigma(a_1, \dots, a_n)$ has infinite order in the homology cobordism group. If all the a_i are odd, the V–manifold W is not spin, and one should alter the construction. In fact, one can construct two different spin plumbed V–manifolds with boundary $\Sigma(a_1, \dots, a_n)$, one with $b_+ = 1$ and $b_- = n-2$, and the other with $b_+ = n-2$ and $b_- = 1$. Theorem 7.31 provides two sets of bounds on the respective w–invariants. Since the two w–invariants are equal to each other by Theorem 7.32, both sets of bounds can be applied to each of them, proving that they vanish. A complete proof can be found in [273].

Corollary 7.34. *The $\bar{\mu}$–invariant of Neumann and Siebenmann is a homology cobordism invariant for all Seifert fibered homology spheres.*

The modulo 2 reduction of the $\bar{\mu}$–invariant coincides with the Rokhlin invariant. In particular, if the Rokhlin invariant of $\Sigma(a_1, \dots, a_n)$ is non-trivial then $\Sigma(a_1, \dots, a_n)$ cannot have order two in $\Theta^3_{\mathbb{Z}}$. Thus if one is to look for an element in $\Theta^3_{\mathbb{Z}}$ solving positively the triangulation conjecture, see Section 7.3.1, all Seifert fibered homology spheres, as well as all homology spheres homology cobordant to them, should be excluded from consideration.

Remark 7.35. There are many Seifert fibered homology spheres with trivial $\bar{\mu}$–invariant having infinite order in the homology cobordism group. For example, let p, q, r be pairwise relatively prime odd integers of absolute value greater than one. If $pq + qr + pr = -1$ then $\bar{\mu}(\Sigma(|p|, |q|, |r|)) = 0$ but $R(|p|, |q|, |r|) = 1$ where R is the Fintushel–Stern invariant, see Theorem 7.20. Therefore $\Sigma(|p|, |q|, |r|)$ has infinite order in the homology cobordism group by Corollary 7.21.

7.2.5 The Group $\Theta^3_{\mathbb{Z}}$ is Infinitely Generated

The following theorem providing an infinite family of linearly independent generators in the abelian group $\Theta^3_{\mathbb{Z}}$ was first proved by Furuta [114]; another proof was given by Fintushel and Stern [83]. Both proofs rely on the study of the set of values of the Chern-Simons function on its critical points.

Theorem 7.36. *Let $p, q \geq 2$ be relatively prime integers. The Brieskorn homology spheres $\Sigma(p, q, pqk-1)$, $k \geq 1$, are linearly independent over $\mathbb{Z}$ in the homology cobordism group.*

7.3 Applications of the Homology Cobordism Group

The group $\Theta^3_{\mathbb{Z}}$ and the homomorphism μ have a distinguished history in the study of manifolds. We already mentioned some of the applications in Chapter 3; more applications are discussed below.

7.3.1 Triangulating Topological Manifolds

The structure of $\Theta^3_{\mathbb{Z}}$ is closely related to the triangulation problem for topological manifolds in dimensions five and higher. Kirby and Siebenmann [162] showed that there exist topological manifolds in all dimensions greater than or equal to five which are not PL, that is, not homeomorphic to any *combinatorial* simplicial complex. The *triangulation conjecture* states that all topological manifolds in these dimensions are *simplicially triangulable*, that is, homeomorphic to a simplicial complex, not necessarily combinatorial, see [284]. This conjecture is still open. The following result is due to Matumoto [209] and Galewski and Stern [119], see also [121].

Theorem 7.37. *Let M be a topological n–manifold. If $n \geq 7$ ($n \geq 6$ if ∂M is compact or $n \geq 5$ if M is closed) then there exists a well–defined obstruction $\tau_M \in H^5(M, \ker\mu)$, where μ is the Rokhlin homomorphism (7.1), such that $\tau_M = 0$ if and only if M is simplicially triangulable. If $\tau_M = 0$ then there are $|H^4(M, \ker\mu)|$ triangulations up to concordance. Furthermore, $\tau_M = 0$ for all M if and only if there exists an integral homology sphere Σ with $\mu(\Sigma) = 1$ and such that $\Sigma\#\Sigma$ bounds a smooth acyclic 4–manifold.*

In other words, every topological n–manifold of dimension n as indicated in the theorem is simplicially triangulable if and only if there is an element of order two in $\Theta^3_{\mathbb{Z}}$ which does not lie in the kernel of the Rokhlin homomorphism. It is not known if such an element exists. However, it is known that such an element cannot be found easily by finding an amphicheiral Σ and noticing that $\Sigma\#\Sigma = \Sigma\#(-\Sigma)$ bounds a smooth acyclic 4–manifold: the Rokhlin invariant of an amphicheiral homology sphere is always zero, see Section 3.6.2. In fact, it is not known whether $\Theta^3_{\mathbb{Z}}$ has any torsion at all or whether $\Theta^3_{\mathbb{Z}}$ modulo torsion is free abelian. However, certain classes of integral homology spheres are known to have infinite order in the homology cobordism group, as described above. It is also known that $\ker\mu$ is infinite, see Example 7.22, therefore, in every dimension $n \geq 5$, there exist closed topological n–manifolds with infinitely many non-concordant simplicial triangulations.

7.3.2 Knot Concordance Group

Another application is to the classical knot concordance group $\mathcal{C}^{3,1}$. The simplest known obstruction to sliceness is the knot signature: it is known to vanish for slice knots. The knot signature induces a homomorphism of abelian groups $\mathcal{C}^{3,1} \to \mathbb{Z}$ whose image is the set of even integers. In particular, all knots with non-zero signature have infinite order in $\mathcal{C}^{3,1}$. Another obstruction to sliceness is the Alexander polynomial: the Alexander polynomial of a slice knot has the form $\Delta_k(t) = p(t)p(t^{-1})$ where $p(t)$ is a polynomial with integer coefficients, see, for example, [263]. This is enough to show that $\mathcal{C}^{3,1}$ is infinitely generated. In particular, if k is slice then $\Delta_k(-1)$ is a square number. The figure eight knot k has signature zero but is not slice because $\Delta_k(-1) = 5$. For other classical obstructions to sliceness see survey [130].

Given a knot k in S^3, one can perform a $(-1/q)$–surgery along k to obtain an integral homology sphere $S^3 - (1/q)\cdot k$. This construction provides a family of maps $D_q : \mathcal{C}^{3,1} \to \Theta^3_{\mathbb{Z}}$, $q \in \mathbb{Z}$, which were first introduced in [125]. A link between $\mathcal{C}^{3,1}$ and the rational homology spheres is provided by the family of maps $D_{p,q} : \mathcal{C}^{3,1} \to \Theta^3_{\mathbb{Q}}$ sending a knot $k \subset S^3$ into $[S^3 - (p/q)\cdot k] - [L(p,q)]$ where $L(p,q)$ is the lens space obtained by $(-p/q)$–surgery on a trivial knot in S^3. Any information about the groups $\Theta^3_{\mathbb{Z}}$ and $\Theta^3_{\mathbb{Q}}$ yields via the maps D_q and $D_{p,q}$ new information about the group $\mathcal{C}^{3,1}$, and vice versa, see [55] and [87].

Remark 7.38. In general, the maps D_q and $D_{p,q}$ are not homomorphisms. For example, let k_1 and k_2 be the left– and right–handed trefoils, then $D_1(k_1) = \Sigma(2,3,5)$ and $D_1(k_2) = \Sigma(2,3,7)$. The square knot $k_1\#k_2$ is slice, therefore, $D_1(k_1\#k_2)$ represents zero in the homology cobordism group, compare with Example 7.9. On the other hand, $\Sigma(2,3,5)\#\,\Sigma(2,3,7)$ is not homology cobordant to zero, see Example 7.16.

Another link between $\mathcal{C}^{3,1}$ and the homology cobordism groups is established via the branched cover construction, see Section 7.1.4. Let Σ be a double branched covering of S^3 with branch set a knot k. Casson and Gordon showed in [51] that if k is a ribbon knot then $\rho_\alpha(\Sigma) = \pm 1$ for any representation $\alpha : H_1(\Sigma,\mathbb{Z}) \to U(1)$. Their proof proceeds through the interpretation of the ρ–invariant $\rho_\alpha(\Sigma)$ in terms of the rational homology cobordism obtained as a double branched cover over the ribbon, compare with Remark 6.17. Fintushel and Stern [88] used gauge theory to show that in fact the conclusion $\rho_\alpha(\Sigma) = \pm 1$ holds for all slice knots. These results were extended in [207] and [266].

7.3.3 PL–discs in Contractible 4–Manifolds

Every knot $k \subset S^3 = \partial D^4$ bounds a (not necessarily locally flat) PL–disc in D^4, namely, the cone over k. Zeeman [311] conjectured that there exist contractible 4–manifolds W and knots $k \subset \partial W$ which do not bound any PL–discs inside W. This conjecture was proved by Akbulut [3] by providing the following example.

Let W be the contractible 4–manifold obtained by attaching a 2–handle to $S^1 \times D^3$ along its boundary as shown in Figure 7.2 with $p = 0$. Let

$k \subset \partial W$ be coming from the loop $S^1 \times \{\mathrm{pt}\}$ on the boundary of $S^1 \times D^3$, then k does not bound any PL–disc inside W. The proof of this fact in [3] is gauge–theoretical.

7.3.4 Constructing Smooth Manifolds

One can study smooth 4–manifolds by splitting them along integral homology spheres. If a homology sphere Σ bounds a smooth 4–manifolds X with intersection form Q_X, and if $-\Sigma$ bounds another smooth manifold Y with intersection form Q_Y then $X \cup_\Sigma Y$ is a closed smooth manifold whose intersection form is isomorphic to $Q_X \oplus Q_Y$. Conversely, if the intersection form of a smooth closed simply connected 4–manifold W decomposes as $Q_1 \oplus Q_2$, then there is an integral homology 3-sphere Σ in W splitting it into two 4–manifolds X and Y with intersection forms Q_1 and Q_2, respectively, see [100]. This fact is useful in constructing exotic 4–manifolds and group actions.

Example 7.39. It was shown in [81] that $\Sigma(3,5,19)$ bounds a smooth contractible 4–manifold W and that the double $W \cup_\tau (-W)$ along the free involution $\tau(x,y,z) = (-x,-y,-z)$ on $\Sigma(3,5,19)$, compare with (1.1), is diffeomorphic to S^4. Interchanging the two copies of W, we obtain an involution on S^4 which is not in any sense smoothly equivalent to the antipodal map. Thus the quotient of S^4 by this involution is a smooth manifold homotopy equivalent to $\mathbb{R}P^4$ but not s–cobordant to it. Other related constructions can be found in [89].

Let X_0 and X_1 be smooth, closed, oriented, simply connected 4-manifolds. If X_0 is homeomorphic to X_1 then there is a compact contractible 4-manifold $W \subset X_0$ such that by cutting W out of X_0 and regluing it by an involution on its boundary, we get a smooth 4-manifold diffeomorphic to X_1, see [62] and [211]. This contractible piece W is called an *Akbulut cork* corresponding to the pair (X_0, X_1).

Example 7.40. Explicit examples of Akbulut corks were constructed in [1] and [124]. They correspond to pairs $(X_0(n), X_1(n))$ where $X_0(n) = E(n) \# (-\mathbb{C}P^2)$ is a blow up of the elliptic surface $E(n)$, and $X_1(n) = (2n-1) \cdot \mathbb{C}P^2 \# 10n \cdot (-\mathbb{C}P^2)$ with $n \geq 2$. For any given n, manifolds $X_0(n)$ and $X_1(n)$ are homeomorphic but not diffeomorphic because their Donaldson polynomials are different, see Example 6.70 and Example 6.74. Their Akbulut cork W is independent of n and is obtained by attaching a two-handle to $S^1 \times D^3$ along its boundary as shown in Figure 7.2 with $p = 0$. Observe that $\Sigma = \partial W$ is an integral homology sphere obtained by surgery on the link shown in Figure 7.3, and that this link is symmetric with respect to the involution $\tau : \Sigma \to \Sigma$ interchanging the two link components. The manifold $X_1(n)$ is obtained from $X_0(n)$ by cutting out W and regluing it using τ. The action of τ on the Floer homology of Σ was studied in [276].

The involution $\tau : \Sigma \to \Sigma$ extends to a self–homeomorphism but not self–diffeomorphism of W. Let V be the manifold with the smooth structure obtained by pulling back the smooth structure of W via this self–homeomorphism. Then V is homeomorphic but not diffeomorphic to W relative to the boundary Σ.

References

1. S. Akbulut, *A fake compact contractible 4-manifold*, J. Diff. Geom. **33** (1991), 335-356.
2. S. Akbulut, *Lectures on Seiberg-Witten invariants*, Turkish J. Math. **20** (1996), 95–118.
3. S. Akbulut, *A solution to a conjecture of Zeeman*, Topology **30** (1991), 513–515.
4. S. Akbulut, R. Kirby, *Mazur manifolds*, Michigan Math. J. **26** (1979), 259–284.
5. S. Akbulut, J. McCarthy, *Casson's invariant for oriented homology 3-spheres. An exposition.* Princeton Univ. Press, 1990
6. L. Alaniya, *Manifolds of Alexander type*, Russian Math. Surveys **46** (1991), 245–246.
7. B. Apanasov, *Conformal geometry of discrete groups and manifolds.* Walter de Gruyter, 2000
8. C. Arf, *Untersuchungen über quadratische Formen in Körpern der Charakteristik 2. I.* J. Reine Angew. Math. **183** (1941), 148–167.
9. M. Atiyah, *New invariants of 3 and 4-dimensional manifolds.* In: "The Mathematical Heritage of Hermann Weyl", Proc. Sympos. Pure Math. **48** (1988), 285–299.
10. M. Atiyah, R. Bott, V. Patodi, *On the heat equation and the index theorem*, Invent. Math. **19** (1973), 279–330.
11. M. Atiyah, N. Hitchin, I. Singer, *Self-duality in four-dimensional Riemannian geometry*, Proc. Roy. Soc. London Ser. A **362** (1978), 425–461.
12. M. Atiyah, V. Patodi, I. Singer, *Spectral asymmetry and Riemannian geometry. I.* Math. Proc. Cambridge Philos. Soc. **77** (1975), 43–69.
13. M. Atiyah, V. Patodi, I. Singer, *Spectral asymmetry and Riemannian geometry. II.* Math. Proc. Cambridge Philos. Soc. **78** (1975), 405–432.
14. M. Atiyah, V. Patodi, I. Singer, *Spectral asymmetry and Riemannian geometry. III.* Math. Proc. Cambridge Philos. Soc. **79** (1976), 71–99.
15. D. Auckly, *Chern-Simons invariants of 3-manifolds which fiber over* S^1, Internat. J. Math. **5** (1994), 179–188.
16. D. Auckly, *Surgery numbers of 3-manifolds: a hyperbolic example*, Geometric Topology (Athens, GA, 1993), 21–34. Amer. Math. Soc., 1997.
17. D. Auckly, *A topological method to compute spectral flow*, Kyungpook Math. J. **38** (1998), 181–203.
18. D. Auckly, *Topological methods to compute Chern-Simons invariants*, Math. Proc. Cambridge Philos. Soc. **115** (1994), 229–251.
19. D. Austin, *Equivariant Floer groups for binary polyhedral spaces*, Math. Ann. **302** (1995), 295–322.

20. D. Austin, P. Braam, *Morse–Bott theory and equivariant cohomology*, The Floer Memorial Volume, 123–183. Birkhäuser, 1995.
21. D. Austin, P. Braam, *Equivariant Floer theory and gluing Donaldson polynomials*, Topology **35** (1996), 167–200.
22. D. Bar-Natan, *On the Vassiliev knot invariants*, Topology **34** (1995), 423–472.
23. S. Bauer, *Parabolic bundles, elliptic surfaces and $SU(2)$-representation spaces of genus zero Fuchsian groups*, Math. Ann. **290** (1991), 509–526.
24. S. Bauer, C. Okonek, *The algebraic geometry of representation spaces associated to Seifert fibered homology 3–spheres*, Math. Ann. **286** (1990), 45–76.
25. J. Berge, *Some knots with surgeries yielding lens spaces*. Preprint.
26. R. H. Bing, *An alternative proof that 3–manifolds can be triangulated*, Ann. of Math. (2) **69** (1959), 37–65.
27. J. Birman, *New points of view in knot theory*, Bull. Amer. Math. Soc. **28** (1993), 253–287.
28. J. Birman, R. Craggs, *The μ–invariant of 3–manifolds and certain structural properties of the group of homeomorphisms of a closed, oriented 2–manifold*, Trans. Amer. Math. Soc. **237** (1978), 283–309.
29. H. Boden, C. Herald, *A connected sum formula for the $SU(3)$ Casson invariant*, J. Diff. Geom. **53** (1999), 443–464.
30. H. Boden, C. Herald, *The $SU(3)$ Casson invariant for integral homology 3–spheres*, J. Diff. Geom. **50** (1998), 147–206.
31. H. Boden, C. Herald, P. Kirk, *An integer valued $SU(3)$ Casson invariant*. Preprint math.GT/0008085.
32. H. Boden, C. Herald, P. Kirk, E. Klassen, *Gauge theoretic invariants of Dehn surgeries on knots*, Geom. Topol. **5** (2001), 143–226.
33. H. Boden, A. Nicas, *Universal formulae for $SU(n)$ Casson invariants of knots*, Trans. Amer. Math. Soc. **352** (2000), 3149–3187.
34. R. Bott, *Morse theory indomitable*, Inst. Hautes Études Sci. Publ. Math. **68** (1989), 99–114.
35. S. Boyer, D. Lines, *Surgery formulae for Casson's invariant and extensions to homology lens spaces*, J. Reine Angew. Math. **405** (1990), 181–220.
36. S. Boyer, A. Nicas, *Varieties of group representations and Casson's invariant for rational homology 3–spheres*, Trans. Amer. Math. Soc. **322** (1990), 507–522.
37. P. Braam, *Floer homology groups for homology three-spheres*, Adv. in Math. **88** (1991), 131–144.
38. P. Braam, S. Donaldson, *Floer's work on instanton homology, knots and surgery*, In: The Floer Memorial Volume, 195–256. Birkhäuser, 1995.
39. P. Braam, S. Donaldson, *Fukaya–Floer homology and gluing formulae for polynomial invariants*. In: The Floer Memorial Volume, 257–281. Birkhäuser, 1995.
40. E. Brieskorn, *Beispiele zur Differenzialtopologie von Singularitäten*, Invent. Math. **2** (1966), 1–14.
41. W. Browder, *Surgery on simply–connected manifolds*. Springer-Verlag, 1972.
42. K. Brown, *Cohomology of groups*. Springer-Verlag, 1982.
43. G. Burde, *$SU(2)$–representation spaces for two-bridge knot groups*, Math. Ann. **288** (1990), 103–119.
44. G. Burde, H. Zieschang, *Knots*. Walter de Gruyter, 1985.
45. S. Cairns, *The manifold smoothing problem*, Bull. Amer. Math. Soc. **67** (1961), 237–238.

46. S. Cappell, R. Lee, E. Miller, *A perturbative $SU(3)$ Casson invariants.* Preprint math.DG/0006018.
47. S. Cappell, R. Lee, E. Miller, *Self-adjoint elliptic operators and manifold decompositions. I. Low eigenmodes and stretching*, Comm. Pure Appl. Math. **49** (1996), 825–866.
48. S. Cappell, R. Lee, E. Miller, *Self-adjoint elliptic operators and manifold decompositions. II. Spectral flow and Maslov index*, Comm. Pure Appl. Math. 49 (1996), 869–909.
49. S. Cappell, R. Lee, E. Miller, *A symplectic geometry approach to generalized Casson's invariants of 3–manifolds*, Bull. Amer. Math. Soc. **22** (1990), 269–275.
50. S. Cappell, J. Shaneson, *Invariants of 3–manifolds*, Bull. Amer. Math. Soc. **81** (1975), 559-562
51. A. Casson, C. Gordon, *Cobordism of classical knots.* In: À la recherche de la topologie perdue, 181–199. Birkhäuser, 1986.
52. A. Casson, J. Harer, *Some homology lens spaces which bound rational homology balls.* Pacific J. Math. **96** (1981), 23–36.
53. W. Chen, *Casson's invariant and Seiberg-Witten gauge theory*, Turkish J. Math. **21** (1997), 61–81.
54. S. S. Chern, J. Simons, *Characteristic forms and geometric invariants*, Ann. of Math. **99** (1974), 48–69.
55. T. Cochran, R. Gompf, *Applications of Donaldson's theorems to classical knot concordance, homology 3–spheres and property P*, Topology **27** (1988), 495–512.
56. O. Collin, *Equivariant Casson invariant for knots and the Neumann-Wahl formula*, Osaka J. Math. **37** (2000), 57–71.
57. O. Collin, N. Saveliev, *Equivariant Casson invariants via gauge theory*, J. Reine Angew. Math. **541** (2001), 143–169.
58. O. Collin, N. Saveliev, *A geometric proof of the Fintushel-Stern formula.* Adv. in Math. **147** (1999), 304–314.
59. O. Collin, B. Steer, *Instanton Floer homology for knots via 3–orbifolds.* J. Diff. Geom. **51** (1999), 149–202.
60. M. Culler, C. Cordon, J. Luecke, P. Shalen, *Dehn surgery on knots*, Ann. of Math. **125** (1987), 237–300.
61. C. Curtis, *Generalized Casson invariants for $SO(3)$, $U(2)$, $Spin(4)$, and $SO(4)$*, Trans. Amer. Math. Soc. **343** (1994), 49–86.
62. C. Curtis, M. Freedman, W. Hsiang, R. Stong, *A decomposition theorem for h-cobordant smooth simply-connected compact 4-manifolds*, Invent. Math. **123** (1996), 343–348.
63. M. Daniel, P. Kirk, *A general splitting formula for the spectral flow*, Michigan Math. J. **46** (1999), 589–617.
64. A. Davidow, *Casson's invariant and twisted double knots*, Topology Appl. **58** (1994), 93–101.
65. J. Dean, *Hyperbolic knots with small Seifert–fibered Dehn surgeries*, Ph. D. Thesis, University of Texas, Austin, 1996.
66. I. Dolgachev, *Weighted projective varieties*, Lect. Notes Math. **956**, 34–71. Springer, 1982.
67. S. Donaldson, *An application of gauge theory to four-dimensional topology*, J. Diff. Geom. **18** (1983), 279–315.

68. S. Donaldson, *The orientation of Yang–Mills moduli spaces and 4-manifold topology*, J. Diff. Geom. **26** (1987), 397–428.
69. S. Donaldson, *Polynomial invariants for smooth four-manifolds*, Topology **29** (1990), 257–315.
70. S. Donaldson, *Floer homology groups in Yang–Mills theory*, Cambridge University Press, 2002.
71. S. Donaldson, P. Kronheimer, *The geometry of four-manifolds*, The Clarendon Press, Oxford University Press, 1990.
72. S. Dostoglou, D. Salamon, *Self-dual instantons and holomorphic curves*, Ann. of Math. **139** (1994), 581–640.
73. N. Duchon, *Involutions of plumbed manifolds*, Ph.D. Thesis, University of Maryland, College Park, 1982.
74. W. Ebeling, C. Okonek, *Donaldson invariants, monodromy groups, and singularities*, Int. J. Math. **1** (1990), 233–250.
75. J. Eells, N. Kuiper, *An invariant for certain smooth manifolds*, Ann. Mat. Pura Appl. (4) **60** (1962), 93–110.
76. D. Eisenbud, W. Neumann, *Three-dimensional link theory and invariants of plane curve singularities.* Princeton Univ. Press, 1985.
77. R. Fenn, C. Rourke, *On Kirby's calculus of links*, Topology **18** (1979), 1–15.
78. R. Fintushel, R. Stern, *The blowup formula for Donaldson invariants*, Ann. of Math. **143** (1996), 529–546.
79. R. Fintushel, R. Stern, SO(3)*-connections and the topology of 4-manifolds*, J. Diff. Geom. **20** (1984), 523–539.
80. R. Fintushel, R. Stern, *Constructing lens spaces by surgery on knots*, Math. Z. **175** (1980), 33–51.
81. R. Fintushel, R. Stern, *An exotic free involution on* S^4, Ann. of Math. **113** (1981), 357–365.
82. R. Fintushel, R. Stern, *Homotopy* $K3$ *surfaces containing* $\Sigma(2,3,7)$, J. Diff. Geom. **34** (1991), 255–265.
83. R. Fintushel, R. Stern, *Instanton homology of Seifert fibred homology three spheres*, Proc. London Math. Soc. **61** (1990), 109–137.
84. R. Fintushel, R. Stern, *Integer graded instanton homology groups for homology three-spheres*, Topology **31** (1992), 589–604.
85. R. Fintushel, R. Stern, *Invariants for homology 3-spheres.* In: Geometry of low-dimensional manifolds 1, 125–148, Cambridge Univ. Press, 1990.
86. R. Fintushel, R. Stern, *A* μ*-invariant one homology 3-sphere that bounds an orientable rational ball*, Contemp. Math. **35** (1984), 265–268.
87. R. Fintushel, R. Stern, *Pseudofree orbifolds*, Ann. of Math. **122** (1985), 335–364.
88. R. Fintushel, R. Stern, *Rational cobordisms of spherical space forms*, Topology **26** (1987), 385–393.
89. R. Fintushel, R. Stern, *Seifert fibered 3-manifolds and nonorientable 4-manifolds*, Contemp. Math. **20** (1983), 103–119.
90. R. Fintushel, R. Stern, *Surgery in cusp neighborhoods and the geography of irreducible 4-manifolds*, Invent. Math. **117** (1994), 455–523.
91. R. Fintushel, R. Stern, *Using Floer's exact triangle to compute Donaldson invariants.* In: The Floer Memorial Volume, 435–444. Birkhäuser, 1995.
92. E. Flapan, *Necessary and sufficient conditions for certain homology 3-spheres to have smooth* Z_p*-actions*, Pacific J. Math. **117** (1985), 255–265.

93. A. Floer, *An instanton-invariant for 3-manifolds*, Comm. Math. Phys. **118** (1988), 215–240.
94. A. Floer, *Instanton homology, surgery, and knots.* In: Geometry of low-dimensional manifolds, 1 (Durham, 1989), 97–114, London Math. Soc. Lecture Note Ser., **150**, Cambridge Univ. Press, 1990.
95. A. Floer, *Morse theory for Lagrangian intersections*, J. Diff. Geom. **28** (1988), 513–547.
96. A. Fomenko, S. Matveev, *Algorithmic and computer methods for three-manifolds.* Translated from the 1991 Russian original. Kluwer Academic Publishers, 1997.
97. D. Freed, K. Uhlenbeck, *Instantons and four-manifolds.* Springer-Verlag, 1984.
98. M. Freedman, *The topology of four-dimensional manifolds*, J. Diff. Geom. **17** (1982), 357–453.
99. M. Freedman, R. Kirby, *A geometric proof of Rochlin's theorem*, Algebraic and geometric topology, Part 2, pp. 85–97, Proc. Sympos. Pure Math., XXXII, Amer. Math. Soc., 1978
100. M. Freedman, L. Taylor, *Λ-splitting 4-manifolds*, Topology **16** (1977), 181–184.
101. C. Frohman, E. Klassen, *Deforming representations of knot groups in $SU(2)$*, Comment. Math. Helv. **66** (1991), 340–361.
102. C. Frohman, D. Long, *Casson's invariants and surgery on knots*, Proc. Edinburgh Math. Soc. **35** (1992), 383–395.
103. C. Frohman, A. Nicas, *An intersection homology invariant for knots in a rational homology 3-sphere*, Topology **33** (1994), 123–158.
104. K. Frøyshov, *Equivariant aspects of Yang-Mills Floer theory*, Topology **41** (2002), 525–552.
105. K. Frøyshov, *The Seiberg-Witten equations and four-manifolds with boundary*, Math. Res. Lett. **3** (1996), 373–390.
106. G. Fujita, *A splicing formula for Casson-Walker's invariant*, Math. Ann. **296** (1993), 327–338.
107. K. Fukaya, *Floer homology for oriented 3-manifolds.* In: Aspects of low-dimensional manifolds, 1–92, Adv. Stud. Pure Math., 20, Kinokuniya, Tokyo, 1992.
108. K. Fukaya, *Floer homology of connected sum of homology 3-spheres*, Topology **35** (1996), 89–136.
109. S. Fukuhara, N. Maruyama, *A sum formula for Casson's λ-invariant*, Tokyo J. Math. **11** (1988), 281–287.
110. S. Fukuhara, Y. Matsumoto, K. Sakamoto, *Casson's invariant of Seifert homology 3-spheres*, Math. Ann. **287** (1990), 275–285.
111. Y. Fukumoto, *On an invariant of plumbed homology 3-spheres*, J. Math. Kyoto Univ. **40** (2000), 379–388.
112. Y. Fukumoto, M. Furuta, *Homology 3-spheres bounding acyclic 4-manifolds*, Math. Res. Lett. **7** (2000), 757–766.
113. Y. Fukumoto, M. Furuta, M. Ue, *W-invariants and Neumann-Siebenmann invariants for Seifert homology 3-spheres*, Topology Appl. **116** (2001), 333–369.
114. M. Furuta, *Homology cobordism group of homology 3-spheres*, Invent. Math. **100** (1990), 339–355.

115. M. Furuta, *Monopole equation and the 11/8-conjecture*, Math. Res. Lett. **8** (2001), 279–291.
116. M. Furuta, B. Steer, *Seifert fibred homology 3-spheres and the Yang-Mills equations on Riemann surfaces with marked points*, Adv. in Math. **96** (1992), 38–102.
117. Y. Fukumoto, M. Furuta, *Homology 3-spheres bounding acyclic 4-manifolds.* Preprint, 1998.
118. D. Gabai, *On the geometric and topological rigidity of hyperbolic 3-manifolds*, J. Amer. Math. Soc. **10** (1997), 37–74.
119. D. Galewski, R. Stern, *Classification of simplicial triangulations of topological manifolds*, Ann. of Math. **111** (1980), 1–34.
120. D. Galewski, R. Stern, *Orientation-reversing involutions on homology 3-spheres*, Math. Proc. Cambridge Philos. Soc. **85** (1979), 449–451.
121. D. Galewski, R. Stern, *A universal 5-manifold with respect to simplicial triangulations*, Geometric topology (Proc. Georgia Topology Conf., 1977), pp. 345–350. Academic Press, 1979.
122. S. Garoufalidis, A. Kricker, *Finite type invariants of cyclic branched covers.* Preprint math.GT/0107220
123. W. Goldman, *The symplectic nature of fundamental groups of surfaces*, Adv. in Math. **54** (1984), 200–225.
124. R. Gompf, A. Stipsicz, *4-manifolds and Kirby calculus.* Amer. Math. Soc. 1999
125. F. Gonzáles-Acuna, *Dehn's construction on knots*, Bol. Soc. Mat. Mexicana **15** (1970), 58–79.
126. C. Gordon, *Knots whose branched cyclic coverings have periodic homology*, Trans. Amer. Math. Soc. **168** (1972), 357–370
127. C. Gordon, *Knots, homology spheres, and contractible 4-manifolds*, Topology **14** (1975), 151–172.
128. C. Gordon, R. Litherland, *On the signature of a link*, Invent. Math. **47** (1978), 53–69.
129. C. Gordon, *Ribbon concordance of knots in the 3-sphere*, Math. Ann. **257** (1981), 157–170.
130. C. Gordon, *Some aspects of classical knot theory*, Knot Theory (Proc. Sem. Plans-sur-Bex, 1977), Lecture Notes in Math. **685** (1978), 1–60.
131. C. Gordon, J. Luecke, *Knots are determined by their complements*, J. Amer. Math. Soc. **2** (1989), 371–415.
132. L. Guillou, A. Marin (eds.), *Á la recherche de la topologie perdue.* Birkhäuser, 1986.
133. L. Guillou, A. Marin, *Notes sur l'invariant de Casson des sphères d'homologie de dimension trois*, Enseign. Math. (2) **38** (1992), 233–290.
134. L. Guillou, A. Marin, *Une extension dún théorème de Rohlin sur la signature.* In : À la recherche de la topologie perdue, 97–118. Birkhäuser, 1986.
135. N. Habegger, *Une variété dimension 4 avec forme d'intersection paire et signature -8*, Comm. Math. Helv. **57** (1982), 22–24.
136. C. Herald, *Existence of irreducible representations for knot complements with nonconstant equivariant signature*, Math. Ann. 309 (1997), 21–35.
137. C. Herald, *Flat connections, the Alexander invariant, and Casson's invariant*, Comm. Anal. Geom. **5** (1997), 93–120.
138. C. Herald, *Legendrian cobordism and Chern-Simons theory on 3-manifolds with boundary*, Comm. Analysis and Geom. **2** (1994), 337–413.

139. C. Herald, *Chern-Simons gauge theory on 3-manifolds*. The Proceedings of the Conference on Geometric Structures on Manifolds (Seoul, 1997), 83–109, Lecture Notes Ser., **46**, Seoul Nat. Univ., Seoul, 1999
140. M. Heusener, *$SO_3(\mathbf{R})$–representation curves for two-bridge knot groups*, Math. Ann. **298** (1994), 327–348.
141. M. Heusener, J. Kroll, *Deforming abelian $SU(2)$–representations of knot groups*, Comment. Math. Helv. **73** (1998), 480–498.
142. H. Hilden, *Every closed orientable 3–manifold is a 3–fold branched covering space of S^3*, Bull. Amer. Math. Soc. **80** (1974), 1243-1244.
143. H. Hilden, *Representations of homology 3-spheres*, Pacific J. Math. **94** (1981), 125–129.
144. H. Hilden, M. Lozano, J. Montesinos, *On knots that are universal*, Topology **24** (1985), 499–504.
145. M. Hirsch, *Obstruction theories for smoothing manifolds and maps*, Bull. Amer. Math. Soc. **69** (1963), 352–356.
146. F. Hirzebruch, W. Neumann, S. Koh, *Differentiable manifolds and quadratic forms*. Marcel Dekker, 1971.
147. F. Hirzebruch, D. Zagier, *The Atiyah-Singer theorem and elementary number theory*. Publish or Perish, 1974.
148. J. Hoste, *A formula for Casson's invariant*, Trans. Amer. Math. Soc. **297** (1986), 547–562.
149. W. C. Hsiang, P. S. Pao, *The homology 3-spheres with involutions*, Proc. Amer. Math. Soc. **75** (1979), 308–310.
150. J. Igusa, *On a property of commutators in the unitary group*, Mem. Coll. Sci. Univ. Kyoto Ser. A. **26** (1950), 45–59.
151. K. Ishibe, *The Casson-Walker invariant for branched cyclic covers of S^3 branched over a double knot*, Osaka J. Math. **34** (1997), 481–495.
152. W. Jaco, P. Shalen, *Seifert fibered spaces in 3–manifolds*, Mem. Amer. Math. Soc. **21** (1979).
153. K. Johannson, *Homotopy equivalences of 3–manifolds with boundaries*, Lect. Notes Math. **761**. Springer-Verlag, 1979
154. D. Johnson, *A survey of the Torelli group*, Contemp. Math. **20** (1983), 165–179.
155. V. Jones, *Hecke algebra representations of braid groups and link polynomials*, Ann. of Math. **126** (1987), 335–388.
156. S. Kaplan, *Constructing framed 4–manifolds with given almost framed boundaries*, Trans. Amer. Math. Soc. **254** (1979), 237–263.
157. A. Kawauchi, *Floer homology of topological imitations of homology 3–spheres*, J. Knot Theory Ramifications **7** (1998), 41–60.
158. M. Kervaire, J. Milnor, *On 2–spheres in 4–manifolds*, Proc. Nat. Acad. Sci. USA **47** (1961), 1651–1657.
159. R. Kirby, *A calculus for framed links in S^3*, Invent. Math. **45** (1978), 35–56.
160. R. Kirby, *Problems in low-dimensional topology*, Geometric topology. Proceedings of the 1993 Georgia International Topology Conference. American Mathematical Society, International Press, 1993.
161. R. Kirby, M. Scharlemann, *Eight faces of the Poincaré homology 3-sphere*, Geometric topology (Proc. Georgia Topology Conference, 1977), 113–146. Academic Press, 1979.
162. R. Kirby, L. Siebenmann, *Foundational essays on topological manifolds, smoothings, and triangulations*. Princeton Univ. Press, 1977.

163. P. Kirk, *Mutations of homology spheres and Casson's invariant*, Math. Proc. Cambridge Philos. Soc. **105** (1989), 313–318.

164. P. Kirk, *$SU(2)$ representation varieties of 3–manifolds, gauge theory invariants, and surgery on knots*, Proceedings of GARC Workshop on Geometry and Topology, 137–176, Seoul Nat. Univ., Seoul, 1993

165. P. Kirk, E. Klassen, *Chern-Simons invariants of 3–manifolds and representation spaces of knot groups*, Math. Ann. **287** (1990), 343-367.

166. P. Kirk, E. Klassen, *Chern–Simons invariants of 3–manifolds decomposed along tori and the circle bundle over the representation space of T^2*, Comm. Math. Phys. **153** (1993), 521–557.

167. P. Kirk, E. Klassen, *Computing spectral flow via cup products*, J. Diff. Geom. **40** (1994), 505–562.

168. P. Kirk, E. Klassen, *Representation spaces of Seifert fibered homology spheres*, Topology **30** (1991), 77–95.

169. P. Kirk, E. Klassen, *The spectral flow of the odd signature operator and higher Massey products*, Math. Proc. Cambridge Philos. Soc. **121** (1997), 297–320.

170. P. Kirk, E. Klassen, D. Ruberman, *Splitting the spectral flow and the Alexander matrix*, Comment. Math. Helv. **69** (1994), 375–416.

171. E. Klassen, *Representations in $SU(2)$ of the fundamental groups of the Whitehead link and of doubled knots*, Forum Math. **5** (1993), 93–109.

172. E. Klassen, *Representations of knot groups in $SU(2)$*, Trans. Amer. Math. Soc. **326** (1991), 795–828.

173. M. Kontsevich, *Vassiliev's knot invariants*, I. M. Gelfand Seminar, 137-150, Adv. Soviet Math. **16**, Part 2, Amer. Math. Soc., 1993.

174. P. Kronheimer, M. Larsen, J. Scherk, *Casson's invariant and quadratic reciprocity*, Topology **30** (1991), 335–338.

175. P. Kronheimer, *Embedded surfaces and gauge theory in three and four dimensions*, Surveys in differential geometry, Vol. III (Cambridge, MA, 1996), 243–298, Int. Press, 1998.

176. P. Kronheimer, T. Mrowka, *Embedded surfaces and the structure of Donaldson's polynomial invariants*, J. Diff. Geom. **41** (1995), 573–734.

177. P. Kronheimer, T. Mrowka, *The genus of embedded surfaces in the projective plane*, Math. Res. Letters **1** (1994), 797–808.

178. P. Kronheimer, T. Mrowka, *Recurrence relations and asymptotics for four–manifold invariants*, Bull. Amer. Math. Soc. **30** (1994), 215–221.

179. S. Kwasik, K. B. Lee, *Locally linear actions on 3–manifolds*, Math. Proc. Cambridge Philos. Soc. **104** (1988), 253–260.

180. R. Landvoy, *The Jones polynomial of pretzel knots and links*, Topology Appl. **83** (1998), 135–147.

181. Thang T. Q. Le, *An invariant of integral homology 3–spheres which is universal for all finite type invariants*, Solitons, geometry, and topology: on the crossroad, 75–100, Amer. Math. Soc. Transl. **179**, Amer. Math. Soc. 1997.

182. C. Lescop, *Global surgery formula for the Casson-Walker invariant.* Princeton Univ. Press, 1996.

183. C. Lescop, *On the kernel of the Casson invariant*, Topology **37** (1998), 25–37.

184. C. Lescop, *A sum formula for the Casson-Walker invariant*, Invent. Math. **133** (1998), 613–681.

185. C. Lescop, *Un calcul élémentaire de l'invariant de Casson des sphères d'homologie entière fibrées de Seifert à trois fibres exceptionnelles*, Enseign. Math. (2) **38** (1992), 276–289.

186. R. Lee, *Semicharacteristic classes*, Topology **12** (1973), 183–199.
187. R. Lee, W. Li, *Floer homologies for rational homology 3-spheres.* MSRI Preprint # 04394.
188. R. Lee, W. Li, *Floer homology for Lagrangian intersections and instantons.* Preprint, Oklahoma State University, 1995.
189. R. Lee, W. Li, *Symplectic Floer homology of $SU(2)$ representation spaces.* Preprint, Oklahoma State University, 1995.
190. J. Levine, *Link concordance*, Algebra and topology 1988 (Taejon, 1988), 57–76, Korea Inst. Tech., Taejon, 1988.
191. W. Li, *Casson-Lin's invariant and Floer homology*, J. Knot Theory Ramifications **6** (1997), 851–877.
192. W. Li, *Connected sums, Floer homologies and spectral sequences.* Preprint, Oklahoma State University.
193. W. Li, *Equivariant knot signatures and Floer homologies*, J. Knot Theory Ramifications **10** (2001), 687–701.
194. W. Li, *Floer homology for connected sums of homology 3-spheres*, J. Diff. Geom. **40** (1994), 129–154.
195. W. Li, *Künneth formula for $SO(3)$ Floer homology*, Topology **38** (1999), 1209–1237.
196. W. Li, J. H. Rubinstein, *Casson invariant is a homotopy invariant.* Preprint, Oklahoma State University.
197. W. B. R. Lickorish, *A representation of orientable combinatorial 3-manifolds*, Ann. of Math. **76** (1962), 531–540.
198. Y. Lim, *The equivalence of Seiberg–Witten and Casson invariants for homology 3-spheres*, Math. Res. Lett. **6** (1999), 631–643.
199. X.-S. Lin, *A knot invariant via representation spaces*, J. Diff. Geom. **35** (1992), 337–357.
200. P. Lisca, *Computations of instanton invariants using Donaldson–Floer theory*, Invent. Math. **119** (1995), 347–359.
201. C. Livingston, *Homology cobordisms of 3-manifolds, knot concordances, and prime knots*, Pacific J. Math. 94 (1981), 193–206.
202. R. Lockhart, R. McOwen, *Elliptic differential operators on noncompact manifolds*, Ann. Scuola Norm. Sup. Pisa Cl. Sci. (4) **12** (1985), 409–447.
203. R. Mandelbaum, *Four-dimensional topology: an introduction*, Bull. Amer. Math. Soc. **2** (1980), 1–159.
204. K. B. Marathe, G. Martucci, M. Francaviglia, *Gauge theory, geometry and topology*, Confer. Sem. Math. Univ. Bari **262**, 1995.
205. M. Marcolli, B.-L. Wang, *Exact triangles in Seiberg–Witten Floer theory.* Preprint math.DG/0009157.
206. K. Masataka, *Casson's knot invariant and gauge theory*, Topology Appl. **112** (2001), 111–135.
207. G. Matic, *$SO(3)$-connections and rational homology cobordisms*, J. Diff. Geom. **28** (1988), 277–307.
208. Y. Matsumoto, *An elementary proof of Rochlin's signature theorem and its extension by Guillou and Marin.* In : À la recherche de la topologie perdue, 119–139. Birkhäuser, 1986.
209. T. Matumoto, *Triangulation of manifolds*, Algebraic and geometric topology (Stanford, 1976), Part 2, 3–6, Proc. Sympos. Pure Math. **32** (1978).
210. S. Matveev, *Generalized surgeries of three-dimensional manifolds and representations of homology spheres*, Math. Notes **42** (1987), 651–656.

211. R. Matveyev, *A decomposition of smooth simply-connected h-cobordant 4-manifolds*, J. Diff. Geom. **44** (1996), 571–582.
212. B. Mazur, *A note on some contractible 4-manifolds*, Ann. of Math. **73** (1961), 221–228.
213. G. Meng, C. Taubes, *$\underline{SW}$ = Milnor torsion*, Math. Res. Lett. **3** (1996), 661–674.
214. J. Milnor, *Differentiable manifolds which are homotopy spheres.* Mimeographed notes, Princeton, 1959.
215. J. Milnor, *Groups which act on S^n without fixed points*, Amer. J. Math. **79** (1957), 623–630.
216. J. Milnor, *Lectures on the h-cobordism theorem.* Princeton Univ. Press, 1965.
217. J. Milnor, *Morse Theory.* Princeton Univ. Press, 1963.
218. J. Milnor, *On the 3-dimensional Brieskorn manifolds $M(p,q,r)$.* In: Knots, groups, and 3-manifolds (Papers dedicated to the memory of R. H. Fox), 175–225. Princeton Univ. Press, 1975.
219. J. Milnor, *Singular points of complex hypersurfaces.* Princeton Univ. Press, 1968.
220. J. Milnor, *A unique decomposition theorem for 3-manifolds*, Amer. J. Math. **84** (1962), 1–7.
221. J. Milnor, M. Kervaire, *Bernoulli numbers, homotopy groups, and a Theorem of Rohlin*, Proc. Internat. Congress Math. 1958, 454–458, Cambridge Univ. Press, 1960.
222. E. Moise, *Affine structures in 3-manifolds. V. The triangulation theorem and Hauptvermutung*, Ann. of Math. **56** (1952), 96–114.
223. J. Montesinos, *Classical tessellations and three-manifolds.* Springer-Verlag, 1987.
224. J. Montesinos, *A representation of closed orientable 3-manifolds as 3-fold branched coverings of S^3*, Bull. Amer. Math. Soc. **80** (1974), 845-846.
225. J. Montesinos, *Surgery on links and double branched covers of S^3.* In: Knots, groups, and 3-manifolds (Papers dedicated to the memory of R. H. Fox), 227–259. Princeton Univ. Press, 1975.
226. J. Montesinos, *Variedades de Seifert que son recubridores ciclicos ramificados de dos hojas*, Bol. Soc. Math. Mexicana **18** (1973), 1–32.
227. J. Morgan, *An introduction to gauge theory.* In: Gauge theory and the topology of four-manifolds (Park City, UT, 1994), 51–143. American Math. Society, 1998.
228. J. Morgan, *The Seiberg-Witten equations and applications to the topology of smooth four-manifolds.* Princeton Univ. Press, 1996.
229. J. Morgan, T. Mrowka, D. Ruberman, *The L^2-moduli space and a vanishing theorem for Donaldson polynomial invariants.* International Press, 1994.
230. S. Morita, *Casson's invariant for homology 3-spheres and characteristic classes of surface bundles. I.* Topology **28** (1989), 305–323.
231. L. Moser, *Elementary surgery along a torus knot*, Pacific J. Math. **38** (1971), 737–745.
232. T. Mrowka, *A local Mayer-Vietoris principle for Yang-Mills moduli spaces*, Ph. D. Thesis, UC Berkeley, 1988.
233. D. Mullins, *The generalized Casson invariant for 2-fold branched covers of S^3 and the Jones polynomial*, Topology **32** (1993), 419–438.
234. V. Muñoz, *Fukaya-Floer homology of $\Sigma \times S^1$ and applications*, J. Diff. Geom. **53** (1999), 279–326.

235. V. Muñoz, *Ring structure of the Floer cohomology of* $\Sigma \times S^1$, Topology **38** (1999), 517–528.
236. R. Myers, *Homology cobordisms, link concordances, and hyperbolic 3–manifolds*, Trans. Amer. Math. Soc. **278** (1983), 271–288.
237. R. Myers, *Homology 3–spheres which admit no PL involutions*, Pacific J. Math. **94** (1981), 379–384.
238. E. Nasatyr, *Seifert–fibred homology 3–spheres, V–surfaces and the Floer index*, Math. Proc. Cambridge Philos. Soc. **111** (1992), 461–485.
239. A. Némethi, *Casson invariant of cyclic coverings via eta–invariant and Dedekind sums*, Topology Appl. **102** (2000), 181–193.
240. A. Némethi, L. Nicolaescu, *Seiberg-Witten invariants and surface singularities.* Preprint math.AG/0111298
241. W. Neumann, *An invariant of plumbed homology spheres*, Topology Symposium, Siegen 1979, 125–144. Lect. Notes in Math. **788**. Springer, 1980.
242. W. Neumann, F. Raymond, *Seifert manifolds, plumbing,* μ*-invariant and orientation reversing maps*, Algebraic and geometric topology (Santa Barbara, 1977), 163–196. Lecture Notes in Math. **664**. Springer, 1978.
243. W. Neumann, J. Wahl, *Casson invariant of links of singularities*, Comm. Math. Helv. **65** (1990), 58–78.
244. W. Neumann, D. Zagier, *A note on an invariant of Fintushel and Stern*, Lect. Notes in Math. **1167** (1985), 241–244.
245. P. Newstead, *Characteristic classes of stable bundles of rank 2 over an algebraic curve*, Trans. Amer. Math. Soc. **169** (1972), 337–345.
246. P. Newstead, *Topological properties of some spaces of stable bundles*, Topology **6** (1967), 241–262.
247. L. Nicolaescu, *Finite energy Seiberg–Witten moduli spaces on 4–manifolds bounding Seifert fibrations*, Comm. Anal. Geom. **8** (2000), 1027–1096.
248. L. Nicolaescu, *Lattice points, Dedekind–Rademacher sums and a conjecture of Kronheimer and Mrowka.* Preprint, 1997.
249. L. Nicolaescu, *The Maslov index, the spectral flow, and decompositions of manifolds*, Duke Math. J. **80** (1995), 485–533.
250. T. Ohtsuki, *Finite type invariants of integral homology 3–spheres*, J. Knot Theory Ramifications **5** (1996), 101–115.
251. P. Orlik, *Seifert manifolds*, Lecture Notes in Math. **291**. Springer-Verlag, 1972.
252. P. Ozsváth, Z. Szabó, *Holomorphic discs and topological invariants for rational homology three–spheres.* Preprint math.SG/0105202
253. P. Ozsváth, Z. Szabó, *The symplectic Thom conjecture*, Ann. of Math. **151** (2000), 93–124.
254. P. Ozsváth, Z. Szabó, *The theta divisor and the Casson–Walker invariant.* Preprint math.GT/0006194
255. M. Postnikov, *Three–dimensional spherical forms.* Proc. Steklov Inst. Math. 1992, no. 4, 129–161.
256. H. Rademacher, E. Grosswald, *Dedekind sums.* The Carus Math. Monographs, **16**, Mathematical Association of America, 1972.
257. F. Raymond, *Classification of the actions of the circle on 3–manifolds*, Trans. Amer. Math. Soc. **131** (1968), 51–78.
258. K. Reidemeister, *Homotopieringe und Linsenräume*, Abh. Math. Sem. Univ. Hamburg **11** (1935), 102–109.
259. K. Reidemeister, *Zur dreidimensionalen Topologie*, Abh. Math. Sem. Univ. Hamburg **9** (1933), 189-194.

260. V. Rokhlin, *New results in the theory of four-dimensional manifolds* (Russian), Doklady Acad. Nauk SSSR **84** (1952), 221–224.
261. V. Rokhlin, *Proof of a conjecture of Gudkov* (Russian), Funkcional. Anal. i Priložen. **6** (1972), 62–64.
262. V. Rokhlin, *Two-dimensional submanifolds of four-dimensional manifolds* (Russian), Funkcional. Anal. i Priložen. **5** (1971), 48–60.
263. D. Rolfsen, *Knots and Links.* Publish or Perish, 1990.
264. D. Ruberman, *Mutation and gauge theory. I. Yang-Mills invariants*, Comment. Math. Helv. **74** (1999), 615–641.
265. D. Ruberman, *Mutation and volumes of knots in S^3*, Invent. Math. **90** (1987), 189–215.
266. D. Ruberman, *Rational homology cobordisms of rational space forms*, Topology **27** (1988), 401–414.
267. D. Ruberman, *Seifert surfaces of knots in S^4*, Pacific J. Math. **145** (1990), 97–116.
268. O. Saeki, *On 4-manifolds homotopy equivalent to the 2-sphere.* In: Aspects of low-dimensional manifolds, 301–330, Adv. Stud. Pure Math. **20**, Kinokuniya, Tokyo, 1992.
269. I. Satake, *The Gauss-Bonnet theorem for V-manifolds*, J. Math. Soc. Japan **9** (1957), 464–492.
270. N. Saveliev, *Dehn surgery along torus knots*, Topology Appl. **83** (1998), 193-202.
271. N. Saveliev, *Floer homology and invariants of homology cobordism*, Internat. J. Math. **9** (1998), 885–919.
272. N. Saveliev, *Floer homology of Brieskorn homology spheres*, J. Diff. Geom. **53** (1999), 15–87.
273. N. Saveliev, *Fukumoto-Furuta invariants of plumbed homology 3-spheres*, Pacific J. Math. (to appear)
274. N. Saveliev, *An involution permuting Floer homology*, J. Knot Theory Ramifications **9** (2000), 543–556.
275. N. Saveliev, *Lectures on the topology of 3-manifolds. An introduction to the Casson invariant.* Walter de Gruyter, 1999.
276. N. Saveliev, *A note on Akbulut corks.* Preprint, 2000.
277. N. Saveliev, *Notes on homology cobordisms of plumbed homology 3-spheres*, Proc. Amer. Math. Soc. **126** (1998), 2819–2825.
278. N. Saveliev, *Representation spaces of Seifert fibered homology spheres*, Topology Appl. (to appear)
279. N. Saveliev, *A surgery formula for the $\bar{\mu}$-invariant*, Topology Appl. **106** (2000), 91–102.
280. S. Schröer, *Floer homology and homotopy planes*, Math. Z. **227** (1998), 279–294.
281. M. Schwarz, *Morse homology.* Birkhäuser, 1993.
282. J. P. Serre, *A course in arithmetic.* Springer-Verlag, 1973.
283. K. Shoda, *Einige Sätze über Matrizen*, Japan J. Math. **13** (1936–37), 361–365.
284. L. Siebenmann, *Are nontriangulable manifolds triangulable ?* In: Topology of Manifolds, 77 – 84, Markham, Chicago, 1970.
285. L. Siebenmann, *Disruption of low-dimensional handlebody theory by Rohlin's theorem.* In: Topology of Manifolds, 57–76, Markham, Chicago, 1970.

286. L. Siebenmann, *On vanishing of the Rohlin invariant and nonfinitely amphicheiral homology 3-spheres*, Topology Symposium, Siegen 1979, 172–222. Lect. Notes in Math. **788**. Springer, 1980.
287. J. Singer, *Three-dimensional manifolds and their Heegaard diagrams*, Trans. Amer. Math. Soc. **35** (1933), 88–111.
288. R. Stern, *Computing Donaldson invariants.* In : Gauge theory and the topology of four-manifolds (Park City, UT, 1994), 145–194, American Math. Society, 1998.
289. R. Stern, *Some more Brieskorn spheres which bound contractible manifolds.* Notices Amer. Math. Soc. **25** (1978), A448
290. A. Stipsicz, Z. Szabó, *Floer homology groups of certain algebraic links.* In : Low dimensional topology (Knoxville, 1992), 173–185. Internat. Press, 1994.
291. D. Sullivan, *On the intersection ring of compact three manifolds*, Topology **14** (1975), 275–277.
292. C. Taubes, *Casson's invariant and gauge theory*, J. Diff. Geom. **31** (1990), 547–599.
293. C. Taubes, *Gauge theory on asymptotically periodic 4-manifolds*, J. Diff. Geom. **25** (1987), 363–430.
294. C. B. Thomas, *A reduction theorem for free actions by the group* $Q(8n; k, l)$ *on* S^3, Bull. London Math. Soc. **20** (1988), 65–67.
295. W. Thurston, *Three-dimensional manifolds, Kleinian groups and hyperbolic geometry*, Bull. Amer. Math. Soc. **6** (1982), 357–381.
296. A. Tristram, *Some cobordism invariants for links*, Math. Proc. Cambridge Philos. Soc. **66** (1969), 251–264.
297. V. Turaev, *Classification of oriented Montesinos links via spin structures*, Lecture Notes in Math. **1346**, 271–289. Springer, 1988.
298. V. Turaev, *Quantum invariants of knots and 3-manifolds.* Walter de Gruyter, 1994.
299. K. Uhlenbeck, *Connections with* L^p *bounds on curvature*, Comm. Math. Phys. **83** (1982), 31–42.
300. V. Vassiliev, *Cohomology of knot spaces.* In : Theory of singularities and its applications, 23–69, Adv. Soviet Math., 1, Amer. Math. Soc., 1990.
301. O. Viro, *Branched coverings of manifolds with boundary, and invariants of links* (Russian), Izv. Akad. Nauk SSSR Ser. Mat. **37** (1973), 1241–1258.
302. K. Walker, *An extension of Casson's invariant.* Princeton Univ. Press, 1992.
303. A. D. Wallace, *Modifications and cobounding manifolds.* Canad. J. Math. **12** (1960), 503–528.
304. J. H. C. Whitehead, *On incidence matrices, nuclei and homotopy types*, Ann. of Math. **42** (1941), 1197–1239.
305. W. Wieczorek, *Immersed spheres and finite type for Donaldson invariants.* Preprint math.DG/9811116.
306. E. Witten, *Topological quantum field theory*, Comm. Math. Phys. **117** (1988), 353–386.
307. J. Wolf, *Spaces of constant curvature.* Publish or Perish, 1984.
308. T. Yoshida, *A computational algorithm of spectral flow in Floer homology.* In : Topology '90 (Columbus, OH, 1990), 445–457. Walter de Gruyter, 1992.
309. T. Yoshida, *Floer homology and splittings of manifolds*, Ann. of Math. **134** (1991), 277–323.

310. T. Yoshida, *A formulation of symplectic Floer homology for Heegaard splittings.* In : Topology, geometry and field theory, 237–254. World Sci. Publishing, 1994.
311. E. C. Zeeman, *On the dunce hat*, Topology **2** (1964), 341–358.

Index

Printed in the United Kingdom by
Lightning Source UK Ltd., Milton Keynes
139467UK00004B/15/A

9 783540 437963